Useful relations

$$R\bar{T} = 2.4790 \text{ kJ mol}^{-1}$$
$$\bar{T} = 298.15 \text{ K}$$
$$k\bar{T}/hc = 207.226 \text{ cm}^{-1}$$

T/K	100.00	298.15	500.00	1000.0	1500.0	2000.0
$(kT/hc)/\text{cm}^{-1}$	69.50	207.22	347.51	659.03	1042.5	1390.1

$$hc = 1.986\,45 \times 10^{-23} \text{ J cm}$$
$$hc/k = 1.438\,77 \text{ cm K}$$
$$h/k = 4.799\,22 \times 10^{-11} \text{ K Hz}^{-1}$$
$$8\pi hc = 4.992\,49 \times 10^{-24} \text{ J m s}^{-2}$$

Greek alphabet

A	alpha	α	N	nu	ν	
B	beta	β	Ξ	xi	ξ	
Γ	gamma	γ	O	omicron	o	
Δ	delta	δ	Π	pi	π	
E	epsilon	ε	P	rho	ρ	
Z	zeta	ζ	Σ	sigma	σ	
H	eta	η	T	tau	τ	
Θ	theta	θ	Y	upsilon	υ	
I	iota	ι	Φ	phi	ϕ	
K	kappa	κ	X	chi	χ	
Λ	lambda	λ	Ψ	psi	ψ	
M	mu	μ	Ω	omega	ω	

Prefixes

f	p	n	μ	m	c	d	k	M	G
femto	pico	nano	micro	milli	centi	deci	kilo	mega	giga
10^{-15}	10^{-12}	10^{-9}	10^{-6}	10^{-3}	10^{-2}	10^{-1}	10^{3}	10^{6}	10^{9}

Quantity	Symbol	Value
Speed of light†	c	$2.99792458 \times 10^8 \text{ m s}^{-1}$
Elementary charge	e	$1.602177 \times 10^{-19} \text{ C}$
Faraday constant	$F = eN_A$	$9.6485 \times 10^4 \text{ C mol}^{-1}$
Boltzmann constant	k	$1.38066 \times 10^{-23} \text{ J K}^{-1}$
Gas constant	$R = kN_A$	$8.31451 \text{ J K}^{-1} \text{ mol}^{-1}$ 8.20578×10^{-2} $\text{dm}^3 \text{ atm K}^{-1} \text{ mol}^{-1}$ $62.364 \text{ L Torr K}^{-1} \text{ mol}^{-1}$
Planck constant	h $\hbar = h/2\pi$	$6.62608 \times 10^{-34} \text{ J s}$ $1.05457 \times 10^{-34} \text{ J s}$
Avogadro constant	N_A	$6.02214 \times 10^{23} \text{ mol}^{-1}$
Atomic mass unit	u	$1.66054 \times 10^{-27} \text{ kg}$
Mass of electron	m_e	$9.10939 \times 10^{-31} \text{ kg}$
proton	m_p	$1.67262 \times 10^{-27} \text{ kg}$
neutron	m_n	$1.67493 \times 10^{-27} \text{ kg}$
Vacuum permeability†	μ_0	$4\pi \times 10^{-7} \text{ J s}^2 \text{ C}^{-2} \text{ m}^{-1}$ $4\pi \times 10^{-7} \text{ T}^2 \text{ J}^{-1} \text{ m}^3$
Vacuum permeativity†	$\varepsilon_0 = 1/c^2\mu_0$ $4\pi\varepsilon_0$	$8.85419 \times 10^{-12} \text{ J}^{-1} \text{ C}^2 \text{ m}^{-1}$ $1.11265 \times 10^{-10} \text{ J}^{-1} \text{ C}^2 \text{ m}^{-1}$
Bohr magneton	$\mu_B = e\hbar/2m_e$	$9.27402 \times 10^{-24} \text{ J T}^{-1}$
Nuclear magneton	$\mu_N = e\hbar/2m_p$	$5.05079 \times 10^{-27} \text{ J T}^{-1}$
Electron g value	g_e	2.00232
Bohr radius	$a_0 = 4\pi\varepsilon_0/\hbar^2 m_e \varepsilon_0$	$5.29177 \times 10^{-11} \text{ m}$
Rydberg constant	$R_\infty = m_e e^4/8h^3 c\varepsilon_0^2$	$1.09737 \times 10^5 \text{ cm}^{-1}$
Fine structure constant	$\alpha = \mu_0 e^2 c/2h$	7.29735×10^{-3}
Gravitational constant	G	$6.67259 \times 10^{-11} \text{ N m}^2 \text{ kg}^{-2}$
Standard acceleration of free fall†	g	9.80665 m s^{-2}

† Exact (defined) values

Quanta

Quanta

A Handbook of Concepts

SECOND EDITION

P. W. Atkins

University Lecturer and
Fellow of Lincoln College, Oxford

Oxford   New York   Tokyo
OXFORD UNIVERSITY PRESS
1991

Oxford University Press, Walton Street, Oxford OX2 6DP

Oxford New York Toronto
Delhi Bombay Calcutta Madras Karachi
Petaling Jaya Singapore Hong Kong Tokyo
Nairobi Dar es Salaam Cape Town
Melbourne Auckland
and associated companies in
Berlin Ibadan

Oxford is a trade mark of Oxford University Press

Published in the United States
by Oxford University Press, New York

First edition 1974
Second edition 1991

British Library Cataloguing in Publication Data

Atkins, P. W. (Peter William), 1940–
Quanta. – 2nd ed.
1. Chemistry. Quantum theory
I. Title
541.28
ISBN 0–19–855572–5
ISBN 0–19–855573–3 (pbk.)

Library of Congress Cataloging-in-Publication Data

Atkins, P. W. (Peter William), 1940–
Quanta: a handbook of concepts/P. W. Atkins.—2nd ed.
Includes index.
1. Quantum chemistry.   2. Quantum theory.   I. Title.
QD462.A846   1991   541.2'8—dc20   90–21970
ISBN 0–19–855572–5
ISBN 0–19–855573–3 (pbk.)

Typeset by H. Charlesworth & Co Ltd, Huddersfield
Printed in Great Britain by
Butler & Tanner Ltd, Frome, Somerset

Preface

Quanta contains my visualizations and interpretations of the concepts of quantum theory. The pictures and descriptions, like all analogies and models, are only a partial representation of the mathematics, but I hope they contain the heart of the matter, and enable the reader to understand each idea at a physical, rather than mathematical, level.

My aim in this completely rewritten edition is to convey qualitative, pictorial interpretations of quantum concepts, particularly the concepts that are used in chemistry. The entries are often based on discussions that I have had with my students or are like fragments of my lectures. In all of them I have sought to impart insight into the subject and to make it accessible to those who are deterred by the mathematics. I hope it will also act as a succinct reminder of a once learned and now half remembered idea.

I agreed in the first edition, when seeking to disarm an imagined critic, and continue to agree in this, that quantum theory has an inescapable mathematical structure. I agree that the full richness of the theory can be conveyed only in the language of mathematics, and that pictures and words have connotations that can mislead. Nevertheless, I remain convinced that not everyone seeks uninterpreted mathematics for their understanding. Most chemists, at least, seek visualizations and verbal interpretations of quantum concepts, and that is what I have stretched myself to provide.

I have written *Quanta* with a broad selection of people in mind. Those who are *not* the target are the professional quantum chemists: this is not a detailed, exhaustive account of technical details about quantum mechanics and the computation of molecular structures. The readers that I do have in mind are principally students of chemistry, who at all stages of their studies encounter unfamiliar, little understood or half forgotten concepts. I hope they will find some enlightenment— and perhaps be brought to the edge of deep understanding—if they read through the appropriate entry and look at the illustrations. I also have in mind that other class of students, the teachers of students, who are pressed for explanations by their students or look for presentations that they can use in lectures. Finally, I think of practising chemists everywhere—in all the branches, roots, and twigs of the subject—who may need to be reminded of a concept. In summary, I like to think of the book as a gentle, largely non-technical, accessible resource of concepts, pictures, and elucidations.

Each entry is intended to explain, in plain language, the physical content of its topic. Where the entry draws on information contained elsewhere in the book, or where further development of a topic is desirable, or where the reader might be felt to require more background, I have labelled a word with an unobtrusive ° to signify that it is treated in its own entry. Unlike most dictionaries this has an index: some concepts are buried inside others. If my explanation of a topic is inadequate, the bibliography will direct the dissatisfied along a trail of others' explanations.

Where I feel it desirable that some mathematics be available (for example, to be complete, or to provide the background of a remark), I have used a system of Boxes. The Tables contain numerical data to convey an idea of the sizes of physical quantities and their trends.

Each entry concludes with *Further information*. This is an annotated guide to the literature, and in it will be found references to books and articles to turn to for more information, the development of the topic, and the missing or emasculated mathematics. This section is not exhaustive (it contains works I have found helpful) but may be used to force an entry into the literature. In this section I have tried to remember to list the books in order of complexity. The exception to this organization is the reference to MQM, the second edition of my *Molecular quantum mechanics* (Oxford University Press, 1983). Many of the topics are treated in more detail in that book, and I give a reference to it at the start of each *Further information* section.

I have received a great deal of help in preparing this edition. I am particularly grateful to

Dr N. L. Allan, University of Bristol

Dr. N. C. Handy, University of Cambridge

Professor S. F. A. Kettle, University of East Anglia

Professor R. G. Parr, University of North Carolina, Chapel Hill

Dr J. N. L. Connor, University of Manchester

Professor L. Pederson, University of North Carolina, Chapel Hill

who read the first draft, gave me a lot of advice about the entries, and provided me with additional current and key references. I am equally grateful to my students, who in tutorials have forced me to explain what previously I had thought I had understood. My publishers, Oxford University Press, were as helpful as any author could wish.

P.W.A.

Oxford
August 1990

Ab initio

The small amount of quantum chemistry still conducted in Latin is the province of °computational chemists. *Ab initio,* roughly translated, means 'from scratch'. The term is applied to calculations of molecular structure that use no input other than the °Schrödinger equation, the values of the fundamental constants (specifically e, m_e, and $\hbar$), and the atomic numbers of the atoms present.

Further information

For an introduction to *ab initio* procedures, see Richards and Cooper (1983), Hehre *et al.* (1986), and Hinchcliffe (1987, 1988). Their application to the calculation of molecular properties is summarized by Hinchcliffe (1987). A collection of thumbnail sketches of the landmarks in the development of *ab initio* methods has been compiled by Schaefer (1984). For an analysis of the links between *ab initio* and semi-empirical procedures, see Freed (1983). Perhaps the first use of the term '*ab initio*' is in Parr, Craig, and Ross (1950).

Absorption coefficient

The **Beer–Lambert law** states:

> The intensity of radiation decreases exponentially as it passes through an absorbing medium:

$$I = I_0 10^{-\varepsilon[J]l}$$

where I_0 is the incident intensity, l is the path length through the sample of molar concentration [J] in the absorbing species J, and ε is the **molar absorption coefficient** (formerly, the 'extinction coefficient'). The dimensionless quantity $\varepsilon[J]l$ is called the **absorbance**, A, (formerly the 'optical density') of the sample, and I/I_0 is the **transmittance**, T. The intensity decays exponentially because the probability that absorption will occur at any stage along the path is proportional to the intensity of the surviving radiation (see °Einstein coefficients).

The molar absorption coefficient depends on the frequency of the incident light, and the **integrated absorption coefficient** $\mathscr{A}$ is the integral of A over an entire absorption band. The integrated absorption coefficient is a measure of the true intensity of the transition, and is related to the °oscillator strength of the transition and the °transition moment.

Further information

For an introduction to measures of spectroscopic intensity see Atkins (1990) and MQM Chapter 12 and Appendix 17: that appendix establishes the connections between the various measures of absorption intensity. See also Struve (1989) for a more detailed description.

Acronyms

Scientific disciplines collect acronyms like combs collect hairs, and quantum chemistry is no exception. Some of the more common are listed in Box A.1. Where there is further information in this text the appropriate entry has been specified. The list is by no means exhaustive, but it includes the more commonly encountered terms.

Box A.1 Acronyms and abbreviations in quantum chemistry and related fields

AM1	Austin method 1, a °semiempirical procedure
AO	°Atomic orbital
CI	°Configuration interaction
CNDO	Complete neglect of differential overlap, a °semiempirical procedure
DZ	Double zeta, a basis set consisting of two STOs for each atomic orbital
EH	Extended Hückel theory
EPR	Electron paramagnetic resonance (see ESR)
ESCA	Electron spectroscopy for chemical analysis (see °photoelectron spectroscopy)
ESR	°Electron spin resonance
FEMO	Free-electron molecular orbitals
GTO	°Gaussian type atomic orbital
HF	Hartree–Fock method for °self-consistent fields
HMO	°Hückel molecular orbitals
HOMO	°Highest occupied molecular orbital
INDO	Intermediate neglect of differential overlap, a °semiempirical procedure
IR	Infrared radiation, as in °vibrational spectroscopy
JWKB	Jeffreys–Wentzel–Kramers–Brillouin, a semiclassical approximation to quantum mechanics
LCAO	°Linear combination of atomic orbitals
LUMO	Lowest unoccupied molecular orbital (see °HOMO)
MINDO	Modified intermediate neglect of differential overlap, a °semiempirical procedure
MNDO	Modified neglect of differential overlap, a °semiempirical procedure
MO	°Molecular orbital
NMR	°Nuclear magnetic resonance
PES	°Photoelectron spectroscopy
PPP	Pariser–Parr–Pople method, a °semiempirical procedure
RHF	Restricted Hartee–Fock method (for open-shell molecules)
SCF	°Self-consistent field
STO	°Slater type orbital
STO–NG	A Slater type orbital expressed as the sum of N gaussian orbitals
TFD	Thomas–Fermi–Dirac method, a statistical treatment of electron density
UHF	Unrestricted Hartree–Fock method for SCF calculations on open-shell molecules
UV	Ultraviolet radiation
VB	°Valence bond theory
VSEPR	°Valence-shell electron pair repulsion, a theory of molecular geometry
WKB	Wentzel–Kramers–Brillouin method, a semiclassical approximation to quantum mechanics
ZDO	Zero differential overlap, a °semiempirical approximation.

Adiabatic process

In quantum mechanics an **adiabatic process** is a process in which, although a system is undergoing a change that modifies its energy levels, it remains in a single, definite state (as defined, for example, by a set of quantum numbers). In a **non-adiabatic process** (or 'diabatic process'), the system may be found in a range of different states after the change has occurred. An adiabatic process generally occurs very slowly, whereas a non-adiabatic process is sudden, or impulsive.

A model of an adiabatic process is the very slow compression of a square well containing a single particle. If the particle is in the nth level of the original well, it will remain in the nth level of the new, shorter well (Fig. A.1). However, if the compression is impulsive, and the walls of the well are suddenly pushed closer together, the particle will be found in any one of a range of states of the shorter well. Another example is a hypothetical °hydrogenic atom with a variable nuclear charge. If the electron is initially in a $1s$ orbital and the nuclear charge is increased from $Z=1$ extremely slowly, the electron will stay in that shrinking orbital, and when $Z=2$ the system will have become a ground state He^+ ion. If the nuclear charge is increased suddenly, the electron will be unable to follow the rapidly changing potential and will be found in any of a range of orbitals of the newly formed He^+ ion.

In an adiabatic process the wavefunction adapts to the slowly changing parameters that define the system (such as the length of the well or the nuclear charge in the foregoing examples): the wavefunction follows the slow remoulding of the system. In an impulsive, non-adiabatic process, the change in the parameter occurs so quickly that the system momentarily retains its initial wavefunction. However, since that initial wavefunction can usually be expressed as a °superposition of the wavefunctions of the newly formed system, we interpret the final state as a mixture of wavefunctions of the new system. Hence we conclude that the system has made a transition to a range of states.

Further information

An account of adiabatic processes in terms of °perturbation theory will be found in Davydov (1976), who derives the condition that a process is adiabatic if the rate of change of the perturbation V is very slow compared to the separation ΔE of the initial state from neighbouring states, in the sense that $\mathrm{d}V/\mathrm{d}t \ll (\Delta E)^2/\hbar$. See also §50 of Bohm (1951) for a nice discussion with straightforward mathematics. The connection between the quantum mechanical and thermodynamic aspects of adiabaticity has been explored by Boulil, Deumié, and Henri-Rousseau (1987).

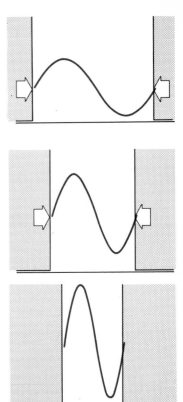

Fig. A.1 An adiabatic process: as the walls are pressed together very slowly (top to bottom), the wavefunction changes smoothly into a single eigenstate of the new system.

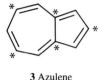

1 Propene   2 Benzene

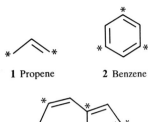

3 Azulene

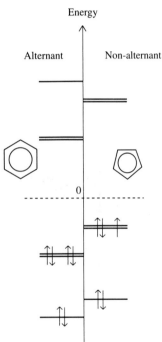

Fig. A.2 The molecular orbital energy levels of alternant (left) and non-alternant (right) molecules in the °Hückel approximation. Note the symmetrical array of the energy levels of the alternant molecule.

Alternant hydrocarbon

A conjugated hydrocarbon is **alternant** if its C atoms can be labelled alternately by a star, no star, a star, and so on, such that no two starred or unstarred atoms are neighbours when the labelling is complete. Two examples are propene (**1**) and benzene (**2**); an example of a **non-alternant hydrocarbon**, a hydrocarbon for which the labelling results in two adjacent starred or unstarred atoms, is azulene (**3**). An **even-alternant hydrocarbon** is an alternant hydrocarbon with an even number of C atoms; an **odd-alternant hydrocarbon** is an alternant hydrocarbon with an odd number of C atoms. Benzene is even-alternant; propene is odd-alternant.

Within the °Hückel approximation, the π systems of alternant hydrocarbons have the following characteristics:

- The bonding and antibonding orbitals are arranged symmetrically about zero energy (Fig. A.2).

Thus, each bonding orbital of energy $-E$ has an °antibonding partner of energy $+E$.

- In an even-alternant hydrocarbon with N C atoms there are usually $\frac{1}{2}N$ bonding orbitals and $\frac{1}{2}N$ antibonding orbitals.

It may happen that the numbers of starred and unstarred atoms differ by 2, in which case there are two nonbonding orbitals in place of one pair of bonding and antibonding orbitals.

- In an odd-alternant hydrocarbon there is usually one more starred atom than unstarred atom, and consequently there is one nonbonding orbital.

This nonbonding orbital lies between the symmetrical array of bonding and antibonding orbitals (as in propene).

- The distribution of electrons is more uniform in alternant than in non-alternant hydrocarbons.

This uniformity is expressed by the **Coulson–Rushbrooke theorem**:

The π-electron °charge density on each C atom in the ground state of an alternant hydrocarbon is unity.

That is, each C atom has a net ownership of one π electron in an alternant hydrocarbon.

The stabilities of even-alternant hydrocarbons (such as °benzene) can be rationalized in terms of the preceding properties. In particular, in an N-atomic even-alternant hydrocarbon, each C atom provides one π electron. According to the °Pauli principle, all N electrons can be accommodated in the $\frac{1}{2}N$ bonding π orbitals, and no antibonding π orbitals are occupied. The resulting structural stability is consolidated kinetically by the uniformity of the charge distribution, which provides no special centres of attraction for nucleophilic or electrophilic reagents.

Further information

A helpful elementary account of alternant hydrocarbons will be found in Pilar (1968), Coulson, O'Leary, and Mallion (1978), and Yates (1978). The book by Pauncz (1967) considers them in detail and the spectroscopic properties of alternant molecules are described by Murrell (1971). For a proof of the Coulson–Rushbrooke theorem, see Coulson and Rushbrooke (1940) and a review article by Coulson (1970).

Angular momentum

The **angular momentum** is the momentum associated with the rotational motion of an object. The greater the angular momentum of an object, the stronger the braking force (more precisely, the braking torque) needed to bring it to rest.

In classical mechanics the magnitude of the angular momentum of a rotating object is given by $I\omega$, where I is the moment of inertia of the object (see °rotational motion) and ω is its angular velocity. For a point mass m moving with a speed v on a circular path of radius r the magnitude of the angular momentum is $I = mvr$. A large, heavy object (with a large moment of inertia) need rotate only slowly (have small angular velocity) to achieve the same angular momentum as a small object rotating rapidly. In classical mechanics an object may be accelerated to any angular momentum by choosing the torque and the time for which that torque is applied, and the angular momentum of a body around each of three perpendicular axes may take any value consistent with the magnitude of the angular momentum. Thus, if we were to think of a satellite in orbit around a planet, the plane of the orbit could lie at any angle and we could know the orientation of the orbit to arbitrary precision.

The angular momentum of a body is usually represented by an arrow — a vector (strictly, an axial vector) — lying perpendicular to the plane of rotation, with a length proportional to the magnitude of the momentum. In classical mechanics the arrow can be of any length (Fig. A.3) and lie at any angle (Fig. A.4a). The projection of the arrow on each axis corresponds to the angular momentum of the body around that axis. According to classical mechanics, we may know the projections on all three axes.

According to quantum mechanics, the magnitude of the angular

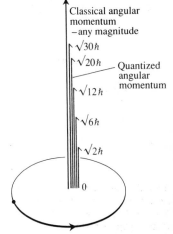

Fig. A.3 According to classical mechanics, the magnitude of the angular momentum (represented by the length of the arrow) may take any value (that is, is continuously variable). According to quantum mechanics, the magnitude of the angular momentum is quantized and limited to the discrete values shown to the right.

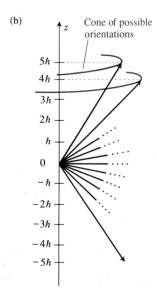

(a)

All orientations
permitted

(b)

Cone of possible
orientations

z

$5\hbar$
$4\hbar$
$3\hbar$
$2\hbar$
$\hbar$
0
$-\hbar$
$-2\hbar$
$-3\hbar$
$-4\hbar$
$-5\hbar$

Fig. A.4 (a) According to classical mechanics, the orientation of the angular momentum is continuously variable. Note how the direction of the vector is related to the direction of motion. (b) According to quantum mechanics, the orientation of the angular momentum is quantized, and only certain projections on the z-axis are permitted. The projections on the x- and y-axes cannot be specified.

momentum is limited to the values $\{j(j+1)\}^{1/2}\hbar$ where j is a non-negative integer (or, under certain circumstances, a half-integer; see °spin). Thus, instead of being continuously variable, the magnitude of the angular momentum of a rotating object can be only 0, $2^{1/2}\hbar$, $6^{1/2}\hbar$, and so on. If we could make very precise observations, we would find that we could accelerate the rotation of a ball only in steps, not smoothly and continuously. The length of the angular momentum vector can therefore in fact take only certain values (Fig. A.3) and is not continuously variable.

According to quantum mechanics, only *one* component of the angular momentum may have a definite, discrete value (Fig. A.4b). This restriction of the component of angular momentum to discrete values is called **space quantization**. The favoured axis (which may be defined by an externally applied electric or magnetic field) is normally identified as the z-axis, and for a given value of j (that is, for a given magnitude of angular momentum), the component of angular momentum of the object around the z-axis is limited to the $2j+1$ values $m_j\hbar$, with $m_j = 0, \pm 1, \pm 2, \dots \pm j$. Positive values of m_j correspond to clockwise rotation about the z-axis (as seen from below) and negative values correspond to counterclockwise rotation about the z-axis (Fig. A.4b).

According to quantum mechanics, if we specify the z-component of angular momentum, we cannot specify the components of angular momentum about the x- and y-axes. Hence we cannot specify the orientation of the motion of the body other than in very general terms. Thus, if m_j is close to its maximum or minimum values ($+j$ or $-j$ respectively), we know only that the motion is largely equatorial (in the xy plane); if $m_j = 0$, the motion is largely polar.

We can judge the magnitude and orientation of the orbital angular momentum of a particle qualitatively by counting the number of angular nodes in its °wavefunction: the greater the number of angular nodes, the greater the angular momentum (Fig. A.5). Similarly, the greater the number of angular nodes in the xy plane, the greater the component of angular momentum around the z-axis (Fig. A.6). The physical reason in each case is that an angular momentum wavefunction can be pictured as a wave wrapped round a sphere, and the greater the number of °nodes, the shorter its wavelength; hence, by the °de Broglie relation, the greater the momentum.

The angular momentum quantum number is denoted l (in place of the more general symbol j) for *orbital* angular momentum (the angular momentum arising from motion around a central point in space). The quantum number l may have only non-negative integral values ($l = 0$, 1, 2,...; see °atomic orbitals for more information). The °spin quantum number, the quantum number that specifies the *intrinsic* angular momentum of a particle, is denoted s or (for nuclei) I. Both s and I may

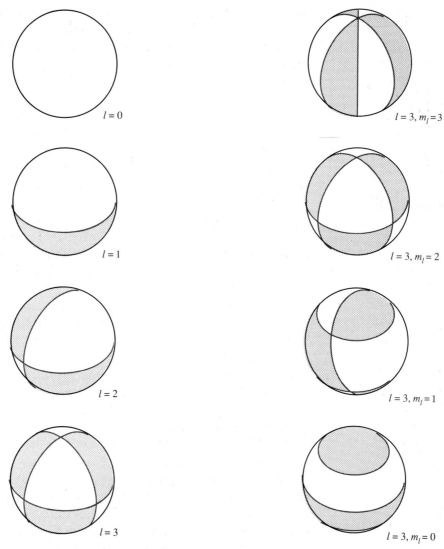

Fig. A.5 The magnitude of the angular momentum increases with the number of angular °nodes in the wavefunction. Note that the number of angular nodes is equal to the quantum number l.

Fig. A.6 The angular momentum around the z-axis increases as the number of nodes that cut through the equator increases, which is equal to the value of the quantum number m_l. These wavefunctions are all for $l = 3$.

have either integral or half-integral non-negative values depending on the identity of the fundamental particle. The quantum numbers for the z-component of angular momentum are m_l and m_s respectively (instead of the more general m_j).

If a system contains several particles with angular momentum, the total angular momentum is also quantized. The composite angular momentum is in general denoted J, but it is denoted L if only orbital angular momenta contribute and S if only spins contribute. The values that J (or L and S) may take are determined by the °Clebsch–Gordan series. Once they have been obtained, the magnitude of the total angular momentum is given by $\{J(J+1)\}^{1/2}\hbar$ and its $2J+1$ permitted components on the z-axis are given by $M_J\hbar$ in the usual way.

In quantum mechanics, an angular momentum is most generally defined in terms of a set of °commutation relations. *Any* set of observables represented by three operators that satisfy the commutation rule $[j_x, j_y] = i\hbar j_z$ (and cyclic permutations of the subscripts) is called an angular momentum. The properties of this more general 'angular momentum' are the same as those outlined above since the properties we have described are a direct consequence of the commutation relations.

Further information

See MQM Chapter 6 for a detailed discussion of angular momentum. An old but interesting account, which emphasizes the connection between the classical and quantum pictures of angular momentum, will be found in Kauzmann (1957). Since the quantum theory of angular momentum is of the greatest importance in a number of contexts, including spectroscopy, scattering, electronic structure, and the properties of fundamental particles, there are many books that describe it in detail. See in particular Zare (1987), Brink and Satchler (1968), Rose (1957), Edmonds (1974), and Biedenharn and Louck (1981). For details of the representation of angular momenta in terms of orientated arrows, see the °vector model and Altmann (1991a).

Anharmonicity

°Harmonic oscillation occurs when a particle experiences a restoring force that is proportional to its displacement. An oscillator is **mechanically anharmonic** if the restoring force is not strictly proportional to the displacement. A harmonic oscillator is characterized by a parabolic potential energy (in which the potential energy is proportional to the square of the displacement); an anharmonic oscillator has a nonparabolic potential energy (Fig. A.7).

If the restoring force weakens at large displacements (as is typical of a chemical bond), the potential becomes less confining as the displacement increases. Correspondingly, the energy levels become closer together than in a harmonic oscillator (Fig. A.7) at energies at which the oscillator can reach such large displacements that the weakened restoring force can be experienced. As a result, spectroscopic transitions to the upper vibrational levels converge. Mechanical anharmonicity also weakens the °selection rules for vibrational transitions, for the wavefunctions are no longer those of a truly harmonic oscillator. As a result, previously forbidden transitions become allowed, and **harmonics** of the fundamental transitions are observed in which the vibrational quantum

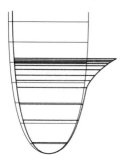

Fig. A.7 The potential energy of a harmonic oscillator is parabolic (black) but the potential energy of a molecule (blue) is more complicated and the vibrations are strongly anharmonic at high excitations. As the potential becomes less confining the levels come closer together (see °Morse potential).

number changes by $+2, +3$, etc. These transitions increase in intensity with the extent of anharmonicity in the potential.

Mechanical anharmonicity can mix together different °normal modes of vibration. Normal modes may not be independent when the potential is not parabolic, and the motion of one mode may stimulate another mode into vibration. That is, in the presence of anharmonicity the wavefunction of a normal mode can mix with, and therefore acquire some of the characteristics of, some of the other normal modes.

An example of a spectroscopic consequence of anharmonicity is the phenomenon of **Fermi resonance**. Fermi resonance is a mechanism whereby the simultaneous excitation of two modes is permitted because nearby (in energy) there is a fundamental excitation frequency of another, allowed mode (Fig. A.8). The anharmonicity in the molecular motion gives the mixture of vibrational modes some of the characteristics of the allowed fundamental, and so their joint excitation becomes allowed.

The intensities of transitions are also affected by **electrical anharmonicity**, which occurs if the °electric dipole moment of a molecule varies nonlinearly with the displacement. Since a dipole moment is proportional to the separation of the charges constituting the dipole, such a nonlinear variation occurs if the charge distribution in a molecule changes as it is distorted (Fig. A.9). The nonlinear variation of the dipole moment can result in the modification of the selection rules (which suppose that the oscillator is both mechanically and electrically harmonic), and transitions can occur with $\Delta v = \pm 2, \pm 3$, etc. Electrical anharmonicity does not affect the energy levels themselves; it affects only the selection rules for transitions between them.

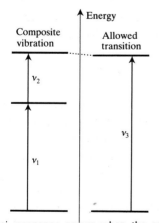

Fig. A.8 Fermi resonance occurs when the energy of joint excitation of two modes of vibration matches the energy of a single excitation of an allowed transition. The forbidden transitions acquire intensity because they acquire some of the character of the allowed transition.

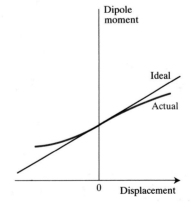

Fig. A.9 Electrical anharmonicity occurs when the electric dipole moment does not vary linearly with the displacement (because the charge distribution is changing too).

Further information

See MQM Chapter 11 for a discussion of anharmonicity and also for a discussion of the role of symmetry in governing what modes may interact by Fermi resonance. For the spectroscopic consequences of anharmonicity, see Hollas (1983, 1987) and Gribov and Orville-Thomas (1988). For a useful model of a mechanically anharmonic oscillator see the entry on the °Morse potential.

Antibonding

An **antibonding orbital** is a °molecular orbital that, when occupied, increases the energy of a molecule relative to the separated atoms.

Imagine an orbital on each of two atoms at a separation R (Fig. A.10), and let the orbitals have opposite phases (opposite signs, as shown by the tinting). As the separation decreases, the region of positive amplitude of one orbital moves into the region of negative amplitude of the other. The two wavefunctions interfere destructively, and the total wave amplitude in the region of overlap is decreased and may become zero over a °nodal surface. Since the square of the total amplitude gives the probability of finding the electrons in a particular region, the effect of bringing the orbitals together with opposite phases is to reduce the electron density in the internuclear region. This reduction has an adverse effect on the energy of the molecule because the internuclear region is the best place to put the electrons, for then their interaction with the two nuclei is most favourable. Hence, if the orbital is occupied the molecule will have an energy greater than that of the two atoms at infinite separation.

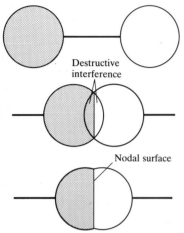

Fig. A.10 Destructive interference occurs when the atomic orbitals on two atoms approach each other with opposite phases. As a result, electrons are excluded from the internuclear region and the energy of the molecule is higher than that of the widely separated atoms.

Further information

The role of antibonding orbitals in the determination of molecular structure is elucidated in Coulson (1982), DeKock and Gray (1980), Gimarc (1979), and Verkade (1986). Orchin and Jaffé (1967) deal with antibonding orbitals exclusively. For some pictorial representations of antibonding orbitals see Gimarc (1979), Verkade (1986), and Tedder and Nechvatal (1985). See also Hout, Pietro, and Hehre (1984) for a computer graphics approach. The book by Albright, Burdett, and Whangbo (1985) is a virtuoso deployment of the concepts of antibonding as well as bonding orbitals. For more information, see °molecular orbitals.

Aromaticity

The property of **aromaticity** applies to molecules that are cyclic, usually planar, and conjugated (possessing alternating single and double bonds), and which have a lower energy than would be expected for a molecule with so many double bonds. Aromaticity is normally identified with enhanced thermodynamic stability, low reactivity, and magnetic properties that indicate the presence of a significant ring current. It has been established that all three indexes of aromaticity correlate with an additional property, the relative °hardness of the molecule in question. °Benzene is the archetype of aromatic molecules.

The additional lowering of energy that is often characteristic of an aromatic molecule is due to °resonance (in °valence bond language) or delocalization (in °molecular orbital language). The extent of energy lowering relative to a hypothetical reference state is called the °resonance energy or delocalization energy. The tricky and somewhat arbitrary problem of what reference state to use is mentioned in the entry on °resonance energy. The advantage of the identification of aromaticity in terms of relative hardness has the advantage that it is independent of the choice of a reference structure.

A rule of thumb that indicates whether a cyclic, planar, conjugated molecule is likely to be aromatic is the **Hückel $4n+2$ rule**, which states that aromatic molecules have $4n+2$ electrons, where n is an integer. Benzene has $n = 1$; the simplest aromatic molecule of all is the cyclopropene cation, with $n = 0$. The large molecule [18]-annulene, consisting of 18 conjugated C atoms in a planar ring, is also aromatic, with $n = 4$.

The origin of the $4n+2$ rule may be understood by considering the energy levels of cyclic hydrocarbons. From N atoms in a ring, N molecular orbitals may be formed, $\frac{1}{2}N$ of which are bonding and $\frac{1}{2}N$ antibonding (one may be nonbonding if N is odd). The lowest orbital is nondegenerate, and the remaining bonding orbitals are all doubly degenerate (Fig. A.11). Two electrons may occupy the lowest energy

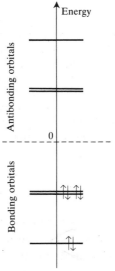

Fig. A.11 The π molecular orbital energy level diagram of the typical aromatic molecule benzene. Note that only the bonding orbitals are occupied.

orbital, and each doubly degenerate pair can accommodate four electrons. Therefore, the bonding orbitals will be full when there are $4n + 2$
electrons present in the molecule, where n is the number of degenerate
pairs of bonding orbitals. An aromatic molecule is therefore the analogue of a closed-shell species, and with only bonding orbitals occupied
it has a low energy.

Cyclic conjugated molecules with $4n$ electrons show an enhanced
thermodynamic instability and are termed **antiaromatic**. Antiaromatic
molecules include the cyclopropenyl anion ($n = 1$). However, a problem
with the concept of antiaromaticity is that since the delocalized system
has an energy greater than that of the corresponding delocalized system, the molecule will distort and eliminate the uniformity of the bonds
in the ring unless the σ bonds are so rigid that they maintain the
uniform structure. As a result, antiaromaticity is a much less useful
concept than aromaticity.

Further information

See Dewar (1969) and Coulson (1979) for aromaticity in the context of chemical
bonding, and Lewis and Peters (1975) and Glidewell and Lloyd (1986) for
more explicit surveys. The best modern survey of aromaticity is that by Gerratt
(1986). Bergmann and Pullman (1971) have edited a collection of papers on
aromaticity and related subjects. See °benzene for additional information and
a different viewpoint. For a discussion of hardness in the context of aromaticity,
see Zhou and Parr (1989). For the extension of the Hückel rule to three
dimensions, see Haddon (1988). The view that aromaticity might have nothing
to do with delocalization may be explored by referring to °benzene.

Atomic orbital

An **atomic orbital** is the °wavefunction of an electron in an atom. Its
square (its square modulus if the wavefunction is °complex) gives the
probability of finding the electron at each point. An electron described
by a particular orbital is said to **occupy** that orbital. According to the
°Pauli exclusion principle, no more than two electrons can be described
by the same orbital: we say that an orbital may **accommodate** up to
two electrons. An atomic orbital has a well defined significance only
for °hydrogenic atoms; in many-electron atoms there are electron–
electron interactions that destroy the validity of discussing atoms in
terms of orbitals. However, hydrogenic orbitals are still employed in
the widely adopted °orbital approximation.

In the °Bohr model of the hydrogen atom it was supposed that an
electron orbited a central nucleus in a definite path. The introduction
of quantum mechanics, and in particular of the °uncertainty principle,
showed that the concept of a definite trajectory was untenable on an

atomic scale, and hence that an electron's orbit cannot in fact be specified. Quantum mechanics replaced the precise trajectory, the *orbit* of the electron, by a mathematical expression, its *orbital*, which tells us the *probability* that an electron will be found at each point near a nucleus. Specifically, in the ground state of the H atom, the electron occupies an orbital that decays exponentially with distance from the nucleus, but which is independent of angle around the nucleus. Hence, the probability of finding the electron is greatest at the nucleus and then declines exponentially with distance at the same rate in all directions (Fig. A.12).

In general, the amplitude of an atomic orbital varies with distance from the nucleus in a more complicated way than a simple exponential decay. Close to the nucleus the wavefunction may pass through zero several times before rising to a maximum; however, at large distances from the nucleus all orbitals decay exponentially towards zero (Fig. A.13). Atoms also have orbitals with a complicated angular dependence, and in general an atomic orbital can be pictured as a series of lobes centred on the nucleus, each lobe representing a region of space where an electron that occupies the orbital is likely to be found. The lobes have amplitudes that may be positive (white in the illustrations in this book) or negative (tinted).

An orbital is often represented by a **boundary surface** within which there is a specified probability (typically 90 per cent), of finding the electron (Fig. A.14). The lobes of nonspherical orbitals are separated by planes of zero amplitude called **angular nodes** (Fig. A.14). The greater the number of angular nodes, the greater the orbital °angular momentum. In general (with the exception of °hydrogenic atoms), the greater the number of nodes of a given type, the higher the energy.

One of the principal applications of atomic orbitals in quantum chemistry is to the construction of °molecular orbitals as °linear combinations of atomic orbitals.

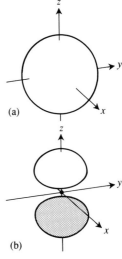

Fig. A.14 (a) All *s* orbitals have spherical boundary surfaces and (b) all *p* orbitals have two lobes separated by a nodal plane. The different shading represents opposite signs of the wavefunction.

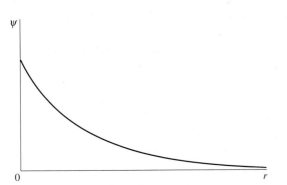

Fig. A.12 The lowest-energy atomic orbital of a °hydrogenic atom decays exponentially with distance.

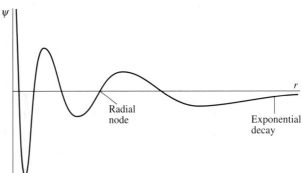

Fig. A.13 In general, atomic orbitals oscillate through zero several times before approaching zero exponentially.

Further information

See °s orbital, °p orbital, °d orbital, and °f orbital for details; orbitals are also examined in detail in the entry on °hydrogenic atoms. For details of orbitals, see MQM Chapter 4 and Gerloch (1986). A convenient source of mathematical expressions for hydrogenic orbitals is the ancient but classic text by Pauling and Wilson (1935); the atomic orbitals of many-electron atoms are often represented by °Slater orbitals. For the extension of the concept of atomic orbitals to molecules, see °molecular orbitals. Orbital diagrams are given by Salem and Owens (1973).

Atomic spectra

An atomic spectrum is produced when an atom makes transitions between states that differ in energy and emits or absorbs the difference in energy as electromagnetic radiation. In **atomic emission spectroscopy** the final state has a lower energy than the initial state, and the atom emits radiation in the transition. In **atomic absorption spectroscopy** the energy of the final state is greater than that of the initial state, and the atom absorbs radiation. If the two states differ in energy by ΔE, the frequency v of the radiation emitted or absorbed is given by the °Bohr frequency condition $\Delta E = hv$.

The observed transitions can normally be ascribed to °electric dipole transitions, and the lines that may appear in the spectrum are governed by °selection rules. The appearance of the spectrum is modified by the application of a strong magnetic field (the °Zeeman effect) or a strong electric field (the °Stark effect). The spectral lines generally have a °fine structure, which arises from °spin–orbit coupling, and a °hyperfine structure, which is due to the interaction of the electrons with the °magnetic dipole and electric °quadrupole moments of the nucleus.

The observation of discrete spectral lines was a historically important clue to the existence of quantized energy levels, and the interpretation of the spectrum of atomic °hydrogen was a major triumph of early quantum theory. Similarly, the °Pauli principle (in chemistry, the key to understanding the Periodic Table) emerged from a study of the spectrum of helium and the puzzling observation that not all the possible states of the atom are allowed.

Atomic spectroscopy is important in part because of its applications to analysis (including that of the composition of inaccessible samples, such as stellar atmospheres). Each element has a characteristic spectrum and its presence can be recognized from this spectral 'fingerprint'. The study of atomic spectra yields information on the energy levels of atoms and their °ionization energies and °electron affinities. The interpretation of photochemical processes (such as those in the upper atmosphere), the selection of °lasers, and the interpretation of the properties of complex metal ions depend crucially on a knowledge of atomic energy levels.

Further information

See MQM Chapter 9 for a description of the interactions that modify the appearance of atomic spectra. Introductory texts on atomic spectra include those by Richards and Scott (1976), Hollas (1983, 1987), Corney (1977), and Cowan and Duane (1981). An advanced text on the interpretation of atomic spectra is the one by Condon and Odabaşi (1980) and the use of the data for complex metal ions is described in Griffith (1964). The original compilation of atomic energy levels by Moore has been reissued (1971), revised (1965 onwards), and extended by Martin, Zalubas, and Hagan (1978). For the graphical representation of atomic energy levels, see the entry on °Grotrian diagrams and the compilation by Bashkin and Stoner (1976).

Atomic units

The appearance of many equations in quantum mechanics (as in other fields) may be considerably simplified if the physical quantities that occur are expressed as multiples of fundamental constants and various combinations of them. These constants and multiples are not strictly 'units' (they are physical quantities), but are often treated as though they were true units. Thus, in the system of atomic units:

- mass is expressed as a multiple of the electron mass m_e
- charge is expressed as a multiple of the elementary charge e
- length is expressed as a multiple of the °Bohr radius a_0
- action (energy × time) is expressed as a multiple of $\hbar$.

These definitions, and some of their consequences, are summarized in Table A.1. Note that it is improper to express a physical quantity as so many 'atomic units' (a.u.), such as expressing an energy as -0.5 a.u. The 'unit' given in the table should always be used, and in this example the energy would be reported as $-0.5\, E_h$.

It is common practice in quantum chemistry to express physical quantities as ratios of the actual physical quantity to the atomic unit of that quantity. Thus, distances r may be expressed in terms of the dimensionless number $\rho = r/a_0$ and energies E may be expressed as the dimensionless number $\varepsilon = E/E_h$. For example, the Schrödinger equation for a hydrogenic atom changes from

$$-\frac{\hbar^2}{2m_e}\nabla^2\psi - \frac{Ze^2}{4\pi\varepsilon_0 r}\psi = E\psi$$

to

$$-\tfrac{1}{2}\nabla^2\psi - \frac{Z}{\rho}\psi = \varepsilon\psi$$

Table A.1 Atomic units

Physical quantity	Name of unit	Symbol	Value
Mass	electron rest mass	m_e	9.105×10^{-31} kg
Charge	elementary charge	e	1.602×10^{-19} C
Action	Planck's constant/2π	$\hbar$	1.055×10^{-34} J s
Length	bohr	a_0	$4\pi\varepsilon_0\, \hbar^2/m_e\, e^2$ 5.292×10^{-11} m
Energy	hartree	E_h	$\hbar^2/m_e\, a_0^2 = 2hcR_\infty$ 4.360×10^{-18} J
Time	au of time	$\hbar/E_h$	$m_e\, a_0^2/\hbar$ 2.419×10^{-17} s
Velocity	au of velocity	$a_0\, E_h/\hbar$	αc 2.188×10^6 m s^{-1}
Momentum	au of momentum	$\hbar/a_0$	1.993×10^{-24} kg m s^{-1}
Dipole moment: Electric	au of electric dipole	ea_0	8.478×10^{-30} C m
Magnetic	au of magnetic dipole	$e\hbar/m_e$	$2\mu_B$ 1.855×10^{-23} J T^{-1}

and the ground state energy is found to be $\varepsilon = -\frac{1}{2}$, corresponding to $E = -0.5\, E_h$.

Further information

The most authoritative account of atomic units, their values, and how they are correctly employed, is to be found in the IUPAC publication edited by Mills (1988).

Aufbau principle

See °building-up principle.

Auger effect

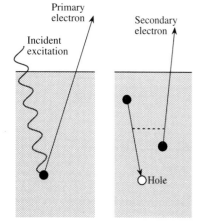

Fig. A.15 In the Auger effect, a primary electron is ejected from a solid by radiation (which may be electromagnetic or a particle beam). An electron falls into the hole left by the ejected primary electron, and its excess energy expels a secondary electron, the Auger electron.

The **Auger effect** (pronounced 'oh-zhay effect', and named in recognition of the French physicist Pierre Auger) is the emission of secondary electrons from an irradiated sample. It occurs when an electron in a high energy level falls into a vacancy left by the emission of an electron, and the energy released in this step ejects a second electron (Fig. A.15). The secondary electrons are called **Auger electrons**, and the measurement of their energies and intensities is called **Auger spectroscopy**. The latter is an important method of studying surfaces of solid samples and identifying adsorbates on them.

Further information

The modern applications of Auger spectroscopy are best explored in books on surface science: see Roberts and McKee (1978). An account of the Auger effect in atomic spectroscopy can be found in Kuhn (1962); for molecular spectroscopy see Rye and Houston (1984).

Band theory

One theoretical description of the structure and properties of metals (which is applicable to all solids) is based on the view that a metal is composed of an array of cations held together by a surrounding sea of electrons. The electrons occupy orbitals that spread throughout the metal and span a nearly continuous band of energies, with gaps between neighbouring bands.

Bands that are incompletely filled (Fig. B.1a) are termed **conduction bands**. This name recognizes that the highest-energy electrons in unfilled bands can adjust rapidly to an applied electric field and give rise to the properties characteristic of metals. These properties include high electrical conductivity, thermal conductivity, reflectivity, malleability, and ductility. A full band is called a **valence band**, and its electrons are not mobile. A solid that has unoccupied conduction bands at $T=0$ (Fig. B.1b) is an electrical insulator. Insulators become °semiconductors at temperatures above $T=0$.

The formation of bands may be described in two ways. In the **tight-binding approximation**, molecular orbitals are formed from the 'tightly bound' atomic orbitals of the individual atoms. In the **nearly free electron approximation** it is supposed that a good first approximation is that the electrons can travel freely through the lattice.

The starting point of the tight-binding approximation is to regard a solid as a huge molecule of almost infinite extent, and then to apply

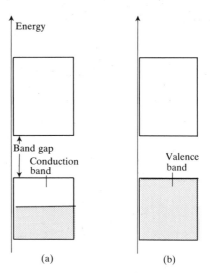

Fig. B.1 (a) According to the band theory of solids, a metal results when a band of orbitals (the conduction band) is incompletely full, for the electrons close to the uppermost occupied level are mobile. (b) If a band of orbitals is full (and hence is a valence band), the solid is an insulator at $T=0$ and a semiconductor at higher temperatures.

°molecular orbital theory. The 'molecule' is composed of atoms in a regular lattice, and the °atomic orbitals on neighbouring atoms overlap each other. We initially suppose that each atom has an *s* orbital available for bond formation, and for simplicity consider a linear chain of *N* atoms.

If *N* = 2, the *s* orbitals of the two atoms overlap and form a bonding and an antibonding pair of molecular orbitals with an energy separation determined largely by the overlap. If a third atom is added to the lattice, the three atomic orbitals overlap to form three molecular orbitals, one of which is bonding, one nonbonding, and the third antibonding. The addition of atoms may be continued, and Fig. B.2 shows that the process gives rise to a set of molecular orbitals that spans a range of energies. When *N* is very large, the *N* molecular orbitals span a virtually continuous band of energies. Since the band has been formed from *s* orbitals it is called an **s band**.

We now suppose that each atom contributes one valence electron (the atoms might be Na atoms in a line). Since each molecular orbital may accommodate two electrons (according to the °Pauli principle), the band will be half full of electrons (Fig. B.1a). Since the electrons near the top of the band can be promoted to nearby empty orbitals very easily, they are mobile and can contribute to the formation of an electric current through the solid. If each atom provides two electrons, the band is full (Fig. B.1b) and the electrons are immobile. (Why net motion cannot occur in this case is described in more detail below.)

The *p* orbitals of the valence shells of the atoms may also overlap to form a **p band** (Fig. B.3). Since the valence *p* orbitals of the atoms

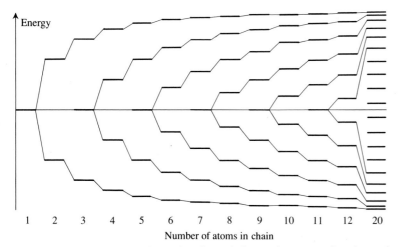

Fig. B.2 The energy levels of *N* atoms in a chain on the assumption that only nearest-neighbour interactions contribute to bonding. (The °Hückel approximation has been used for the calculation.) Note how the density of energy levels in the band increases but that the width remains finite.

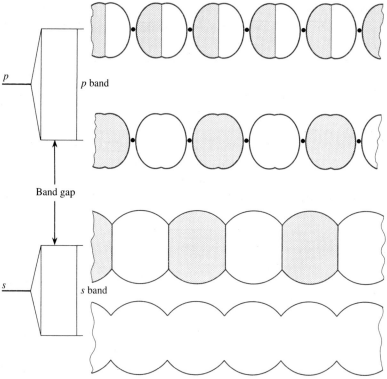

Fig. B.3 If an atom has both *s* and *p* orbitals for bond formation, it may form both *s* and *p* bands. The uppermost level of each band is the combination with the maximum number of internuclear nodes and hence is maximally antibonding. The lowest level of each band is maximally bonding. The size of the band gap depends on the separation of the *s* and *p* orbitals in the atoms and on the strength of the overlap interaction in the solid (which determines the width of the bands).

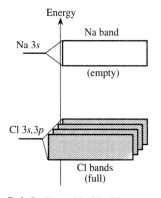

Fig. B.4 Sodium chloride (like any nonmetallic solid) may be discussed in terms of band theory. The Cl atoms form four narrow bands that can accommodate all the valence electrons. The Na atoms form another narrow band, but it is of higher energy and is empty at $T=0$.

lie above the valence *s* orbitals, there may be a **band gap** between the highest energy orbital of the *s* band and the lowest energy orbital of the *p* band. The size of the gap depends on the strength of bonding between the atoms and the separation of their *s* and *p* orbitals. If the separation is large, there may be a large gap; if it is small, the higher levels of the *s* band might coincide with the lower levels of the *p* band.

Band theory is applicable to all types of solid, not only to metals. Thus, in sodium chloride we can distinguish two sets of narrow bands, one formed largely from the 3*s* orbitals of the Na atoms and the other from the 3*s* and 3*p* orbitals of the Cl atoms (Fig. B.4) with the Cl *s* and *p* bands below the Na *s* band in energy. All the valence electrons (7 from each Cl atom and 1 from each Na atom, giving 8 for each NaCl unit) can be accommodated in, and fill, the four Cl bands. Since the Cl bands are separated from the Na band by a substantial gap, sodium chloride is an insulator at $T=0$. Moreover, four full Cl bands are equivalent to a collection of individual Cl$^-$ ions (see °localized and delocalized orbitals), and an empty Na band is equivalent to an array of Na$^+$ ions. Hence, we recover the conventional picture of the compound as an ionic solid.

The starting point of the nearly-free electron approximation is the wavefunction of a freely moving electron. Since the solid is of almost infinite length, the energies available to free electrons take an almost continuous range of values (see °particle in a box). A wavefunction of linear momentum $k\hbar$ has an energy proportional to k^2 (Fig. B.5). Adding electrons to this near-continuum of energy levels fills them up to some energy E_F, the **Fermi energy**, and (at $T=0$) leaves vacant all higher energy levels. Note that at this point we might conclude that all solids would be metals, for the 'band' has no upper limit and so can never be filled.

The periodic lattice imposes a band structure. To see how it arises, we note that a free electron with momentum $k\hbar$ has a definite wavelength $\lambda = 2\pi/k$ (which follows by combining the °de Broglie relation $\lambda = h/p$ with $p = k\hbar$). Waves for which $\lambda \gg a$ (the lattice spacing) propagate freely through the lattice. However, when λ is comparable to a, the waves are diffracted. When $\lambda = 2a$, a pattern of standing waves is set up because a wave that begins moving to the right is reflected by the lattice and moves to the left, where it is reflected to the right, and so on. These standing waves, with wavelengths close to $2a$, occur either with amplitude maxima at the positions of the lattice points (the sites of the positive ions forming the background positive charge of the solid) or with their maxima between the lattice points. Since the two kinds of standing wave have different potential energies, there is an energy gap at $k = \pm\pi/a$ (Fig. B.6). Another gap occurs at $k = \pm 2\pi/a$ and so on. Thus we see that the periodic lattice splits the free-electron near continuum of energy levels into a series of bands separated from each other by gaps (Fig. B.6). The energy bands are widely called **Brillouin zones** (but, strictly, that name relates to an aspect of the translational symmetry of the solid). The available electrons occupy the bands, with the same consequences as in the tight-binding approach.

The relation of the nearly free electron bands to the tight-binding orbital bands can be appreciated by comparing the form of the orbitals at the edges of the s and p bands with the waves at the edges of the first and second Brillouin zones (Fig. B.7). We see that the nodal structure is the same. Only the details of the electron distributions differ, particularly in the regions close to the nuclei. Well below the Brillouin zone boundary (for levels for which $|k| \ll \pi/a$ in the first zone) the states of the electron may be regarded as running waves (as distinct from the standing waves forced on the system at the zone edge). In one dimension the waves run with momentum $k\hbar$ to the right and $-k\hbar$

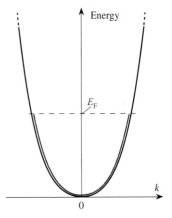

Fig. B.5 The parabolic dependence of the energy of free electrons on the linear momentum (specifically, the wavevector k). At $T=0$, the electrons fill the near-continuum of states up to the Fermi energy.

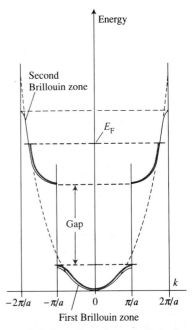

Fig. B.6 The effect of a periodic lattice on the parabolic energy dependence of Fig. B.5 is to split it into a series of zones separated by gaps. The gaps denote energies that an electron cannot have in the periodic lattice.

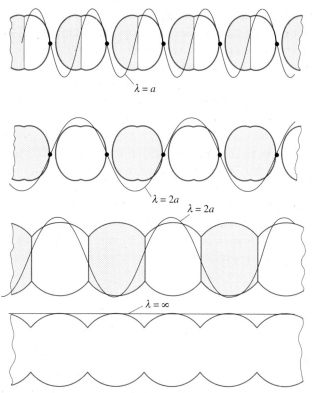

Fig. B.7 The nodal structure of the molecular orbitals of the tight-binding approximation reflects that of the free-electron waves. The states shown are for the bottom and top of an *s* band (lower pair) and of a *p* band (upper pair).

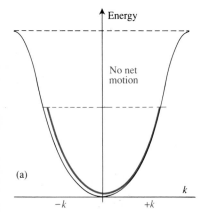

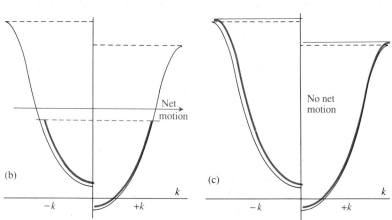

Fig. B.8 (a) In the absence of an electric field the states with positive and negative momenta are degenerate, and their populations are identical. (b) In the presence of an electric field, the states corresponding to motion in one direction (to the right, here) have a lower energy than those travelling to the left, and if the band is not full, there are more electrons travelling to the right: there is an electric current. (c) If the band is full, the populations cannot adjust, and the numbers of electrons travelling to the left and the right remain equal: there is no net current in this insulator.

to the left; since the energy of the waves is proportional to k^2, these two running waves have the same energy. Therefore, in a solid there are equal numbers of electrons running to the left and to the right (Fig. B.8a) and in the absence of external stimuli no current flows. However, when a potential difference is applied, the energy of the electrons running to the right may differ from that of the electrons running to the left (Fig. B.8b). The electrons will redistribute themselves to attain the lowest energy; as a result, more electrons occupy states with momentum to the right, and there is a steady flow of current.

If the band is full, the electrons cannot reorganize themselves to give a net direction of travel (Fig. B.8c) and no current flows. Such a material is an insulator (at $T = 0$). A very strong field might so affect the energies that an empty zone is brought down to the filled one. Then **dielectric breakdown** may occur and a current may flow, but often at the cost of disintegrating the material.

Further information

A simple account of the structures of solids and the interpretation of their properties in terms of band theory will be found in Shriver, Atkins, and Langford (1990). See also Cox (1987), Ladd (1979), and Stiddard (1975) for introductory accounts; the first two have a specific chemical flavour, as has Hoffmann (1988). Duffy (1990) applies band theory specifically to inorganic solids. For more advanced treatments, written in the context of the physical properties of metals, see Anderson (1984), Hall (1974), Kittel (1987), and Moruzzi, Janak, and Williams (1978). A useful source guide is Parker (1987). The population of bands at temperatures $T > 0$ is described by the °Fermi–Dirac distribution. For band theory from the viewpoint of symmetry see Altmann (1991b).

Basis functions

See °linear combination of atomic orbitals.

Benzene

Benzene (C_6H_6, **1**) is the archetype of °aromatic molecules. In the °valence bond theory of molecular structure, the enhanced stability of the π electrons in benzene is ascribed to °resonance. In the °molecular orbital theory, the explanation is expressed in terms of delocalization. These descriptions are not mutually exclusive, but merge into one when they are refined.

The first point to note in the °molecular orbital description of such aromatic molecules is that benzene, an even °alternant hydrocarbon, may be described in terms of the symmetrical array of orbitals and

1 Benzene

energies shown in Fig. B.9. Note that the energy of the lowest orbital (-2β) is substantially lower than that expected for a single localized CC bond ($-\beta$), largely because each atomic orbital overlaps with both its neighbours and the orbital exhibits bonding character between all neighbouring pairs of atoms. That is, the orbital has a very low energy because **delocalization** results in enhanced electron density in the favourable internuclear regions throughout the ring.

The six electrons that are to be accommodated in the delocalized π orbitals can all enter the three orbitals with net bonding character, including the very deep lowest energy orbital, and hence all six electrons contribute to the stability of the molecule.

So far we have ignored the σ electrons, yet they have an important role to play in both the valence bond and the molecular orbital descriptions. Thus, six sp^2-hybridized C atoms (with lobes at 120°) form a six-membered ring (with angle 120°) without strain. Each of the six C atoms has its full share of valence electrons (that is, zero °formal charge), and so not only is there no energy arising from an imbalance of charge, but there are also no centres of charge excess or deficiency to provide a reactive site (see °alternant hydrocarbon). The hexagonal ring of carbon atoms is a very well poised system.

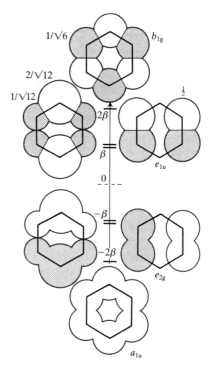

Fig. B.9 The molecular orbitals of benzene with their group theoretical designations and coefficients. White represents positive signs of the combination coefficients, grey represents negative signs.

The conventional °valence bond description of benzene starts out from the classical **Kekulé structure (2)**, a hexagon of alternating single and double bonds. This structure is described by a °valence-bond wavefunction, and its energy can be calculated. However, there is an equally valid description of the molecule in terms of a second Kekulé structure (**3**) and its corresponding wavefunction, in which the positions of the single and double bonds are interchanged. Whenever there is more than one valid description of a molecule, the principles of quantum mechanics tell us to form a °superposition of the structures and to regard the true structure as a blend of these contributions. A blend of wavefunctions has a lower energy than either structure alone (see °resonance). Hence, the π electron energy of benzene is lower than that calculated from a single Kekulé structure.

In conventional valence bond theory, Kekulé structures contribute about 80 per cent of the true total structure, the remaining 20 per cent being due largely to **Dewar structures (4, 5,** and **6)**. The more extensive blending that occurs as more contributing structures are added lowers the energy still further.

According to the modern version of °valence bond theory, it is *not* appropriate to regard benzene as a delocalized system. An analysis of its structure in terms of spin-coupled °valence bond theory leads to a wavefunction that corresponds to *localized* bonds between neighbouring C atoms. The stability of the structure, and perhaps its °aromatic properties in general, stem not from spatial delocalization but from the symmetrical coupling of the electron spins round the ring (analogous to the coupling of spins in magnetic solids). The calculations that have been carried out using this approach appear to suggest that aromatic character is a more deeply quantum mechanical phenomenon than molecular orbital theory has led us to believe. The jury is still out on this centrally important question.

Further information

See MQM Chapter 10 for a simple account of the molecular orbital structure of benzene. For a description of the valence bond approach see Gerratt (1986) and Cooper *et al.* (1986, 1989). More conventional approaches will be found in Pilar (1990), Dewar (1969), and McGlynn *et al.* (1972). For a review of the electronic spectroscopy of benzene see Kitao and Nakatsuji (1988). Other relevant entries are °alternant hydrocarbon, °aromaticity, °Hückel method, and °resonance.

2 Kekulé benzene (1)

3 Kekulé benzene (2)

4 Dewar benzene (1)

5 Dewar benzene (2)

6 Dewar benzene (3)

Birefringence

A substance is **birefringent** if its refractive index depends on the polarization of the incident light. Birefringence may be a natural property of

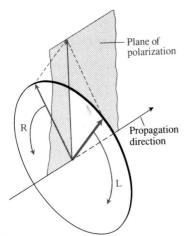

Fig. B.10 A plane polarized electromagnetic wave can be regarded as a superposition of two counter-rotating components of equal amplitude. The relative phases of the rotating components determine the plane of polarization of the resultant. Here the resultant lies in a vertical plane at all times.

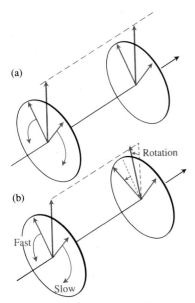

Fig. B.11 (a) If the two circularly polarized components travel at the same speed through a medium, their resultant remains in the same plane. (b) If the two components travel at different rates, the plane of their resultant rotates in the direction of the faster component.

isolated molecules and of a particular aggregate of molecules, or it may be induced electrically, magnetically, and mechanically.

An important example of natural birefringence is **optical activity**, in which the plane of polarization of light is rotated as it passes through a °chiral medium. Optical activity is an example of **circular birefringence** because the effect depends on the different refractive indices of left and right circularly polarized light. Thus, a plane polarized ray of light may be regarded as a superposition of left and right circularly polarized components (Fig. B.10), with the plane of polarization defined by the resultant of the two counter-rotating components. If the medium through which the beam is passing is circularly birefringent, the two components travel at different speeds, and when they emerge the plane of their resultant has rotated (Fig. B.11).

The **Faraday effect** is the production of circular birefringence by a magnetic field parallel to the direction of propagation of the light. The induced angle of optical rotation of a plane polarized beam of light is proportional to the strength of the applied field, the constant of proportionality being the **Verdet constant**. All substances exhibit a Faraday rotation in a magnetic field.

In the **Cotton–Mouton effect**, a magnetic field that is applied perpendicular to the propagation direction of the light induces a **linear birefringence** in the sample. In linear birefringence, one plane-polarized component of the incident light travels faster than its perpendicular partner. A linearly birefringent sample produces an ellipticity in an initially plane polarized beam (in the sense that the electric vector of the field sweeps out an ellipse as it propagates through space). The analogous phenomenon induced by a transverse electric field is called the **Kerr effect**.

Birefringence may result from fluid flow if the molecules are sufficiently anisotropic. Then the alignment of long thin molecules introduces anisotropy into the optical properties of the medium and the refractive indices depend on the orientation of the polarization of the beam. This mechanically induced anisotropy is called **streaming birefringence**.

Further information

See MQM Chapter 13 for a discussion of the quantum mechanical explanation of optical activity. Three thorough reviews of birefringence are those of Caldwell and Eyring (1971), Charney (1979), and Barron (1983). For an account that concentrates on the birefringence of solids, particularly in terms of their symmetries, see Wooster (1973). A standard work on classical optics is that by Born and Wolf (1989).

Black-body radiation

A **black body** is one that absorbs and emits all radiation without favouring any particular frequency. **Black-body radiation** is electromagnetic radiation in thermal equilibrium with a black body at a specific temperature.

A practical realization of a black body is a pinhole in an otherwise sealed container. The hole behaves as a black body because all radiation incident on it from outside passes through and, once in, cannot immediately escape through the vanishingly small hole. Inside the container the radiation experiences an indefinitely large number of absorptions and emissions which bring the electromagnetic field inside the container into thermal equilibrium with the walls. Thus, inside the container the electromagnetic field is at the same temperature as the walls. The presence of the hole enables a tiny proportion of the radiation inside the container to escape and be examined.

Examination of the escaping radiation shows that different frequencies of radiation are present with different intensities (Fig. B.12). Although the hole looks black at low temperatures, a sensitive detector shows that a small amount of low frequency (long wavelength) radiation is present. At higher temperatures the energy emitted from the hole (and therefore present in the electromagnetic field inside the container) is much greater, and is most intense in the infrared. At progressively higher temperatures the pinhole glows dull red, white, and then blue, and the total energy present (and emitted) increases strongly (as T^4). As the temperature is raised further, the radiation becomes most intense in the ultraviolet, although the temperatures at

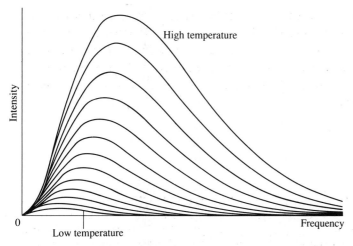

Fig. B.12 The observed intensity of black-body radiation increases with temperature and the maximum shifts towards higher frequencies. The area under the curve grows as T^4.

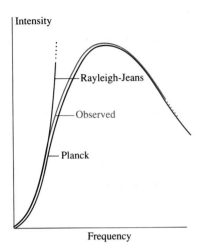

Fig. B.13 The Rayleigh–Jeans law leads to the ultraviolet catastrophe, an infinite accumulation of intensity at high frequencies. The Planck distribution fits the observed intensity at all frequencies.

which this occurs are too great to be conveniently accessible. The increase in the total energy present at thermal equilibrium is summarized quantitatively by the °Stefan–Boltzmann law, and the shift in the radiation peak as the temperature is raised is summarized by °Wien's displacement law.

The historical notoriety of black-body radiation lies in its role in the overthrow of classical mechanics and the formulation of quantum theory. The attempt to account for the distribution of radiation in a black body in terms of classical mechanics culminated in the **Rayleigh–Jeans law** (1900). Lord Rayleigh (1842–1919) began his calculation by counting the number of oscillators of each frequency that could be found per unit volume of a cavity. Sir James Jeans (1877–1946) has his name attached to the law by virtue of his remark, made in 1905, that 'It seems to me that Lord Rayleigh has introduced an unnecessary factor 8 by counting negative as well as positive values of his integers', and hence reducing Rayleigh's expression by that factor. Rayleigh, meanwhile, had used the °equipartition theorem to ascribe an average energy kT to each mode of oscillation. In this way he arrived at the expression shown in Box B.1 for the energy per unit volume of oscillators in a specified range of frequencies (Fig. B.13).

Unfortunately (for Rayleigh), his result agrees with none of the observations mentioned above. Since his expression increases without limit as the frequency increases, it predicts an **ultraviolet catastrophe**, an enormous energy density at high frequencies: according to classical mechanics, even a lighted match should emit γ-rays intensely. It also predicts an infinite energy in any enclosure (the integral of the energy over all frequencies diverges), in conflict with the Stefan–Boltzmann law as well as common sense. Moreover, there is no maximum in the distribution, and so the formula is also unable to account for the observation of one.

The German physicist Max Planck (1858–1947) tackled the problem of black-body radiation from the viewpoint of thermodynamics. He was forced to the conclusion that he needed a single distribution law that yielded one radiation law at low frequencies and another at high frequencies. He therefore looked for a *single* formula — an interpolation formula — which gave the two distributions at low and high frequencies respectively. The formula he derived is shown in Box B.1 and plotted in Fig. B.13.

Planck came to the conclusion that his formula could be deduced from the assumption that a radiation mode of a specified frequency v could be excited only in steps of energy of magnitude hv, where h is a universal constant now known as **Planck's constant**. It is found experimentally that $h = 6.626 \times 10^{-34}$ J Hz^{-1}. The discrete amounts of energy that may be transferred to an object are called **quanta**. It follows that

Box B.1 The Planck black-body radiation law

The **energy density**, $d\mathcal{U}$, the energy per unit volume, in the range v to $v + dv$ or the range λ to $\lambda + d\lambda$ is

$$d\mathcal{U} = \rho(v)\,dv \qquad d\mathcal{U} = \rho(\lambda)\,d\lambda$$

The **Planck distribution** at the temperature T is

$$\rho(v) = \frac{8\pi h v^3}{c^3}\,\frac{1}{e^{-hv/kT} - 1}$$

$$\rho(\lambda) = \frac{8\pi hc}{\lambda^5}\,\frac{1}{e^{-hc/\lambda kT} - 1}$$

The **Rayleigh–Jeans law** is the low frequency ($v \ll kT/h$), long wavelength ($\lambda \gg hc/kT$) limit of the Planck distribution:

$$\rho(v) = \frac{8\pi v^2 kT}{c^3} \qquad \rho(\lambda) = \frac{8\pi kT}{\lambda^4}$$

The **Stefan–Boltzmann law** for the total energy density, $\mathcal{U}$, or the **excitance**, M, the power emitted per unit area:

$$\mathcal{U} = \int \rho\,dv = aT^4 \qquad M = \sigma T^4$$

where σ is the **Stefan–Boltzmann** constant:

$$\sigma = \frac{\pi^2 k^4}{60 c^2 h^3} = 5.670 \times 10^{-8}\ \text{W m}^{-2}\,\text{K}^{-4}$$

and $a = 4\sigma/c$.

The **Wien displacement law** for the wavelength of maximum energy density:

$$T\lambda_{max} = \tfrac{1}{5} c_2$$

where c_2 is the **second radiation constant**:

$$c_2 = hc/k = 1.439\ \text{cm K}$$

very high frequency modes are not excited at normal temperatures because the thermal motion of the walls of the container is insufficient to supply them with adequate energy (see °quantum). This elimination of high-frequency oscillations quenches the rise of the Rayleigh–Jeans distribution at high frequencies (short wavelengths) and thus eliminates the ultraviolet catastrophe. It also introduces a maximum into the energy distribution versus wavelength curve, which is in accord with the Wien displacement law at high frequencies.

Planck's introduction of quantization and of the constant h was originally regarded as no more than a calculational device. However, such was the success of the hypothesis that quantization became accepted as physically real, particularly as other evidence of a different kind soon accumulated (see °quantization). Indeed, the entire current structure of °quantum mechanics is based on it.

Further information

See MQM Chapter 1 for the derivation of the classical and Planck distribution laws. Two excellent historical accounts of the grasping towards understanding that took place at the end of the 19th century are those by Jammer (1966) and Kuhn (1978). Numerical values and other properties of the Planck formula will be found in Gray (1972) and Abramowitz and Stegun (1965).

Bohr frequency condition

In the course of his analysis of the structure of the hydrogen atom, the Danish physicist Neils Bohr (1885–1962) proposed that the frequency

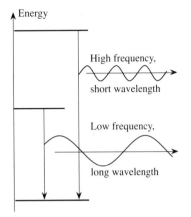

Fig. B.14 According to the Bohr frequency condition, high frequency, short wavelength radiation is emitted when a system undergoes a transition with a large change in energy. Low frequency, long wavelength radiation is obtained from transitions involving only a small change in energy.

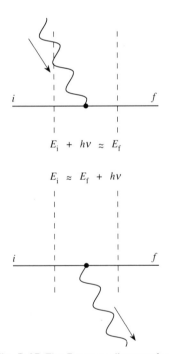

Fig. B.15 The Feynman diagrams for absorption and emission. In each case the overall change in energy is zero when the photon energy matches the transition energy. This is the condition for degeneracy and overall strong coupling.

v of light emitted or absorbed by an atom or molecule when it undergoes a transition between two states differing in energy by ΔE is given by

Bohr frequency condition: $hv = \Delta E$

where h is °Planck's constant. Thus, a transition between two states that are widely separated in energy results in the emission or absorption of light of high frequency (short wavelength).

The Bohr frequency condition is a direct consequence of the conservation of energy and the relation between the energy of a °photon and the frequency of the radiation. A photon of light of frequency v has an energy hv; hence, when it is generated by an atom, that atom must lose an equal amount of energy (Fig. B.14). Likewise, when an atom absorbs a photon, the atom's energy must increase by hv.

A deeper interpretation of the Bohr condition is that it is a consequence of the 'looseness' of degenerate systems. In the entry on °perturbation theory it is explained that a perturbation mixes °degenerate states very effectively. That is, two degenerate states are very 'sloppy', and even a feeble perturbation can push a system entirely from one degenerate state into another. If we draw the °Feynman diagram for the absorption of a photon by a molecule (Fig. B.15), we see that the *overall* 'before' and 'after' states have exactly the same energy when the frequency of the radiation satisfies the Bohr condition. When this is so, the weak interaction between the molecule and the electromagnetic field can result in a high probability that molecular excitation (or de-excitation) will occur, because it is very easy for the overall system to make a transition that involves no net energy change.

Further information

For the original paper on the Bohr frequency condition, see Bohr (1913).

Bohr magneton

The **Bohr magneton** μ_B is the fundamental unit of magnetic dipole moment of an electron:

$$\mu_B = \frac{e\hbar}{2m_e} = 9.274 \times 10^{-24} \text{ J T}^{-1}$$

If the z-component of the angular momentum is $m_l \hbar$, the z-component of the orbital magnetic moment is

$$\mu_z = -m_l \mu_B$$

One source of the magnetic moment of an electron is its orbital angular momentum. The orbital motion of the electron (of charge $-e$) travelling with angular momentum $m_l\hbar$ around a central nucleus is equivalent to an electric current. Hence, the orbiting electron gives rise to a magnetic field (any moving charge generates a magnetic field), and the source of this magnetic field can be represented as a magnetic dipole at the centre of the orbit, with a magnitude proportional to the angular momentum.

The °spin of the electron is also a source of magnetic field (and the only one when all orbital momentum is °quenched). The magnetic moment arising from electron spin is also proportional to the spin angular momentum $m_s\hbar$, but according to the °Dirac equation for the electron, the relation between spin angular momentum and spin magnetic moment is

$$\mu = -gm_s\mu_B$$

The constant g is the °g-factor of the electron, and is almost exactly 2. Thus, an α electron (with $m_s = +\frac{1}{2}$) has a magnetic moment of magnitude μ_B, the same as the orbital magnetic moment of an electron in a p orbital (with $m_l = +1$). ·

Further information

See °magnetogyric ratio for a continuation of this discussion and further references. The magnetic dipole moments of nuclei with spin are discussed in terms of the °nuclear magneton.

Bohr model of the atom

Before the discovery of quantum mechanics, Bohr (see preceding entries) applied the concept of quantization to the problem of the structure of the °hydrogen atom. He accepted the Rutherford model of the nuclear atom and asserted that:

- An electron remains in a **stationary state**, an orbit around the central nucleus, until it undergoes a transition.
- A transition between stationary states differing in energy by ΔE is accompanied by the emission or absorption of radiation with a frequency v determined by the condition $hv = \Delta E$.

The latter assertion is the °Bohr frequency condition.

- The permitted stationary states may be found by balancing the Coulombic attractive force between the electron and nucleus against the centrifugal effect of the angular momentum of the electron in its orbit.

$$E = -hcR/25 \qquad n = 5, r = 25a_0$$

$$-hcR/16 \qquad n = 4, r = 16a_0$$

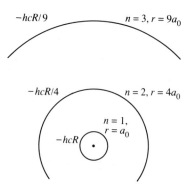

$$-hcR/9 \qquad n = 3, r = 9a_0$$

$$-hcR/4 \qquad n = 2, r = 4a_0$$

$$n = 1, \; r = a_0$$

$$-hcR$$

Fig. B.16 The radii and energies of the first five Bohr orbits of the hydrogen atom. The radius increases as n^2. The Bohr radius, a_0, is equal to 53 pm. Note how the atom rapidly becomes very large as the electron is excited into states of higher energy.

Finally, Bohr brought in Planck's constant (specifically, as $\hbar = h/2\pi$), and introduced a quantization postulate:

● The only orbital angular momenta permitted are integral multiples of $\hbar$.

When Bohr carried through the calculation on the basis of this model, he deduced that the energies of the electron in its permitted orbits (Fig. B.16) are

$$E = -\frac{hcR}{n^2} \qquad R = \frac{m_e e^4}{8h^2 \varepsilon_0^2} \qquad n = 1, 2, \ldots$$

This expression is in excellent agreement with experiment (and in exact agreement with the solutions of the °Schrödinger equation, which came later). In particular, it reproduces the positions of all the series in the spectrum of atomic hydrogen.

The refinement of Bohr's initially promising model proceeded in three steps. The first took into account the fact that the orbital motion occurred about the centre of mass of the system rather than about the nucleus itself. This refinement merely involved replacing the mass of the electron in the Bohr formula by the °reduced mass μ of the proton and the electron.

The second improvement was introduced by Arnold Sommerfeld (1868–1951) in 1915. In the generalization of the Bohr theory known as the **Bohr–Sommerfeld atom** the orbits are allowed to be elliptical (Fig. B.17). The eccentricity (the degree of ellipticity) of the elliptical orbits is determined by a quantum number k, the **azimuthal quantum number**. The energy of an orbit is independent of its eccentricity.

The third improvement was also made by Sommerfeld (again in 1915), who incorporated relativity into the model. Sommerfeld found that the effect of relativity on the rapidly moving electron was to cause a mismatch of the ends of the elliptical orbits, so the electron described an open orbit around the nucleus — a continuously evolving orbit that resembled a rosette (Fig. B.18). The inclusion of relativity caused the energies to depend weakly on k, and remarkable agreement with experiment was obtained. (The numbers obtained are identical to those obtained in the later quantum-relativistic °Dirac theory of the hydrogen atom.)

However, although the numerical agreement is almost exact, the Bohr model of the hydrogen atom is fundamentally wrong. Modern quantum theory shows that

● An electron does not travel in a definite orbit.

Later developments of quantum theory show that it is not possible to regard an electron as travelling along a precise trajectory (see °uncer-

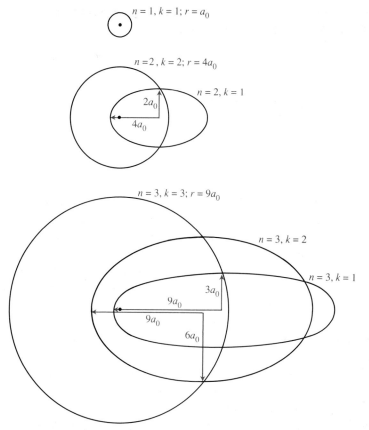

$n = 1, k = 1; r = a_0$

$n = 2, k = 2; r = 4a_0$

$n = 2, k = 1$

$2a_0$

$4a_0$

$n = 3, k = 3; r = 9a_0$

$n = 3, k = 2$

$n = 3, k = 1$

$3a_0$

$9a_0$

$9a_0$

$6a_0$

Fig. B.17 The Bohr–Sommerfeld model of the hydrogen atom led to the introduction of a second quantum number, k, that determines the ellipticity of the orbits and which is limited to integral values from 1 to n. Note that the major diameter of each orbit is independent of k, and that the electron approaches more closely to the nucleus the smaller the value of k.

Fig. B.18 A fragment of the open rosette orbital that results when relativistic effects are taken into account in the Bohr–Sommerfeld model. Orbits with different values of k are now no longer degenerate.

tainty principle and °atomic orbital). That is, the Bohr and Bohr–Sommerfeld orbits are macroscopic concepts that can have at best only approximate validity on the scale of atoms.

- In the ground state of the hydrogen atom, the electron has zero angular momentum around the nucleus.

Hence, it cannot be the centrifugal force that balances the nuclear attraction.

- The electron may be found at the nucleus itself, which is forbidden in the Bohr model.

The Bohr model is also incomplete in the sense that in its postulates it virtually *asserts* the structure of the atom, and no justification is given for the stationarity of states and the quantization of angular momentum.

We have to conclude that the agreement of the calculated energy levels in the Bohr model with experiment is coincidental. This coincidence stems from the special properties of the Coulomb potential, which also turn up in the quantum mechanical theory of the atom (see °hydrogen atom and °degeneracy). Nevertheless, despite the coincidences, Bohr's achievement was considerable, for he applied to a problem in mechanics a theory that had been constructed to describe the behaviour of radiation. His work was an important precursor of the view that optical and mechanical phenomena are essentially identical.

Further information

Good accounts of the Bohr theory of hydrogenic atoms (those with one electron around a nucleus of arbitrary atomic number) will be found in Herzberg (1944) and Pauling and Wilson (1935). For an examination of Bohr's model in a historical perspective, see Jammer (1966); this reference also explains the contributions made by Sommerfeld in a reasonably simple manner.

Bond

Elementary chemistry distinguishes two kinds of bond between atoms:

• The **ionic bond**, in which electrons are transferred between atoms and the bond arises from the Coulombic interaction between the resulting ions.

• The **covalent bond**, in which electron pairs are shared more or less equally between neighbouring atoms.

A covalent bond is denoted by a line between two atoms, as in H—H. In earlier days (and sometimes still), chemists identified a covalent bond as a bond in which each atom contributed *one* electron to the shared pair; they distinguished a covalent bond from a **dative bond**, or **covalent–coordinate bond**, in which one partner donates *both* electrons of the shared pair. The term 'dative bond' and its symbol B→A are now obsolete, and the term 'covalent bond' signifies that a bond has been formed by electron sharing irrespective of the origin of the members of the pair.

A covalent bond may be classified as polar or nonpolar:

• A **polar bond** is a covalent bond in which the electron pair is shared unequally between the two atoms, with the result that the atoms carry partial electric charges.

• A **nonpolar bond** is a covalent bond in which the electron pair is shared equally between the two atoms.

The bond in HCl is polar, with a partial negative charge on Cl and a partial positive charge on H. The bond in N_2 is nonpolar. An ionic bond may be regarded as the limit of a polar bond in which one atom has acquired virtually full possession of the 'shared' electron pair, so becoming an anion, and the other atom has lost possession of the pair, thereby becoming a cation.

A further classification of covalent bonds is as single, double, or triple:

- A **single bond** consists of one shared pair of electrons.
- A **double bond** consists of two shared pairs of electrons.
- A **triple bond** consists of three shared pairs of electrons.

Quadruple bonds (four shared pairs of electrons) are rarely encountered but occur in some metal clusters (see °δ bonds).

The **valence** of an atom is the number of bonds that it may form. Thus, oxygen is typically divalent, since it normally forms two bonds, as in carbon dioxide, $O{=}C{=}O$; carbon is typically tetravalent, since it normally forms four bonds (as in carbon dioxide). The early views on bond formation, particularly the conceptual framework established by G.N. Lewis, correlated an element's valence with the completion of an atom's 'octet' of electrons (see °Lewis structures). According to Lewis, each atom acquired or lost electrons until it had achieved the octet configuration characteristic of a noble gas atom near it in the Periodic Table. The loss or gain of electrons could be by total transfer (with the formation of ionic bonds) or by sharing (with the formation of covalent bonds). For instance, Na, with electron configuration $[Ne]3s^1$, could lose an electron and become the Na^+ ion with a neon-like configuration; Cl, with configuration $[Ne]3s^2 3p^5$, could gain an electron and become a Cl^- ion with an argon-like closed shell configuration. Likewise, two H atoms could share a pair of electrons, and hence both would effectively acquire a helium-like configuration. The view that atoms form bonds until they have attained a noble gas configuration is still a useful rule of thumb, despite the many exceptions that are known (see °hypervalence and °electron-deficient compounds).

Modern theories of chemical bonding have largely dispensed with the distinction between the various types of bond. Instead, they see bonding as a consequence of the lowering of energy that occurs when electrons occupy °molecular orbitals that spread over all the atoms in the molecule. According to molecular orbital theory, the bonding influence of an electron pair may be distributed over many atoms. Molecular orbital theory also shows that an electron pair is not essential to bond formation, but merely represents the maximum number of electrons that can occupy a given orbital and hence contribute to bonding. The bond polarity is a consequence of the distribution of electron probability that molecular orbitals represent, and the extreme case of an ionic bond corresponds to a molecular orbital that is composed almost entirely of atomic orbitals on one atom. Counting the number of bonds between neighbouring atoms is an ill-defined procedure in the molecular orbital theory of polyatomic molecules since the bonding influence of an electron is distributed, but in a diatomic molecule (and, to some degree, in localized fragments of polyatomic molecules, such as the C=C bonds in alkenes) the concept is valid, and is expressed in terms of the °bond order.

Further information

See the entries on the °ionic model, °covalent bonds, °Lewis structures, °valence bond theory, and °molecular orbitals for more detailed information on some of the topics mentioned here. Introductions to the modern description of the chemical bond in terms of the occupation of molecular orbitals will be found in DeKock and Gray (1980), Coulson (1979, 1982), Gimarc (1979), and Webster (1990). Pauling (1960), although expressed in somewhat outmoded terms, is still an excellent illustration of the rationalization of chemical properties in terms of the concepts associated with bond formation. Up-to-date surveys have been published by Baird (1986), DeKock (1987), and Gallup (1988).

Bond order

The **bond order** b is a measure of the net number of bonds between a specific pair of atoms in a molecule. It is defined as

$$b = \tfrac{1}{2}(n - n^*)$$

where n is the number of electrons in bonding orbitals and n^* is the number of electrons in antibonding orbitals. The definition is based on the identification of a pair of electrons in a bonding orbital with a single bond in the °Lewis account of bonding, and the cancellation of the bonding effect of a bonding electron pair by two electrons in an °antibonding orbital. We identify $b = 1$ with a 'single bond', $b = 2$ with a 'double bond', and so on. The definition can lead to bonds having a fractional order, as in NO, for which the bond order is $2\tfrac{1}{2}$. The calculation of the bond order is the molecular-orbital analogue of counting shared electron pairs in °Lewis structures.

The definition of bond order is most directly relevant to diatomic molecules, for their molecular orbitals are bonding or antibonding between the only two atoms in the molecule. To determine the bond order, we write down the electron configuration of the molecule using the °building-up principle and a molecular orbital energy level diagram like that in Fig. B.19a. Thus, the configuration and electron count of O_2 is

$$1\sigma_g^2 2\sigma_u^2 3\sigma_g^2 4\sigma_u^2 5\sigma_g^2 1\pi_u^4 2\pi_g^2$$

$$n = 1 \qquad +2 \qquad +2+4 \qquad = 10$$

$$n^* = \qquad 2 \qquad +2 \qquad +2 = \ 6$$

and $b=2$, which is consistent with the chemical structure O=O for the molecule. When applied to polyatomic molecules, the simple definition of bond order given above gives the average degree of the bonding influence of all the electrons. Thus, the valence electron configuration of SF_6 is illustrated schematically in Fig. B.19b, where the ten molecular orbitals shown in the diagram have been constructed from the four s and p valence orbitals of the S atom and one p orbital on each of the six F atoms (see °hypervalence for more discussion). The $6+6=12$ valence electrons occupy the four bonding orbitals and the two largely nonbonding orbitals, hence the *total* bond order in the molecule is $b=\frac{1}{2}(10-2)=4$. Since this bonding influence is distributed over six SF links, the average bond order of each link is $\frac{2}{3}$.

A more detailed analysis can be established by expressing the bond order between pairs of atoms in a molecule in terms of the coefficients of the atomic orbitals in the occupied molecular orbital. Thus, if the coefficients of the atomic orbitals on atoms A and B are c_A and c_B, then the contribution of that orbital to the bonding will be proportional to the product $c_A c_B$. The AB bond order is then defined as the sum of $c_A c_B$ over all the occupied orbitals. When c_A and c_B are simultaneously large for a particular orbital, a large contribution to the total order results (but it cannot exceed unity). When c_A and c_B have opposite signs, corresponding to °antibonding character between A and B, the product is negative; it subtracts from the overall result and reduces the bond order.

As an example, each carbon–carbon bond in °benzene (Fig. B.9) has a contribution of $\frac{2}{3}$ from the π orbitals, and 1 from the σ orbitals; hence each bond is of order $1\frac{2}{3}$. The carbon–carbon bond in ethane has $b=1$, in ethene $b=2$, and in ethyne $b=3$, in accord with the normal chemical representation of their structures.

One application of the quantitative definition of bond order is to the estimation of bond lengths, especially of related species, such as carbon–carbon bonds in hydrocarbons. An empirical relation has been recognized between the length R and order b of carbon–carbon bonds:

$$R/\text{pm} = 166.5 - 13.98\,(1+b)$$

As the bond order increases, the bond length decreases.

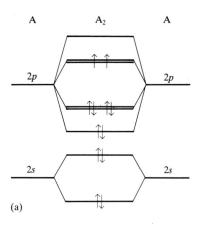

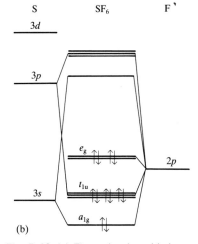

Fig. B.19 (a) The molecular orbital energy level diagram for Period 2 homonuclear diatomics, and the electron configuration of the ground state of O_2. The bond order is 2. (b) The molecular orbital energy level diagram for SF_6. Note that predominantly bonding orbitals are occupied. This is reflected in the value of the bond order, which is $\frac{2}{3}$ for each SF link.

Further information

See DeKock and Gray (1980) and Coulson (1979, 1982) for a further introductory discussion. Some of the correlations between bond order and bond length are illustrated in Shriver, Atkins, and Langford (1990). A helpful and lengthy discussion will be found in Streitwieser (1961).

Born–Oppenheimer approximation

The **Born–Oppenheimer approximation** is the separation of the motion of the electrons in a molecule from the motion of the nuclei. The consequence of the approximation is that the electron wavefunction can be calculated for any specific static nuclear framework.

The approximation is based on the great differences between the masses of electrons and nuclei. The consequence of this difference is that the nuclei move so slowly that the electrons can adjust their distribution instantaneously to take into account the changing potential. The practical effect of the approximation is that arbitrary, fixed locations of the nuclei can be selected and the electronic wavefunctions calculated for each one of them. The nuclei can then be moved to new positions, and the electronic calculation repeated. Thus it is possible in principle to calculate the energy for all possible arrangements of the nuclei, and then to find the one corresponding to the lowest energy, which is identified as the equilibrium shape of the molecule.

The Born–Oppenheimer approximation makes it possible to discuss the shape of a molecule in terms of a **molecular potential energy curve** — the plot of the calculated energy against the relative location of the nuclei (Fig. B.20) — for each of its electronic states. For a diatomic molecule, the curve is a plot of energy against internuclear separation; for a polyatomic molecule the curve is a complicated °potential energy surface of all the nuclear coordinates (Fig. B.21). The plot shows the variation of potential energy of the molecule as the nuclear positions change. The energy is 'potential' in the sense that it is the total energy of the molecule less the kinetic energy of the nuclei (which are assumed stationary at each location). Once the variation of molecular energy with nuclear geometry is known it can be used as the potential energy surface for the °vibrational motion of the molecule.

In practice, the Born–Oppenheimer approximation is imperfect and small spectroscopic consequences of its breakdown can be observed. In each case, the consequences can be traced to a mixing between states that results from the electrons only imperfectly following the slow motion of the nuclei: if an electron lags behind a moving nucleus its wavefunction might resemble another state of the molecule. The consequences of the failure of the approximation include **intensity**

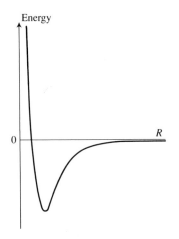

Fig. B.20 A molecular potential energy curve, showing the variation of the molecule's energy (in a given electronic state) for a series of static locations of the nuclei.

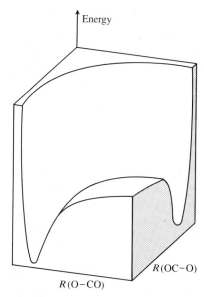

Fig. B.21 For a polyatomic molecule, the potential energy is represented by a surface that varies with the displacements of all the relative locations of the atoms. This surface is a slice through a much higher dimensional surface: it is confined to displacements of the atoms that preserve the linearity of the molecule.

stealing, in which a transition that is expected to be forbidden acquires intensity by virtue of some mixing with states to which transitions are allowed. The consequences also include small shifts in the energies of states and the removal of some degeneracies.

Further information

The original paper in which the approximation is proposed is Born and Oppenheimer (1927). For a discussion and justification of the approximation, see Slater (1963). Some of the consequences of the failure of the approximation are described in King (1964), Hollas (1983), and Struve (1989).

Boson

A **boson** is a particle with integral °spin (including zero). Examples include the deuteron, ^2H ($I = 1$), the ^4He nucleus or α-particle ($I = 0$), the ^4He atom ($I = 0$), the ^{14}N nucleus ($I = 1$) and the °photon ($I = 1$). In general, *fundamental* particles that are bosons are transmitters of forces (for example, the photon is the agent of the electromagnetic force) whereas fundamental particles that are °fermions are constituents of matter.

The °Pauli principle (as distinct from the Pauli *exclusion* principle), requires a wavefunction to be symmetrical under the interchange of any pair of identical bosons. It follows that (in contrast to °fermions), bosons are not restricted by the Pauli exclusion principle, and any number of bosons may occupy a single state. The description of the states that integral-spin particles can occupy was set out by the Indian mathematician Satyendra Bose (1894–1974), and then developed by Einstein, and the name 'boson' recognizes the former's contribution.

The consequences of the ability of many bosons to occupy a single state are the existence of superfluidity and °superconductivity. The operation of °lasers depends on the °photon being a boson, for an intense monochromatic beam of light consists of a large number of photons in the same state. The Planck formula for °black-body radiation can be treated as the outcome of searching for the most probable distribution of particles (photons) that obey the Pauli principle for bosons (that is, obey 'boson statistics').

Further information

See °spin and the °Pauli principle for further information. The table of nuclear properties (Table S.1) shows which nuclei are bosons and which are fermions. For a full discussion of the distinction between fermions and bosons, and of whether all particles are either fermions or bosons, see the article by Peierls in Salam and Wigner (1972). For an authoritative survey, refer to Pauli (1940). For a derivation of the Planck distribution from Bose–Einstein statistics, see Chandler (1987).

Bracket notation

Paul Dirac (1902–1984), one of the intellectually most elegant of theoretical physicists, introduced a fittingly elegant notation for the expression of quantum mechanics. He denoted the state of the system by a quantity called a **ket**, denoted $|n>$, where n stands for the quantum numbers needed to specify the state. The ket corresponds to the °wavefunction ψ_n in the conventional notation. He also denoted the complex conjugate $\psi_n{}^*$ by the **bra** $<n|$. The notation $<n|n>$ is taken to imply not just the multiplication of $\psi_n{}^*$ by ψ_n but integration over the product too. That is,

$$<n|n> = \int \psi_n{}^* \psi_n \mathrm{d}\tau$$

It follows that the °normalization of ψ_n is expressed by

$$<n|n> = 1$$

and the °orthogonality of two states ψ_n and ψ_m is expressed by

$$<n|m> = 0$$

The typical integral that is widely encountered in quantum mechanical calculations, which has the form

$$I = \int \psi_m{}^* \Omega \psi_n \mathrm{d}\tau$$

where Ω is an operator, is written much more succinctly as

$$I = <m|\Omega|n>$$

Thus, untidy integrals are swept off pages, formulae are made to look easier, and the structure of a calculation takes on a much less fearsome appearance. The expression for I shows the origin of the names 'bra' and 'ket', for they are derived from $<\text{bra}|c|\text{ket}>$.

The property of °hermiticity is also easier to express in bracket notation than in conventional terms. Thus, if an operator Ω is hermitian,

$$<n|\Omega|m> = <m|\Omega|n>{}^*$$

Bracket notation is mildly helpful when we write °eigenvalue equations. Thus, in place of $\Omega\psi_n = \omega\psi_n$, we write

$$\Omega|n> = \omega|n>$$

Further information

See MQM Chapter 5 for an account of the notation and Dirac (1958) for an authoritative discussion. A deeper analysis of the significance of bras and kets is given by Salam and Wigner (1972). Writing quantum mechanical states in terms of bras and kets is a component of 'transformation theory'; see Dirac (1958) and Davydov (1976).

Branch

The °rotational transitions that accompany a °vibrational transition of a molecule give rise to lines in the spectrum which can be grouped into **branches**:

- **O branch**: lines arising from $J \rightarrow J - 2$ (Raman spectra)
- **P branch**: lines arising from $J \rightarrow J - 1$ (infrared and Raman spectra)
- **Q branch**: lines arising from $J \rightarrow J$ (if allowed, see below)
- **R branch**: lines arising from $J \rightarrow J + 1$ (infrared and Raman spectra)
- **S branch**: lines arising from $J \rightarrow J + 2$ (Raman spectra)

The Q branch is allowed only for linear molecules that have a component of angular momentum about their internuclear axis because only then can the angular momentum of the incoming photon be transferred to the molecule without causing a change in its rotational state. Very few diatomic molecules give a Q branch; the exceptions are molecules

---

Box B.2 Branches

For the transitions $J' \leftarrow J$ accompanying a vibrational transition of wavenumber $\tilde{v}$:

O branch $(\Delta J = -2, J-2 \leftarrow J)$, $\tilde{v}_O(J) = \tilde{v} - 2B(2J-1)$

P branch $(\Delta J = -1, J-1 \leftarrow J)$, $\tilde{v}_P(J) = \tilde{v} - 2BJ$

Q branch $(\Delta J = 0, J \leftarrow J)$, $\tilde{v}_Q(J) = \tilde{v}$

R branch $(\Delta J = +1, J+1 \leftarrow J)$, $\tilde{v}_R(J) = \tilde{v} + 2B(J+1)$

S branch $(\Delta J = +2, J+2 \leftarrow J)$, $\tilde{v}_S(J) = \tilde{v} + 2B(2J+3)$

More complicated expressions apply when the rotational constants B are different for the two vibrational states.

---

with a component of electronic orbital angular momentum about the internuclear axis (such as NO, which has a Π ground state). The energies of the transitions are given in Box B.2.

A classical analogy that shows why rotational transitions are expected to accompany vibrational excitation is the mechanism by which ice skaters change their rotational states: if they throw out their arms they rotate more slowly, and if they draw them in they rotate more quickly. In much the same way, the sudden lengthening or contraction of a bond when a vibrational transition occurs changes the state of rotation of the molecule.

Further information

See MQM Chapter 11 and Richards and Scott (1985) for an introduction to the branch structure of vibrational transitions. More detailed accounts will be found in Hollas (1983, 1987), Steinfeld (1985), Struve (1989), and Graybeal (1988). The standard reference, which illustrates the depths of scholarship which may be brought to bear on the analysis of spectra, and of their branch structure in particular, is Herzberg (1950). See also Papousek and Aliev (1982) for details.

Building-up principle

The **building-up principle**, or '*Aufbau* principle', is a statement about how electrons should be fed into the orbitals of an atom or molecule to obtain its ground-state electron °configuration. The principle states:

> An electron enters the lowest available orbital consistent with the °Pauli exclusion principle; when degenerate orbitals are available, electrons occupy empty orbitals first before pairing and completely filling each orbital.

Thus, the first electron enters the lowest orbital, the second joins the

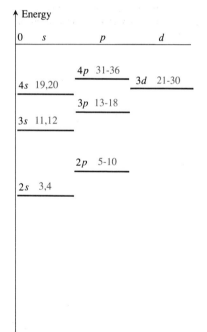

Fig. B.22 The energy levels of a typical many-electron atom and the order in which they are filled as electrons are successively added.

first (but with opposite °spin); the third electron enters the next higher orbital, then the fourth pairs with it and so on (Fig. B.22; see °Hund's rules).

In atoms, the energies of orbitals are determined by the value of the principal quantum number and the effects of °penetration and shielding. They lie in the order

$$1s < 2s < 2p < 3s < 3p < 4s < 4d < 4p \ldots$$

Occupied by up to   2   2   6   2   6   2   10   6 … electrons

The variation in orbital energy with atomic number throughout the Periodic Table is shown in Fig. B.23. The configurations obtained from this diagram follow the format of the Periodic Table, and indeed account for the periodicity of the elements.

One mnemonic for remembering the sequence of occupation is shown in Fig. B.24. A superior alternative is to remember the general

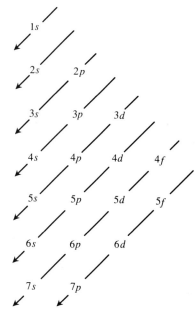

Fig. B.24 In the building-up principle for atoms, the orbitals are occupied in the order shown by the arrows and starting at 1*s*; once the tip of one arrow is reached, move to the arrow on its right.

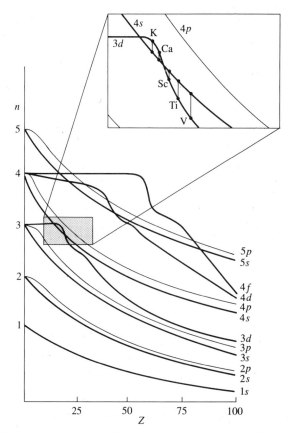

Fig. B.23 The energies of the orbitals in many-electron atoms. The inset shows a magnified view of the change in relative energies of the 4*s* and 3*d* orbitals between Ca and Sc. Note how the energies of the 3*d* orbitals fall away swiftly (after Cu) and are thereafter best regarded as part of the atomic core.

layout of the Periodic Table, and to occupy the orbitals so that the Table is reproduced. It must be noted that the order of occupation is not necessarily the order of orbital energies: the scheme is found empirically to give the *overall* ground configuration of the atoms when all the electrons have been accommodated (see below). For example, the ground configurations of the first few elements are

H: $1s^1$ He: $1s^2$, denoted [He]

Li: [He]$2s^1$ Be: [He]$2s^2$ B: [He]$2s^22p^1$ ... Ne: [He]$2s^22p^6$, denoted [Ne]

Na: [Ne]$3s^1$ ...

A complete list of ground state electron configurations is given in Table B.1.

The criterion for a configuration to be the ground state is that its *overall* energy is the lowest of all configurations. The overall energy takes into account the effects of electron–electron repulsion, and may not correspond to the electrons occupying the orbitals in the order of their energies. In particular, the total energy of a configuration is not the sum of the individual orbital energies, since account must also be taken of the repulsions between the electrons. Thus, it may be an advantage to occupy an orbital of slightly higher energy if in that way the electron repulsion energy can be reduced. When individual orbitals have similar energies, the true ground configuration might differ slightly from that predicted by a straightforward application of the building-up principle.

The variation from a straightforward application of the building-up principle is particularly important when the d orbitals start to be filled, at the element Sc. We should note from Fig. B.23 that at Sc the energy of the $3d$ orbitals falls below that of the $4s$ orbital. However, the two $4s$ electrons that are present (according to the order of occupation specified above) do not transfer to the $3d$ orbitals, for although that would appear to lead to a lower *orbital* energy, the total energy would be higher on account of the increase in electron repulsion that would occur. Thus, the configuration of Sc is [Ar]$3d^14s^2$. At the right of the d block, the energy of the d orbitals has fallen so far that they are now best regarded as part of the core, and the valence shell consists of the $4s$ and $4p$ subshells (and their analogues in the later periods).

For reasons that also stem from the effects of electron–electron repulsions, it turns out that half-full and full subshells (d^5 and d^{10} configurations) have a low energy and may be formed by transfer of an s electron into a d orbital. Thus, the ground configuration of Cr is [Ar]$3d^54s^1$, and not [Ar]$3d^44s^2$ as might have been expected, and Cu is [Ar]$3d^{10}4s^1$, not [Ar]$3d^94s^2$. A similar tendency to form f^7 and f^{14} in the f block is also encountered (see Table B.1).

The building-up principle is also applied to the occupation of °mo-

Table B.1 Ground state electron configurations of the elements

Element	Configuration	Term	Element	Configuration	Term	Element	Configuration	Term
H	$1s^1$	^2S	Kr	$[Ar]3d^{10}4s^24p^6$	^1S	Lu	$[Xe]4f^{14}5d^16s^2$	^2D
He	$1s^2$	^1S	Rb	$[Kr]5s^1$	^2S	Hf	$[Xe]4f^{14}5d^26s^2$	^3F
Li	$[He]2s^1$	^1S	Sr	$[Kr]5s^2$	^1S	Ta	$[Xe]4f^{14}5d^36s^2$	^4F
Be	$[He]2s^2$	^1S	Y	$[Kr]4d^15s^2$	^2D	W	$[Xe]4f^{14}5d^46s^2$	^5D
B	$[He]2s^22p^1$	^2P	Zr	$[Kr]4d^25s^2$	^3F	Re	$[Xe]4f^{14}5d^56s^2$	^6S
C	$[He]2s^22p^2$	^3P	Nb	$[Kr]4d^45s^1$	^6D	Os	$[Xe]4f^{14}5d^66s^2$	^5D
N	$[He]2s^22p^3$	^4S	Mo	$[Kr]4d^55s^1$	^7S	Ir	$[Xe]4f^{14}5d^76s^2$	^4F
O	$[He]2s^22p^4$	^3P	Tc	$[Kr]4d^55s^2$	^6S	Pt	$[Xe]4f^{14}5d^96s^1$	^3D
F	$[He]2s^22p^5$	^2P	Ru	$[Kr]4d^75s^1$	^5F	Au	$[Xe]4f^{14}5d^{10}6s^1$	^2S
Ne	$[He]2s^22p^6$	^1S	Rh	$[Kr]4d^85s^1$	^4F	Hg	$[Xe]4f^{14}5d^{10}6s^2$	^1S
Na	$[Ne]3s^1$	^2S	Pd	$[Kr]4d^{10}$	^1S	Tl	$[Xe]4f^{14}5d^{10}6s^26p^1$	^2P
Mg	$[Ne]3s^2$	^1S	Ag	$[Kr]4d^{10}5s^1$	^2S	Pb	$[Xe]4f^{14}5d^{10}6s^26p^2$	^3P
Al	$[Ne]3s^23p^1$	^2P	Cd	$[Kr]4d^{10}5s^2$	^1S	Bi	$[Xe]4f^{14}5d^{10}6s^26p^3$	^4S
Si	$[Ne]3s^23p^2$	^3P	In	$[Kr]4d^{10}5s^25p^1$	^2P	Po	$[Xe]4f^{14}5d^{10}6s^26p^4$	^3P
P	$[Ne]3s^23p^3$	^4S	Sn	$[Kr]4d^{10}5s^25p^2$	^3P	At	$[Xe]4f^{14}5d^{10}6s^26p^5$	^2P
S	$[Ne]3s^23p^4$	^3P	Sb	$[Kr]4d^{10}5s^25p^3$	^4S	Rn	$[Xe]4f^{14}5d^{10}6s^26p^6$	^1S
Cl	$[Ne]3s^23p^5$	^2P	Te	$[Kr]4d^{10}5s^25p^4$	^3P	Fr	$[Rn]7s^1$	^2S
Ar	$[Ne]3s^23p^6$	^1S	I	$[Kr]4d^{10}5s^25p^5$	^2P	Ra	$[Rn]7s^2$	^1S
K	$[Ar]4s^1$	^2S	Xe	$[Kr]4d^{10}5s^25p^6$	^1S	Ac	$[Rn]6d^17s^2$	^2D
Ca	$[Ar]4s^2$	^1S	Cs	$[Xe]6s^1$	^2S	Th	$[Rn]6d^27s^2$	^3F
Sc	$[Ar]3d^14s^2$	^2D	Ba	$[Xe]6s^2$	^1S	Pa	$[Rn]5f^26d^17s^2$	^4K
Ti	$[Ar]3d^24s^2$	^3F	La	$[Xe]5d^16s^2$	^2D	U	$[Rn]5f^36d^17s^2$	^5L
V	$[Ar]3d^34s^2$	^4F	Ce	$[Xe]4f^26s^2$	^3H	Np	$[Rn]5f^46d^17s^2$	^6L
Cr	$[Ar]3d^54s^1$	^7S	Pr	$[Xe]4f^36s^2$	^4I	Pu	$[Rn]5f^67s^2$	^7F
Mn	$[Ar]3d^54s^2$	^6S	Nd	$[Xe]4f^46s^2$	^5I	Am	$[Rn]5f^77s^2$	^8S
Fe	$[Ar]3d^64s^2$	^5D	Pm	$[Xe]4f^56s^2$	^6H	Cm	$[Rn]5f^76d^17s^2$	^9D
Co	$[Ar]3d^74s^2$	^4F	Sm	$[Xe]4f^66s^2$	^7F	Bk	$[Rn]5f^97s^2$	^6H
Ni	$[Ar]3d^84s^2$	^3F	Eu	$[Xe]4f^76s^2$	^8S	Cf	$[Rn]5f^{10}7s^2$	^5I
Cu	$[Ar]3d^{10}4s^1$	^2S	Gd	$[Xe]4f^75d^16s^2$	^9D	Es	$[Rn]5f^{11}7s^2$	^5I
Zn	$[Ar]3d^{10}4s^2$	^1S	Tb	$[Xe]4f^96s^2$	^6H	Fm	$[Rn]5f^{12}7s^2$	^3H
Ga	$[Ar]3d^{10}4s^24p^1$	^2P	Dy	$[Xe]4f^{10}6s^2$	^5I	Md	$[Rn]5f^{13}7s^2$	^2F
Ge	$[Ar]3d^{10}4s^24p^2$	^3P	Ho	$[Xe]4f^{11}6s^2$	^4I	No	$[Rn]5f^{14}7s^2$	^1S
As	$[Ar]3d^{10}4s^24p^3$	^4S	Er	$[Xe]4f^{12}6s^2$	^3H	Lr	$[Rn]5f^{14}6d^17s^2$	^2D
Se	$[Ar]3d^{10}4s^24p^4$	^3P	Tm	$[Xe]4f^{13}6s^2$	^2F			
Br	$[Ar]3d^{10}4s^24p^5$	^2P	Yb	$[Xe]4f^{14}6s^2$	^1S			

lecular orbitals and the prediction of the ground state electron configurations of molecules. Now the complication is that the order of energies of molecular orbitals depends on the geometry of the molecule. The simplest procedure is to adopt a single molecular geometry and to construct the molecular orbital energy level diagram. The electrons are then fed into the orbitals in accord with the rules of the building-up principle. An elaboration of this procedure is to refer to a °Walsh diagram, in which the molecular geometry is variable, and to use the building-up principle to deduce the equilibrium geometry of the molecule.

Further information

See Puddephatt and Monaghan (1986) for an account of the structure of the Periodic Table and Shriver, Atkins, and Langford (1990) for further information about the consequences of the electron configurations of atoms. A careful analysis of the relative energies of the $3d$ and $4s$ orbitals has been given by Vanquickenborne, Pierloot, and Devoghel (1989). For a discussion of the precise starting point of the f block see Jensen (1982).

Charge density

The **charge density** at a point in an atom, molecule, or solid is the charge per unit volume at each location. For a one-electron system it is equal to the square of the wavefunction at the point of interest multiplied by the electronic charge, $-e$. For a many-electron system, the total charge density is the sum of all the one-electron contributions. The product of the charge density at a location and the volume element $d\tau$ is equal to the charge inside the volume element.

The **atomic charge density** (the charge per atom rather than the charge per unit volume) is often encountered. If the contribution of an atomic orbital to a molecular orbital is c_A, then that orbital contributes $-e|c_A|^2$ to the charge density on atom A if it is occupied by a single electron. The atomic charge density on atom A is the sum of all such terms for the occupied orbitals on the atom, and is normally expressed as a multiple of the elementary charge e. The analysis of the distribution of charge in terms of the population of molecular orbitals and the contributing atomic orbitals is known as a **Mulliken population analysis** after the American theoretical chemist and spectroscopist Robert Mulliken, who introduced the procedure.

There are other ways of establishing the charge distribution. One of the most interesting makes use of the slope (more precisely, the gradient) of the overall wavefunction. The slope is used to divide the molecule up into regions by looking for the minima between regions of charge density and identifying these minima as the 'edges' of the atoms in the molecule. The charge within each region is then summed. See the references to the work by Bader in *Further information*.

The atomic charge density is zero for each atom of a neutral diatomic molecule A_2 because the electrons are distributed symmetrically over them both. In a homonuclear diatomic molecular ion (such as the carbide ion C_2^{2-} or the mercury(I) ion, Hg_2^{2+}), the charge is shared equally, and the atomic charge density is the same on each atom (-1 and $+1$ respectively). The charge density is not symmetrically distributed in a heteronuclear diatomic molecule because electrons may accumulate closer to the more °electronegative atom. The charge density is not necessarily zero on each atom of a homonuclear polyatomic molecule, since atoms in different locations in a molecule may be in different electronic environments. In the ozone molecule, O_3, for instance, the central O atom is distinct from the outer two, and the charge distribution is not uniform.

Further information

See °formal charge for an explanation of how charges are ascribed to atoms for which a Lewis structure is available, and °electronegativity for some information about trying to estimate atomic charge densities (including the pitfalls). The entry on °alternant hydrocarbons states the Coulson–Rushbrooke theorem on the charge density. A discussion of charge densities in molecules will be found in Coulson (1982) and McGlynn *et al.* (1972). For an up-to-date review and illustrations of how charge density is contributing to drug design see Richards (1983, 1989). The topological properties of the gradient of the wavefunction are described by Bader (1985, 1990).

Charge-transfer transition

In a **charge-transfer transition** (a 'CT transition'), an electron migrates from one region of a molecule to another with a net °transition dipole moment. The sudden migration of the electron to a state of lower energy (if the transition is an emission) results in an impulsive shock to the electromagnetic field, and a photon is generated. In the corresponding absorption (which is the more commonly observed type of CT transition), an incoming photon jolts the molecule, and an electron is shaken from one region of the molecule to another.

Charge-transfer transitions in complexes are classified according to the direction of migration of the electron. In a **ligand-to-metal charge-transfer transition** (an LMCT transition) a ligand electron migrates to the central metal ion. An example is the transition responsible for the purple colour of the permanganate ion (MnO_4^-); another example of an LMCT transition is the $Cd^{2+}(5s) \leftarrow S^{2-}(3p\pi)$ transition responsible for the yellow colour of 'cadmium yellow' (CdS). In a **metal-to-ligand charge-transfer transition** (an MLCT transition) the electron migrates from the metal ion to the surrounding ligands. Such transitions are most common for aromatic, carbonyl, and cyano ligands with empty π^* orbitals bonded to a metal atom in a low °oxidation state (because such metal atoms are electron-rich).

Charge-transfer transitions are generally more intense than ligand-field transitions (transitions of the electron from one metal orbital to another). This difference is partly because the electron migrates so far that it may give rise to a large transition dipole moment, and hence may deliver a big dipolar shock to the electromagnetic field. A second reason is that ligand-field transitions are often forbidden, and there is only a very weak impulse given to the electromagnetic field when an electron migrates from one d orbital to another.

Further information

See Shriver, Atkins, and Langford (1990) for a survey of some charge-transfer transitions of complexes. The book by Nassau (1983) gives a helpful introductory survey and describes the range of colours that can be ascribed to CT transitions.

Chiral

The term **chiral** comes from the Greek word for hand, and denotes the 'handedness' of molecules. A molecule is chiral if it is distinguishable from its mirror image (Fig. C.1). This is the case when the mirror image cannot be superimposed on the original molecule by rotation alone. A human hand is chiral because the mirror image of a left hand is a right hand, and no amount of rotation can superimpose the left hand on the right. Chirality provides the criterion for deciding whether or not a substance is °optically active: only chiral molecules and chiral aggregates are optically active.

Any molecule in which four different groups are attached tetrahedrally to a central atom is chiral; an example is tartaric acid (**1** and **2**). A **prochiral** molecule (**3a**) is a molecule that becomes chiral on substitution of one of its ligands (**3b**) and hence is a precursor of a chiral molecule. An example of an **achiral** molecule, one that is not chiral and which has a mirror image that can be superimposed on itself, is NH_3 (out of a myriad of other choices). A molecule and its mirror image partner are called **enantiomers**, from the Greek word for 'both'. Pairs of compounds of composition AB and AB′, where A is chiral and B and B′ are enantiomeric, are called **diastereomers**. There may be a significant energy difference between diastereomers, and corresponding differences in their physical properties (such as solubilities), and it is therefore reasonably simple to separate them by classical

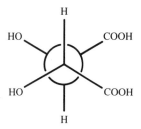

1 D(+)-Tartaric acid

2 L(−)-Tartaric acid

3a Prochiral

$$-X \downarrow +W$$

3b Chiral

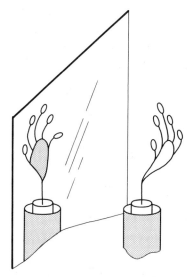

Fig. C.1 To test whether an object is chiral we determine whether it can be superimposed on its mirror image. The illustration shows a stylized human right hand and its mirror image, a left hand. Since the two cannot be superimposed by rotation, a human hand is chiral.

chemical techniques. There is also a very small energy difference between enantiomers, which is termed **chiral discrimination**. This energy difference is a consequence of the weak interaction (one of the fundamental forces) and the lack of symmetry of that force under reflection (see *Further information*).

The formal definition of chirality is that a molecule is chiral if it has no axis of improper rotation, S_n (see °symmetry operation). Since S_1 is a mirror plane and S_2 is a centre of inversion, molecules that have either reflection or inversion symmetry are achiral. However, a molecule might have neither a mirror plane nor a centre of inversion but, because it has an S_4 axis, be achiral. An example of an achiral molecule that has neither a mirror plane nor a centre of inversion is **4**.

Further information

For an account of the R, S notation used to label chiral molecules and distinguish enantiomers see the articles by Huheey (1986) and Ruekberg (1987). For the dependence of molecular properties on their chirality see Brand and Fisher (1987) and Mason (1982). The latter concentrates on chiral discrimination. A very accessible article on chiral discrimination, together with a pictorial interpretation of its origin, is that by Hegstrom and Kondepudi (1990); see also Hegstrom *et al.* (1988). Careless use of group theory can be misleading, as Kettle (1989) points out in his explanation of why ammonia is not chiral. See optical °birefringence for more information about optical activity and further references.

Clebsch–Gordan series

One of the questions that arises in the consideration of the structures of atoms is the magnitude of their total angular momentum. This information has a number of applications, including deciding what °term symbol to use to describe the state of the atom and determining the °selection rules for the transitions. The **Clebsch–Gordan series** is a rule of thumb for deriving the total angular momentum quantum numbers from the angular momentum quantum numbers of the individual electrons. The series was ascribed by Hermann Weyl to the 19th century German mathematicians Paul Clebsch and Rudolf Gordan, but there appears to be no additional evidence that they were the first to formulate it (or that they ever used it).

The physical picture behind the series may be appreciated by considering two orbiting electrons with angular momentum quantum numbers l_1 and l_2. If the two electrons are both orbiting in the same direction, their combined total angular momentum will be the sum of the two individual angular momenta, and so the total orbital angular momentum quantum number will be $L = l_1 + l_2$ (Fig. C.2a). On the

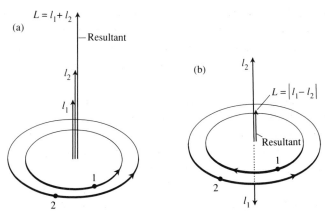

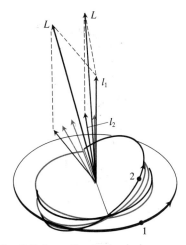

Fig. C.2 (a) The maximum angular momentum of two orbiting particles is obtained when they circulate in the same direction. In quantum mechanical terms, the total angular momentum quantum number is equal to $l_1 + l_2$. (b) The minimum is obtained when they are counter-rotating. In this case, $L = |l_1 - l_2|$.

Fig. C.3 According to classical physics, the total angular momentum is continuously variable between its maximum and minimum values. We can imagine the intermediate values as arising as the plane of the orbit of electron 2 (the inner orbit here) takes arbitrary orientations with respect to the plane of the orbit of electron 1.

other hand, if the two electrons are travelling in opposite directions, their total orbital angular momentum will be the *difference* of the two momenta because the angular momentum of one electron is partially cancelled by that of the other (Fig. C.2b). The minimum total orbital angular momentum quantum number, which must be a positive integer, is therefore $L = l_1 - l_2$ (if $l_1 > l_2$) or $L = l_2 - l_1$ (if $l_2 > l_1$). That is, the minimum value of L is $|l_1 - l_2|$ (the quantum numbers l and L are always non-negative).

According to classical physics, all intermediate values of L are possible, and we can imagine them as arising as the plane of the orbit of electron 2 takes up arbitrary orientations with respect to the plane of the orbit of electron 1 (Fig. C.3). However, according to quantum mechanics the total orbital angular momentum is quantized, and only certain values of the quantum number L are permissible. To put this restriction in classical terms, we would have to say that the orbit of electron 2 can adopt only certain angles to the plane of the orbit of electron 1 (Fig. C.3). The permitted values of L differ from each other in steps of 1, and are

$$L = l_1 + l_2, l_1 + l_2 - 1, \ldots, |l_1 - l_2|$$

This is the Clebsch–Gordan series for the orbital angular momentum of two electrons. For example, a d^2 configuration (with $l_1 = l_2 = 2$) can have $L = 4, 3, 2, 1,$ and 0. One interpretation of the series (and a handy way of employing it) is to note the **triangle rule**, that lines of length l_1, l_2, and L must be able to form a triangle (Fig. C.4).

Similar series apply to the combination of the spins of two electrons and to the combination of spin and orbital angular momenta. If the

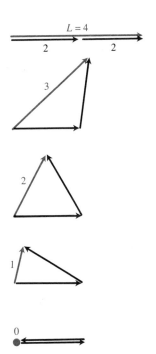

Fig. C.4 The triangle rule for $l_1 = l_2 = 2$. The only (integral) values of L that are allowed are ones that enable a triangle to be drawn.

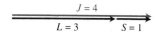

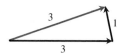

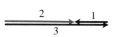

Fig. C.5 An illustration of the triangle rule for coupling L and S. In this case $L=3$ and $S=1$.

spin quantum numbers of the two particles are s_1 and s_2, the total spin quantum number S is one of the values

$$S = s_1 + s_2, s_1 + s_2 - 1, ..., |s_1 - s_2|$$

If an atom has a total orbital angular momentum quantum number L and a total spin quantum number S, the total angular momentum quantum number J is one of the values

$$J = L + S, L + S - 1, ..., |L - S|$$

For example, if $L=3$ and $S=1$, J may take the values 4, 3, and 2 (Fig. C.5).

To couple more than two angular momenta together, we use a Clebsch–Gordan series for each coupling. Suppose we want to know the possible total orbital angular momentum states of three p electrons, we use the Clebsch–Gordan series to derive the total states of two electrons, which gives $L' = 2$, 1, and 0, and then we couple the third electron to each of these states in turn:

coupling $L' = 2$ and $l_3 = 1$ gives $L = 3$, 2, and 1
coupling $L' = 1$ and $l_3 = 1$ gives $L = 2$, 1, and 0
coupling $L' = 0$ and $l_3 = 1$ gives $L = 1$

Thus, the three p electrons can give rise to any of these seven total angular momenta. However, we must be alert to the possibility that the °Pauli exclusion principle may rule out certain combinations. For example, the configuration $2p^1 3p^1 4p^1$ can give rise to a term with $L = 3$ but the analogous configuration $2p^3$ cannot give rise to $L = 3$ because it would require all three electrons to occupy the same orbital.

Which state of total angular momentum has the lowest energy depends on the details of the interactions between the electrons and of their spin–orbit coupling. In some circumstances the state of lowest energy may be predicted using °Hund's rules.

Further information

See MQM Chapter 6 for a discussion of the coupling of angular momenta. See also Zare (1987) for an introductory treatment. More advanced treatments will be found in Brink and Satchler (1968) and in Condon and Odabaşi (1980). The latter includes a report of an attempt to identify the contribution of Clebsch and Gordan. See °angular momentum for more information and references.

Closure approximation

The expressions for the properties of molecules derived using second-order °perturbation theory are sums over all the excited states of the system. Normally we know neither the excited state energies nor the wavefunctions accurately, so the sums cannot be evaluated. However, when only an order of magnitude of a physical quantity is sufficient, or when we are content to see in a general way how the property depends on various aspects of the molecule, we can make an approximation.

We suppose that the actual excitation energy of each state may be replaced by a single *effective* excitation energy Δ that is the same for them all (Fig. C.6). The consequence of replacing the individual excitation energies by Δ is that the sum over the excited states can be evaluated without needing to know the excited state wavefunctions explicitly (see Box C.1). The 'closure' of the sum into a single, compact term is the origin of the name of the approximation. The common energy Δ is interpreted as a parameter that can be varied in a plausible way to simulate the effect of modifications to the molecular structure.

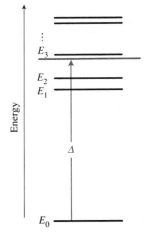

Fig. C.6 In the closure approximation, the true excitation energies ($E_1 - E_0$, $E_2 - E_0$, and so on) are replaced by a single parameter Δ.

Further information

See MQM Chapter 8 for a more thorough description of the approximation and Chapters 13 and 14 for some of its applications to the electric and magnetic properties of molecules. Further applications are outlined by Davies (1967), who underlines (on his p.47) the remark made by McLachlan (1960) about possible limitations of the closure approximation.

Box C.1 The closure approximation

The expression for the second-order correction to the energy obtained from perturbation theory is

$$E^{(2)} = \sum_n{}' \frac{\langle 0|H^{(1)}|n\rangle\langle n|H^{(1)}|0\rangle}{E_0 - E_n}$$

If all energy differences are replaced by Δ, this expression simplifies to

$$E^{(2)} \approx \frac{1}{\Delta} \sum_n{}' \langle 0|H^{(1)}|n\rangle\langle n|H^{(1)}|0\rangle$$

$$\approx \frac{1}{\Delta} \left\{ \sum_n \langle 0|H^{(1)}|n\rangle\langle n|H^{(1)}|0\rangle \right.$$

$$\left. - \langle 0|H^{(1)}|0\rangle\langle 0|H^{(1)}|0\rangle \right\}$$

$$\approx \frac{\langle 0|H^{(1)2}|0\rangle - \langle 0|H^{(1)}|0\rangle^2}{\Delta}$$

which requires no information about excited state wavefunctions.

Chemical shift

In °nuclear magnetic resonance (NMR), radiofrequency radiation is used to observe the energy separations between nuclear spin states in the presence of a magnetic field. However, because the applied field can induce extra fields in the vicinity of the nuclei, the nuclei experience different *local* magnetic fields. Nuclei in different molecular environments experience different but characteristic local fields, and hence give NMR signals that are characteristic of their environments.

The location of the resonance signal in an NMR spectrum is reported as a **chemical shift** from the resonance signal of a standard compound, which in ^{1}H-NMR is the reasonably inert, proton-rich compound tetramethylsilane (TMS, $Si(CH_3)_4$). If the standard gives a signal at a frequency v_0 and the nuclei in question give a signal at a frequency v, the chemical shift is reported on the dimensionless **δ scale**, where

$$\delta = \frac{v - v_0}{v_0} \times 10^6$$

Values of δ range up to about 10 for protons, implying that in a 400 MHz spectrometer, the resonance signal is shifted by up to 4 kHz from the TMS signal. If $\delta > 0$, the resonance is at higher frequency than the standard, implying that the nuclei in the compound being studied experience a stronger local magnetic field than those in the standard compound.

The interpretation of a chemical shift normally concentrates on the **shielding constant** σ, which relates the local field B' to the applied field B through

$$B' = (1 - \sigma)B$$

When $\sigma > 0$, the local field is less than the applied field and the nucleus is **shielded**. When $\sigma < 0$, the local field exceeds the applied field and the nucleus is **deshielded**. The sign and the value of δ in each case depend on the shielding or deshielding relative to that of the standard, and are not easy to predict.

The local contribution $(-\sigma B)$ to the total magnetic field arises from currents induced by the applied field in the atom itself or in neighbouring groups. These currents may be either paramagnetic or diamagnetic (see °magnetic properties), and whether they augment or reduce the local field at the resonant nucleus depends on their position relative to it.

The paramagnetic current (which depends on electrons being forced to travel through the molecular framework by taking advantage of excited states of the molecule, see °magnetic properties) is small when the excited states are high in energy. However, a paramagnetic current cannot be induced around the axis of a linear molecule because the mobility of the electrons in the molecule's degenerate orbitals allows them to escape being forced into excited orbitals. Diamagnetic currents (which arise from the flow of charge through the ground state orbitals of the molecule) are independent of the availability of excited states, and so always contribute to the shielding (but do not always dominate it).

Three examples of the shielding effects caused by groups adjacent to nuclei of interest are illustrated in Fig. C.7 The position of the resonance in the protons near a C=C double bond is determined largely by the magnetic moment induced in the π electrons of the bond. The effect of the field that arises from the largely paramagnetic current is to shield or deshield protons nearby, depending on whether they lie respectively inside or outside a double cone of half-angle 54° 44′ (the 'magic angle' at which $1 - 3\cos^2\theta = 0$). The shifts are opposite near a C≡C triple bond, since there is no paramagnetic current around this linear structure and the current is entirely diamagnetic. In benzene the proton shift to low field arises largely from the ability of the applied field to generate a **ring current**, a current of electrons that circulates around the conjugated ring, when it is perpendicular to the plane of the molecule.

Further information

See MQM Chapter 14 for a more detailed description. The origin and calculation of chemical shifts are described in Harris (1986), Memory (1968), and Sanders and Hunter (1987). An introduction to the theory involved will be found in Carrington and McLachlan (1967) and an excellent account is in Slichter (1988). The role of the ring current is discussed by Carrington and McLachlan and also by Davies (1967); for a more modern account see Haddon (1988).

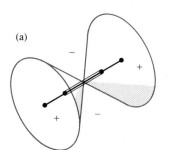 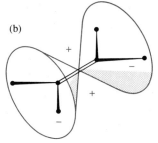 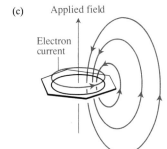

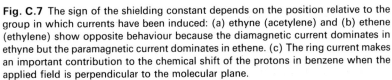

Fig. C.7 The sign of the shielding constant depends on the position relative to the group in which currents have been induced: (a) ethyne (acetylene) and (b) ethene (ethylene) show opposite behaviour because the diamagnetic current dominates in ethyne but the paramagnetic current dominates in ethene. (c) The ring current makes an important contribution to the chemical shift of the protons in benzene when the applied field is perpendicular to the molecular plane.

Coherence

One important characteristic of laser radiation is that it is coherent. An elementary interpretation of coherence is that the peaks and troughs of the radiation emitted by the laser are in step (Fig. C.8). However, there are two types of coherence. By **spatial coherence** is meant the extent to which the peaks and troughs of the waves are in step *across* the beam (Fig. C.9). By **temporal coherence** is meant the extent to which the peaks and troughs remain in step along the path of the beam. The latter is normally expressed in terms of a **coherence length** l_C which is inversely proportional to the span of wavelengths in the beam (the wider the range, the more rapidly the wave peaks get out of step; specifically, $l_C = \lambda^2/2\Delta\lambda$). If the beam were strictly monochromatic, $\Delta\lambda$ would be zero and the waves would remain in step for an infinite distance. If many wavelengths are present the waves get out of step in a short distance (and a correspondingly short time), and the coherence length is short. A typical incandescent source (a lamp bulb) has

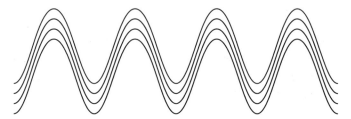

Fig. C.8 In a perfectly coherent beam the waves are in phase at all times.

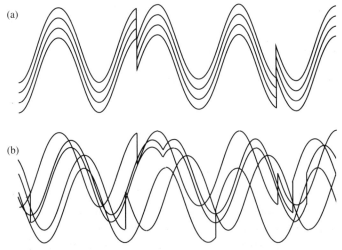

Fig. C.9 (a) Radiation that is spatially coherent because the waves are in phase across the beam, but which is only partially temporally coherent. (b) An incoherent beam in which there is very little spatial and temporal coherence.

$\lambda_C \approx 400$ nm, about the same as the wavelength of the light it emits. A He–Ne laser might have $\lambda_C \approx 10$ cm.

The coherence of radiation is a *wave* property and hence is complementary to the number of photons in the beam, which is a *particle* property. Consequently, we cannot specify the number of photons in a perfectly coherent beam. Another way of stating this complementarity is to note that if one tried to observe the arrival of individual photons in a coherent beam, a completely random distribution would be found. Since radiation from an incandescent source is less coherent, we can begin to make statements about the number of photons present in this case. In particular, the number of coincident arrivals of photons at a point is more orderly than random: it is more likely that two photons will arrive simultaneously at the same point in a noncoherent beam than in a coherent beam of the same intensity.

The coherence of laser radiation is employed in certain types of spectroscopy, particularly coherent anti-Stokes Raman spectroscopy (CARS) and related procedures.

A coherent state of a system, in a broadly defined sense, is in fact its most classical state, or **minimum uncertainty state**. The state has 'minimum uncertainty' in the sense that the expectation values of the momentum and position calculated from it satisfy the *equality* in the °uncertainty relation $\Delta p \Delta q \geq \hbar/2$. Thus, coherent states have been widely studied to explore the connection between quantum mechanics and classical mechanics.

Further information

See Atkins (1990) for an introduction to laser action, and the texts by Hecht (1988), Wilson and Hawkes (1987), and Siegman (1986) for progressively more rigorous treatments of coherence. A collection of papers dealing solely with coherence has been published by Mandl and Wolf (1970). For a discussion of coherent states and their near-classical properties see Louisell (1973), Loudon (1979), and Schlegel (1980). For a novel application of coherence in chemistry, which may point the way to its future role, see Brumer and Shapiro (1989).

Colour

A substance is coloured if it is able to absorb a band of wavelengths from incident white light. Other sources of colour include emission of particular frequencies of light from excited states, and the scattering of light of one frequency more favourably than light of another frequency.

Visible radiation is that part of the °electromagnetic spectrum with wavelengths from about 700 nm (red) through yellow (at about 580 nm), green (530 nm), and blue (470 nm), to violet (420 nm). Since the energies of photons vary from 1.7 to 3.0 eV over this range (see

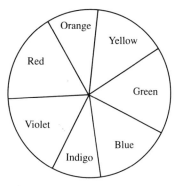

Fig. C.10 Newton's colour wheel. When light of a specific colour is absorbed from white light, the perceived colour is the one diametrically opposite across the wheel.

Table C.1), for a substance to absorb in the visible region, and hence appear coloured, it must possess excited states within that range of energies above the ground state. When red is absorbed (by virtue of the presence of low-lying excited states) the object appears blue; when blue or violet is absorbed it appears red. The perceived colour when a single colour is removed from white light can be worked out from the colour wheel devised by Isaac Newton (Fig. C.10): the perceived colour is diametrically opposite the absorbed colour on the wheel.

The most intense colours (the strongest absorptions and emissions) are due to °electric dipole transitions. The transitions may take place in a localized entity, such as a single atom, molecule, or complex, or may be due to excitation of an electron across the °band gap of a solid. A **chromophore** is a group of atoms that is responsible for the colour of a molecule and which results in absorption in approximately the same spectral region whatever molecule it is in. An important organic chromophore is the C=C double bond, in which the relevant transition is of an electron from the full π orbital to the empty °antibonding π^* orbital (Fig. C.11a), which is called a $\pi^* \leftarrow \pi$ transition (a 'pi-to-pi-star' transition). The carbonyl group (C=O) is also important: here the relevant transition is from a nonbonding °lone pair on the oxygen to the π^* orbital (Fig. C.11b). This $\pi^* \leftarrow n$ transition ('n-to-pi-star' trans-

(a)

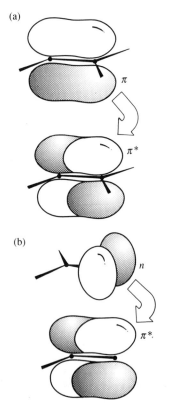

(b)

Fig. C.11 The orbital transitions in (a) a $\pi^* \leftarrow \pi$ transition, as in ethene, and (b) a $\pi^* \leftarrow n$ transition, as in a carbonyl group.

Table C.1 The electromagnetic spectrum

Colour	Frequency $\nu/10^{14}$ Hz	Wavelength λ/nm
X-rays and γ-rays	10^3 and above	3 and below
Ultraviolet radiation	10	300
Visible light:		
Violet	7.1	420
Blue	6.4	470
Green	5.7	530
Yellow	5.2	580
Orange	4.8	620
Red	4.3	700
Infrared radiation	3.0	1000
Microwaves and radiowaves	3×10^{-3} and below	1 mm and above

ition) is electric-dipole forbidden (see °selection rules and °oscillator strength), and so is generally weaker than the $\pi^* \leftarrow \pi$ transition.

The fact that many d-metal complexes are coloured is a consequence of the presence of the d electrons and the small energy splittings arising from the surrounding ligands (see °ligand field theory). The absorption is weak in octahedral complexes because $d \leftarrow d$ transitions are forbidden in centrosymmetric complexes: they are slightly allowed on account of their °vibronic character. More intense colours in the same systems (such as the characteristic intense purple of the permanganate ion, MnO_4^-) are often due to °charge-transfer transitions.

Objects generally remain coloured as they are observed (unless chemical reactions occur) because the molecules that are excited rapidly lose their excess energy to their immediate surroundings (as heat). Thus, the number of molecules available for excitation remains approximately the same even though absorption continues. A technically important exception to the permanence of colour is the use of 'saturable dyes' in lasers, in which the intensity of absorption may be so great that the reservoir of absorbing species is substantially depleted, and the color fades as the exposure continues.

Two examples of colouring arising by scattering may be mentioned briefly. The sky is blue because blue light is scattered more strongly than red; therefore more of the sun's blue radiation is scattered down to us than is its red. In the late evening the sun appears red because people further west are enjoying its blue light for their daytime sky. The reason why clouds (and this paper) appear white even though their presence is also seen by a scattering of incident light is to be found in the size of the scattering particles (see *Further information*). Another example of scattering is the colouring of glass by the precipitation of colloidal gold and other substances. The minute colloidal particles scatter away some components of transmitted light, and impart colour to the glass. Colloidal selenium is used to achieve a variety of colours (which depend on the size of the colloidal particles) including pink, amber, and brown; with cadmium sulfide a rich ruby glass is achieved.

Further information

A very interesting account of the *fifteen* causes of colour (the subtitle of the book) has been published by Nassau (1983). The fifteen of the subtitle is a bit of an exaggeration, since it counts $d \leftarrow d$ transitions, charge-transfer transitions, and the excitations of organic molecules as separate causes; nevertheless, the book gives an impression of the subtle contributions of molecules and solids to the beauty of our environment and is a rich source of references. A starting point for pursuing the fates of electronically excited species is Wayne (1988). For an introduction to colour science see Overheim and Wagner (1982), and for an introduction to the relation between colour and the chemical constitution of organic molecules, see Griffiths (1976). A thorough treatment of the scientific aspects of the reproduction of colour is given in the book by Hunt (1987).

Combination principle

The combination principle proposed by the German spectroscopist Walter Ritz states:

> The wavenumber of any spectroscopic transition in an atom can be expressed as the difference of two terms, each of which contains an integer.

For example, the transitions in the spectrum of the H atom may all be expressed as the difference of two terms, each of which has the form R/n^2, with n an integer.

The combination principle was historically important because it focused attention on the problem of why *integers* should be involved. In due course the puzzle was resolved by interpreting spectra as a result of transitions between quantized energy levels. Thus, each term corresponds to a specific energy level in an atom, and the frequency of the radiation arising from a transition between two terms is given by the °Bohr frequency condition.

Further information

See Jammer (1966) for an account of the combination principle in a historical perspective.

Commutator

A **commutator** of two °operators A and B is denoted $[A,B]$, and is the difference $AB - BA$. The symbol AB means that operation B is performed first, and is followed by operation A; BA implies that A precedes B.

Two operators are said to **commute** if their commutator is zero. If a commutator is nonzero, the final result of performing two operations depends on the order in which the operations are done: operation A followed by operation B results in a different outcome from operation B followed by A. For example, multiplication of a function by x followed by differentiation gives a different outcome from differentiation followed by multiplication by x. A pictorial representation of noncommutation is shown in Fig. C.12, which shows that rotation by $30°$ around the x axis followed by $60°$ around the y axis has a different outcome from the same two rotations done in the opposite order.

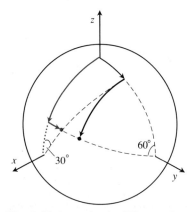

Fig. C.12 A rotation by 30° around an x axis followed by 60° around the y axis has a different effect on the same initial point from the same two rotations done in the opposite order.

The importance of commutators in quantum mechanics comes from the identification of physical observables with °operators, not all of which commute with each other. It turns out that if two operators do not commute with each other, then the observables they represent cannot be determined simultaneously (see °complementarity and the °uncertainty principle). The fact that some operators do not commute with each other is a principal factor underlying the differences between quantum and classical mechanics.

The technical importance of a commutator is that the °eigenfunctions of one operator are also eigenfunctions of any other operator with which it commutes.

Further information

The properties of commutators, and the consequences of noncommutation in quantum mechanics, are described in Chapters 5 and 6 of MQM. The standard texts on quantum mechanics describe the consequences in detail; see, for instance, Davydov (1976) and Bransden and Joachain (1989). For deeper accounts see Jauch (1968) and Salam and Wigner (1972). Take this concept further through the entry on °matrix mechanics.

Complex wavefunctions

That a wavefunction is 'complex' simply means that it has two components, one of which we call its 'real' component and the other its 'imaginary' component. However, both components are equally 'real' in the sense of being present. In the diagrams in this section (but nowhere else), we draw the real component of the wavefunction in black and the imaginary component in blue. In general we should depict a wavefunction using both colours, as in Fig. C.13.

In many cases a spatial wavefunction has only one component, which is usually taken be its 'real' component. We usually regard a spatial wavefunction that has only one component as a 'real' wavefunction. A °particle in a box is described by wavefunctions with a single component (one that is proportional to $\sin x$), and most molecular orbitals are real.

All wavefunctions of definite and nonzero energy are complex if we allow for their time dependence, since a time-dependent wavefunction is the product of a spatial wavefunction ψ and a factor $e^{-iEt/\hbar}$. The rate at which a time-dependent wavefunction changes from real to imaginary is therefore determined by its energy: the higher the energy the faster the wavefunction oscillates between purely real and purely imaginary. In this sense (and perhaps all the other rich, familiar attributes of energy are consequences of this sense), 'energy' is the rate of modulation of a wavefunction from real to imaginary.

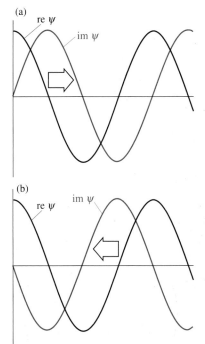

Fig. C.13 The real (black) and imaginary (blue) components of a free particle wavefunction. (a) The wavefunction corresponding to motion towards positive x; (b) the wavefunction corresponding to motion towards negative x. Note how the real component effectively chases the imaginary.

The question we address here, though, is the significance of a *spatial* wavefunction having one or two components and hence being called either 'real' or 'complex'. All physical observables depend on combinations of the form $\psi^*\psi$ (the probability density, for instance is $\psi^*\psi$, and °expectation values are integrals over $\psi^*\Omega\psi$), and hence all physical observables are real even though the wavefunction of the system may be complex.

A spatial wavefunction is complex if the particle it describes has a net motion; a spatial wavefunction is real if the particle has no net motion. For example, the spatial wavefunction (the only component we consider from here on) for a particle with linear momentum $k\hbar$ is

$$\psi = e^{ikx} = \cos kx + i \sin kx$$

The wavefunction is complex, and the particle has a net momentum (to the right, increasing x). The real and imaginary components of ψ are drawn in Fig. C.13a, and we see that the imaginary component precedes the real component in phase (that is, the imaginary component is shifted in the direction of the particle's motion). The wavefunction of a particle travelling with the same momentum in the opposite direction is

$$\psi = e^{-ikx} = \cos kx - i \sin kx$$

Now the imaginary component is shifted to the left of the real component (Fig. C.13b), and so once again its relative location marks the direction of travel.

The wavefunction $\psi = \cos kx$ is real, and corresponds to a standing wave with no net motion in either direction. It can be expressed as a superposition of the wavefunctions for motion to the left and right, because

$$\cos kx = \tfrac{1}{2}(e^{ikx} + e^{-ikx})$$

and the imaginary, direction-indicating component of the wavefunction has cancelled.

The same interpretation can be put on the complex form of atomic orbitals: complex orbitals have a definite angular momentum and real orbitals have zero net angular momentum. For example, the complex form of a p orbital is proportional to

$$\psi = e^{i\phi} = \cos \phi + i \sin \phi \qquad \text{for } m_l = +1$$

$$\psi = e^{-i\phi} = \cos \phi - i \sin \phi \quad \text{for } m_l = -1$$

Both orbitals correspond to motion around the z axis, but in opposite senses. Once again, the sense of motion is carried by the relative phases of the real and imaginary components of the orbitals, and in effect the real component chases the imaginary (Fig. C.14). The real forms of the

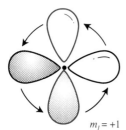

$m_l = +1$

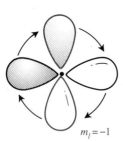

$m_l = -1$

Fig. C.14 A depiction of the real (black) and imaginary (blue) components of a p orbital. The $m_l = \pm 1$ orbitals are distinguished by the relative phases of the real and imaginary components. As for linear motion (Fig. C.13), the real component effectively chases the imaginary component.

°p orbitals that are widely used in chemistry (that is, the p_x and p_y orbitals) are the sums and differences of the two complex forms, and correspond to zero net motion around the z-axis.

Further information

See MQM Chapters 3 and 4 for some more information on complex wave-functions. Dirac took the view that the introduction of i into mechanics was a more profound development than the introduction of noncommuting oper-ators; an analysis of this view, and a survey of why i is so important has been given by Yang (1987), who also quotes from Dirac's lecture on the topic. For a different way of depicting complex wavefunctions (as rotating amplitudes), see Penrose (1989).

Complementarity

In quantum mechanics, **complementarity** refers to the impossibility of specifying simultaneously the wave and corpuscular attributes of a particle. (Etymologically, the wave and corpuscular attributes are both needed to give a 'complete' picture of a particle.) The wave and corpus-cular properties of 'particles' are complementary in the sense that if we specify the precise value of a wavelike property we cannot simultan-eously specify a corpuscular property (see °uncertainty principle).

Further information

For an introduction to the emergence of the concepts mentioned here, see Jammer (1966, 1974). A convenient collection of early papers on quantum theory is the one edited by Wheeler and Zurek (1983). See the Born–Einstein letters (Born, 1970) for an indication of the struggles that even great — but possibly too classical — minds have with the concept of complementarity.

Compton effect

In the **Compton effect**, electromagnetic radiation scattered from elec-trons shows an increase in wavelength: the shift in wavelength is independent of the initial wavelength but characteristic of the angle through which the radiation is deflected (Fig. C.15). A.H. Compton (1892–1962) carried out his original experiments in 1923 using paraffin wax as a concentrated sample of electrons.

The characteristics of the Compton effect strongly suggest that the interaction between the radiation and the electron is like a collision between two particles in which energy and linear momentum are conserved. If it is assumed that a °photon of energy $h\nu$ and momentum h/λ is in collision with a stationary electron, and that both energy and momentum are conserved in the collision of the two particles, then it

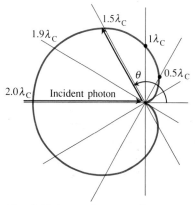

Fig. C.15 In the Compton effect, the shift in wavelength of the scattered photon is determined by the angle through which it is scattered. The illustration shows the shifts in wavelength as multiples of the Compton wavelength of the scattering particle for radiation incident from the left.

is straightforward to deduce the expression $\delta\lambda = (h/m_e c)(1 - \cos\,\theta)$, where $\delta\lambda$ is the wavelength shift (always an increase), and θ is the deflection of the radiation. The value of $\delta\lambda$ is independent of the initial wavelength, and gives a unique value of $\delta\lambda$ for a given θ.

The quantity $h/m_e c = 2.425$ pm is the **Compton wavelength** λ_C of the electron. Therefore, even in the backward scattering direction ($\theta = 180°$) the wavelength shift is only 4.8 pm. The smallness of the shift indicates why it is necessary to use X-rays or γ-rays, for only then is the shift a significant proportion of the wavelength: the effect is independent of the wavelength, but it is easier to detect when the shift is proportionally large.

Further information

Quite a good account of the differences between the classical and quantum versions of the scattering of radiation by electrons may be found in Bohm (1951). More complete treatments are given by Heitler (1954), Jauch and Rohrlich (1955), and Schweber (1961). The original work is reported by Compton (1923) and its historical relevance is explained by Jammer (1966). The most authoritative source on Compton scattering is Williams (1977).

Computational chemistry

Computational chemistry is the branch of chemistry that, by making use of computers and supercomputers, generates information that is complementary to experimental data on the structures, properties, and reactions of substances.

Much of computational chemistry is based on the °Schrödinger equation and its numerical solution for atoms and molecules. Applications include the calculation of electron and charge distributions, molecular geometry in ground and excited states, °potential energy surfaces, rate constants for elementary reactions, and details of the dynamics of molecular collisions. The techniques of computational chemistry are particularly useful for the determination of properties that are inaccessible experimentally and for the interpretation of experimental data.

Computational chemistry also includes the simulation of molecular motion in bulk phases, particularly liquids, in which the trajectories and distributions of molecules are calculated from classical mechanics and used to interpret the thermodynamic and transport properties of fluids. The structures of macromolecules and biomolecules, including the role of solvating water molecules, may also be explored using computational techniques.

Further information

Good introductory surveys of the scope of computational chemistry are those by Clark (1985) and Hirst (1990). A thorough treatment of the techniques of molecular simulation is given by Allen and Tildesley (1987). A source of information about elementary programs that are available for microcomputers is the *Journal of Chemical Education*, which has a regular series of articles on all manner of applications (not all related to quantum chemistry). Collections of the articles are available edited by Batt and Moore (1987).

Configuration

To specify the electronic **configuration** of an atom or molecule we list the orbitals that are occupied in it, as in $1s^2 2s^1$ and $1s^2 2p^1$ for two of the large number of possible configurations of a lithium atom.

The **ground configuration** is the statement of the orbitals that are occupied in the state of lowest energy. It is established by analysis of atomic and molecular spectra and may often be predicted by application of the °building-up principle. For the C atom, for instance, the ground configuration is $1s^2 2s^2 2p^2$. Sometimes the inner complete shells are abbreviated to K, L, M, etc. or to [X], where X is the corresponding noble gas. Thus the ground configuration of Na can be written as $1s^2 2s^2 2p^6 3s^1$, $KL3s^1$, or $[Ne]3s^1$. Likewise, the ground state configuration of H_2 is $1\sigma_g^2$ and that of molecular oxygen is

$$O_2: 1\sigma_g^2 2\sigma_u^2 3\sigma_g^2 4\sigma_u^2 5\sigma_g^2 1\pi_u^4 2\pi_g^2$$

Each configuration that is not a closed shell may give rise to several spectroscopic states. This is because the orbital and spin angular momenta of the electrons in a configuration may take a variety of orientations with respect to each other. The different states group together into a variety of °terms of similar energy.

Further information

Experimentally determined electron configurations of atoms (not only in their ground states) are listed by Moore (1965 onwards, 1971). For the consequences of electron configurations in chemistry see Shriver, Atkins, and Langford (1990). See also Gerloch (1986) for a thoughtful analysis. The ground configurations of atoms are listed in Table B.1.

Configuration interaction

The description of a molecule as a single electron °configuration might not be particularly accurate. For example, the single-configuration description of H_2 has both electrons localized on the same nucleus to too great an extent:

$$\psi(1, 2) \propto \{A(1) + B(1)\}\{A(2) + B(2)\}$$

$$\propto A(1)B(2) + B(1)A(2) + A(1)A(2) + B(1)B(2)$$

$$\uparrow \qquad \uparrow \qquad \uparrow \qquad \uparrow$$

One on A, the other on B   Both on A   Both on B

The description is improved by permitting **configuration interaction** (CI) in which the wavefunction of the molecule is expressed as a superposition of wavefunctions corresponding to different configurations. For instance, the admixture of the excited configuration $2\sigma_u^2$ modifies the ground-state function of H_2 to

$$\psi(1,2) \propto A(1)B(2) + B(1)A(2) + \lambda\{A(1)A(2) + B(1)B(2)\}$$

with $\lambda < 1$. Because it is now less likely that both electrons will be found on the same nucleus, the wavefunction is a better (lower energy) description of the molecule. Thus, CI achieves a certain amount of **charge correlation**, the spatial separation of electrons. Simple molecular orbital theory underestimates electron correlation, and the admission of CI goes some way to repair the deficiency.

The configurations that can be mixed into the ground state must have the same symmetry: for example, the $^1\Sigma$ ground state of H_2 is improved by the admixture of configurations giving rise to $^1\Sigma$ terms rather than $^3\Sigma$ or $^1\Pi$ terms. The second requirement stems from a result known as **Brillouin's theorem**:

Singly excited states do not interact directly with the ground state if °self-consistent field configurations are being considered.

That is, we need not consider the *direct* admixture of a configuration in which only a single electron has been excited.

The simplest way of doing a CI calculation in practice is to calculate the orbitals and their energies, and then feed in electrons to form various configurations. The ground state corresponds to the configuration with the lowest energy. Then the actual wavefunction of the molecule is expressed as a °linear combination of these configurations (using symmetry criteria to decide which configurations to include), and then the variation principle is used to determine the best mixture. Other, more sophisticated procedures are also available.

Further information

A major application of modern computational chemistry is to the treatment of configuration interaction, and so all texts that cover the computation of electron structure deal with the topic. A straightforward account of CI will be found in Richards and Cooper (1983), McGlynn *et al.* (1972), McWeeny (1989), and Naray-Szabo, Surjan, and Angyan (1987). For more complete treatments, see Szabo and Ostlund (1982) and Wilson (1984, 1987). For an intellectual history of the approaches that have been adopted see Schaefer (1984). See the entry on °correlation energy for further information.

Conserved property

A **conserved property** is a physical observable that (in the absence of any time-dependent interaction) does not change with time. One example is the total energy of an isolated atom or molecule; others are the angular momentum of an electron in an atom or the component of angular momentum about the internuclear axis in a linear molecule.

An example of a non-conserved property is the angular momentum about an axis perpendicular to the internuclear axis in a linear molecule. The origin of the changing value can be envisaged classically as the electron beginning its journey about the perpendicular axis, but colliding with a nucleus before it has completed its orbit. The collision changes the value of the angular momentum about the perpendicular axis, and so it is not a conserved quantity.

A conserved property can be discussed quantitatively by defining it as an observable for which the corresponding °operator °commutes with the °hamiltonian of the system. For example, the linear momentum of a system is conserved if the linear momentum operator commutes with the hamiltonian for the system. A physical observable represented by an operator that commutes with the hamiltonian is called a **constant of the motion.**

Further information

See MQM Chapter 5 for the relationship between lack of commutation and time-dependence of measurements. More information will also be found in texts on quantum mechanics, such as Roman (1965), Davydov (1976), and Bransden and Joachain (1989).

Coriolis interaction

The **Coriolis interaction** is the interaction between the rotation and vibration of a molecule.

Consider a diatomic molecule rotating about an axis perpendicular to its internuclear axis. If we treat the problem classically, we may imagine the bond lengthening and shortening as it vibrates. This vibration changes the moment of inertia of the molecule. Since the °angular momentum ($I\omega$) of an isolated molecule is conserved, the angular velocity must increase as the moment of inertia decreases, and vice versa. Therefore, the rotation accelerates and decelerates as the bond vibrates. This link between the rotation and the vibration results in modifications of the molecule's spectral lines.

An important application of the Coriolis interaction is encountered in the case of a vibrating, rotating, linear, triatomic molecule. There are four °normal modes of vibration of this molecule; one is a symmetric stretch in which both A—B bonds vibrate in phase; another (named v_3) is the asymmetric stretch in which as one A—B bond shortens the other lengthens; the other two are the bending motions, which may occur in two perpendicular planes (see °normal modes for pictures).

Consider the effect of the v_3 vibration interacting with the rotation by the Coriolis mechanism (Fig. C.16). As one A—B bond contracts there is a tendency for that half of the molecule to speed up; therefore, that bond tends to bend forward relative to the rest of the molecule. Meanwhile the other A—B bond lengthens, and the Coriolis interaction requires that half of the molecule to decelerate; therefore, it tends to lag behind in the rotation of the molecule. The net effect is that the molecule tends to bend in the plane of the rotation. But now the stretching motion is at the end of its swing, and it begins to swing back: the long bond shortens and the short bond lengthens. The Coriolis interaction comes into operation and the faster rotating bond becomes the laggard and vice versa. This induces the molecule to bend the other way. The net effect of the continuing process is that the bending vibration is stimulated by the combined antisymmetric stretch and the rotation of the molecule. As a result, the two vibrations are not independent and the effect of the Coriolis interaction has been to mix different vibrations together. This mixing appears in the spectrum as a °doubling.

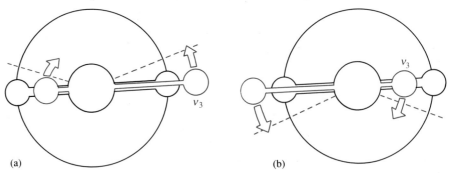

(a)   (b)

Fig. C.16 (a) As one A—B bond contracts that half of the molecule accelerates and the B atom moves forward relative to the rest of the molecule. As the other A—B bond lengthens, that half of the molecule decelerates and B lags behind. The net effect is that the molecule tends to bend in the plane of the rotation. (b) The opposite effects occur as the vibration reverses its swing.

Further information

See MQM Chapter 11 for more information. For the classical origin of the Coriolis effect see Goldstein (1982). Applications to spectroscopy will be found in Hollas (1983, 1987). For an authoritative account see Allen and Cross (1963).

Correlation energy

Even if a °self-consistent field calculation of an atomic or molecular structure in the Hartree–Fock scheme is done exactly, the calculated energy differs from the true energy. A part of the reason for the discrepancy lies in the neglect of relativistic effects, which might be very large for core electrons with high kinetic energies; however, even when relativity is allowed for a discrepancy remains. The magnitude of this residual difference is the **correlation energy,** and its presence reflects the approximations inherent in the Hartree–Fock scheme. (Some care is needed with the precise meaning of correlation energy: physicists use it to signify any contribution to the energy beyond a °Hückel type approximation.)

A basic approximation of the Hartree–Fock scheme is the neglect of the *local* distortion of the distribution of electrons. Instead of an orbital being distorted in the vicinity of another electron, the whole orbital is modified in an averaged way. Therefore the scheme neglects local electron–electron effects — it neglects electron correlations (see °configuration interaction).

Consequences of the inability of the Hartree–Fock scheme to accommodate electron correlation include the inaccurate prediction of dissociation energies and, more generally, the shape of molecular potential energy curves (Fig. C.17). However, the minimum of the Hartree–Fock curve occurs at about the same position as it does in the true curve. Thus the equilibrium geometries are quite well predicted if correlation effects are neglected, but the °force constants, and hence the molecular vibration frequencies, are exaggerated.

Further information

As in the case of the related problem of °configuration interaction, one of the principal targets of computational chemistry has been the treatment of electron correlation. Therefore, see the texts mentioned there. Specifically, see Richards and Cooper (1983) initially, then McGlynn *et al.* (1972), Cook (1978), McWeeny (1989), and Naray-Szabo, Surjan, and Angyan (1987). For more complete treatments, see Szabo and Ostlund (1982) and Wilson (1984, 1987). For an intellectual history of the approaches that have been adopted, see Schaefer (1984). Pauncz (1969) and Sinanoğlu and Brueckner (1970) are further useful references.

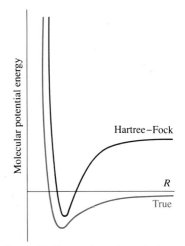

Fig. C.17 The neglect of correlation results in a molecular potential energy curve that differs from the true curve in a variety of ways. Although its minimum is quite close to the true minimum, the curvature is inaccurate close to equilibrium and the dissociation limit is poorly represented.

Correlations

The expression 'correlation' is used in at least three ways in chemistry. Thus, many chemical 'explanations' are actually correlations of properties, either with each other or with numerical parameters. For instance, in inorganic chemistry the behaviour of an element is often correlated with its °electronegativity, its °ionic radius, or its standard reduction potential. Similarly, one property may be correlated with another, as in the correlation between bond length and °bond order, or bond length and force constant.

A second entirely distinct usage refers to the manner in which electrons correlate their distributions so that there is only a low probability of one being at the same location as another. This **charge correlation** is treated in the entries on °configuration interaction and °correlation energy. In addition there is the phenomenon of **spin correlation** in which electrons (and indeed, °fermions in general) of the same spin tend to avoid each other and to cluster together if they have opposite spins (see °Fermi hole).

A third usage relates to the manner in which the states of a system change when a characteristic parameter is modified, and we study the transformation of one set of states into those of the other. An example is the variation of the energies of the molecular orbitals of H_2O as the HOH bond angle is changed from 180° to 90° and the depiction of the correlation as a °Walsh diagram. Another example is the correlation of the states of a free gas-phase atom with those of the atom when it is part of a complex. In this case we may plot the energies of the states as the ligand field strength increases from zero. This is the approach used in the discussion of complexes using °Tanabe–Sugano diagrams.

Correlation diagrams are useful for concentrating attention on two extreme cases (such as the hypothetical 180° and 90° H_2O molecules) and helping to identify the competing effects that favour one state or the other. The actual state of the system (the observed angular H_2O molecule with bond angle 104°, for instance) can then be rationalized in terms of a compromise between these competing influences.

Further information

For a number of the chemical correlations of properties in chemistry, see Shriver, Atkins, and Langford (1990). Other aspects of correlations are treated in several entries; see, for example, °Walsh diagrams, °Woodward–Hoffmann rules, °united atom, and °Tanabe–Sugano diagrams.

Correspondence principle

The **correspondence principle** states that:

At high °quantum numbers, the mean motion of a system becomes identical with its motion calculated using classical mechanics.

The principle implies that quantum mechanics reproduces the structure of classical mechanics when the fine details of the motion are ignored.

An example of an application of the correspondence principle is provided by the construction of a °wave packet to represent the motion of a free particle when the energy or momentum is only coarsely specified. The packet moves along the trajectory that the mass point would have in classical mechanics. However, as the energy becomes more precisely specified, and fewer quantum states contribute to its representation, the distribution becomes less classical and more quantum mechanical. Another example is provided by the Planck distribution of energy in a °black-body: as Planck's constant tends to zero (in a hypothetical classical world) the energy distribution becomes that of a classical system and agrees with the Rayleigh–Jeans distribution. Likewise, the momentum of a °photon (a light quantum) is transmitted to the object that absorbs or reflects it. When a sufficiently large number of photons is involved, this impulse is interpreted as the steady radiation pressure of classical electromagnetic theory.

The correspondence principle was of profound importance in the early days of quantum theory, for it acted as a guide through the exciting gloom of those days: any calculation based on quantum theory had to correspond to a classical result in all details when sufficiently large quantum numbers were involved and quantum fluctuations ignored. As an example of this kind of development we may consider the radiation emitted by a harmonically oscillating electron: the existence of an array of equally spaced quantized energy levels in a °harmonic oscillator suggests that a very wide range of frequencies could be emitted because of the indefinitely large number of different energy separations that may be obtained. Nevertheless, a classical oscillator, to which the quantum harmonic oscillator must correspond, emits only a single frequency, that of its natural classical motion. To reproduce this result it is necessary to impose quantum mechanical restrictions on the transitions that can occur; thus the correspondence principle leads to the °selection rule that an oscillator may make a transition only to a neighbouring level. Such rules emerged naturally from quantum mechanics, but even there the correspondence limit is often a very good check on the validity of a calculation. Finally it should be noted that purely quantum phenomena disappear in the correspondence limit; in particular, all effects due to °spin are eliminated.

Further information

For a historical perspective, and an account of the ways in which the principle was used to disentangle the confusion of the old quantum theory and construct quantum mechanics, see Jammer (1966). Applications of the principle will be found in Kemble (1958), Kramers (1964), and Crawford (1989a).

Coulomb integral

The **Coulomb integral**, J, is the contribution of the classical Coulombic interaction between charge distributions to the total energy of an atom or molecule.

Consider an electron distributed in an °atomic orbital ψ_a, and hence having °charge density $-e\psi_a{}^2$ at a certain point. The charge in a region of volume $d\tau_1$ at that point is $-e\psi_a{}^2 d\tau_1$. Another electron present in the atom may occupy the orbital ψ_b, and the charge in a volume at another point is $-e\psi_a{}^2 d\tau_2$. If these two charges are separated by a distance r, the potential energy of their interaction is the product of the two charges divided by their separation. The total contribution to the energy is then obtained by summing over all the volume elements $d\tau_1$ and $d\tau_2$. The procedure leads to the Coulomb integral (Fig. C.18). As the charges are the same, J is positive, and so leads to an increase in the energy of the atom.

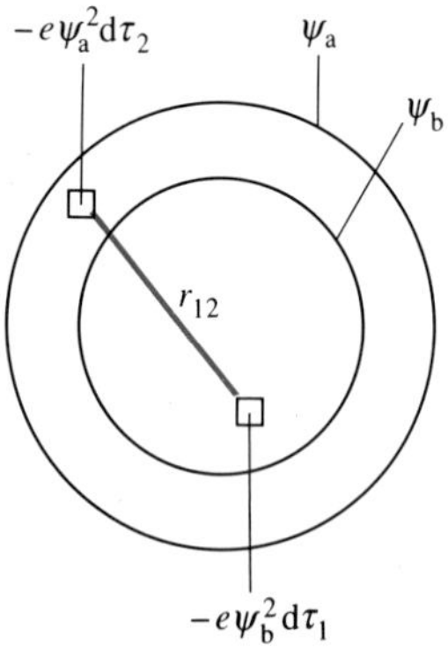

Fig. C.18 The Coulomb integral is the total electrostatic interaction between two electrons. It is calculated by summing the interactions between the electron densities in two volume elements: in this case, electron 1 occupies orbital ψ_a and electron 2 occupies orbital ψ_b (which might be the same orbital).

Further information

See MQM Chapter 10 for an introduction to the role of the Coulomb integral in molecular structure calculations. More thorough but still accessible accounts will be found in Richards and Cooper (1983) and McGlynn et al. (1972). The latter describes in some detail how they are calculated. For thorough treatments see Szabo and Ostlund (1982) and McWeeny (1989).

Covalent bond

A **covalent bond** consists of a pair of electrons shared by two neighbouring atoms. It is denoted by a line, as in C—H and N—O; if more than one pair of electrons is shared by two atoms, we speak of a 'double bond' (as in C=O) or a 'triple bond' (as in N≡N). The classification of covalent bonds as polar or nonpolar is described in the entry on the chemical °bond. Here we concentrate on the origin of the strength of the bond.

In °molecular orbital terms, a covalent bond consists of a pair of electrons that occupy a bonding orbital that spreads over both atoms (and possibly over other atoms too). In the °LCAO approximation the wavefunction of the bond A—B is {A(1)+B(1)}{A(2)+B(2)}, where A and B are the atomic orbitals on the two linked atoms. Two electrons are characteristic of such a bond because that is the maximum number allowed to occupy a single orbital (by the °Pauli exclusion principle), and hence a pair of electrons gives rise to the strongest bonding influence. In °valence bond terms, a covalent bond consists of two paired electrons in orbitals on neighbouring atoms. The valence-bond wavefunction of the bond A—B is $A(1)B(2) + A(2)B(1)$.

To discover why an electron in a bonding molecular orbital has such a low energy, we need to analyze its potential and kinetic energies. However, the contributions of all the changes are very difficult to assess, and it is often presumed that the dominant effect is the lowering of the potential energy that occurs when electrons are moved to the location between nuclei. Nevertheless, ignoring the kinetic energy is dangerous and makes the description of chemical bonding look simpler than it really is, as the following remarks will show.

The *conventional* description of bonding in the simplest molecule of all, the one-electron hydrogen molecule-ion H_2^+, is that the electron occupies the internuclear region and experiences there a favourable Coulombic potential energy by virtue of its interaction with both nuclei. However, the true source of bonding appears to be much more subtle. Thus, as H^+ and H are brought together, the electron is indeed shifted into the internuclear region, but in fact this shift leads to an *increase* in its potential energy because it is moved slightly away from the nuclei.

As the electron accumulates in the internuclear region, the molecular orbital contracts slightly near each nucleus. This contraction is a contribution that lowers the potential energy of the electron because it can now be found closer to the nuclei than in the free atom. The contribution from orbital contraction is considerable and is the dominant change in the potential energy. However, there is a price to pay, because as the electron is confined to a smaller region, its °kinetic energy rises and almost cancels the decrease in potential energy. The cancellation is not complete, however, partly because although it is squeezed closer to the nuclei, the electron also has more freedom to move along the length of the molecule, so its kinetic energy parallel to the bond falls. Finally, there is also a shift of electron density from regions outside the nuclei into the bonding region, and this shift reduces the potential energy and increases slightly the kinetic energy.

The net effect of the changes is a large decrease in the potential energy, which is dominated (at the equilibrium bond length) by the orbital contraction, and a large increase in the kinetic energy, which is also dominated by the orbital contraction. However, the increase in kinetic energy does not quite win, and the overall effect is that H_2^+ has a lower energy than the separated H and H^+.

We stress that this complicated story has been elucidated on the basis of a careful study of the H_2^+ wavefunction, and might need to be modified for more complex species. But it is an excellent example of the importance of detailed and accurate calculations in discovering the true nature of the chemical bond.

Further information

See MQM Chapter 10 for a discussion of bonding, and Coulson (1979) for a more thorough treatment (but Coulson subscribes to the 'electron pile' description). The close analysis on which the preceding remarks on bonding in H_2^+ are based is given by Ruedenberg (1962) and Feinberg, Ruedenberg, and Mehler (1970). See also the more accessible articles by Baird (1986), DeKock (1987), and Gallup (1988).

Covalent radius

The **covalent radius** of an atom is its contribution to a °covalent bond length. If the covalent radii of two atoms A and B are r_A and r_B, the A—B bond length is $r_A + r_B$ (Fig. C.19a). The covalent radius of a singly bonded atom is greater than that of the same multiply bonded atom.

Covalent radii are little more than crude parametrizations of the contributions of atoms to bonds, and may not have an absolute significance: their numerical values are chosen so that reasonably reliable bond lengths can be predicted in a number of molecules. At the most pessimistic level, all we can be sure about is that if two atoms have large covalent radii, then they are likely to form long bonds. To a certain extent they may be interpreted as the distance of closest approach allowed by the contact of the cores of the atoms (Fig. C.19b), but this interpretation probably should not be taken too seriously.

Actual bond lengths depend on the details of molecular structure, bond angles, hybridization, and the identities of the other atoms that may also be attached to the bonded atoms. The presence of ionic character in the bond also reduces its length; an approximate way of taking the effect into account is to use the **Schomaker–Stevenson equation**:

$$r = r_A + r_B - |\chi_A - \chi_B| \times 9 \text{ pm}$$

where the χs are the °electronegativities of the two atoms. Some covalent radii are listed in Table C.2, and an example of their application is shown in Fig. C.20.

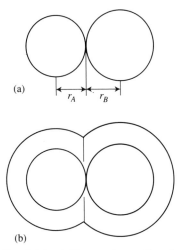

(a)

(b)

Fig. C.19 (a) The covalent radius of an atom is that atom's contribution to the length of a covalent bond. (b) The covalent radius is broadly an indication of the radius of the inner core of the atom; the cores may be thought of as in contact at the equilibrium bond length.

Table C.2 Covalent radii, R_{cov}/pm

H 37			
C 77 (1)	N 74 (1)	O 74 (1)	F 72
67 (2)	65 (2)	57 (2)	
60 (3)			
	P 110	S 104 (1)	Cl 99
		95 (2)	
	As 121	Se 117	Br 114
	Sb 141	Te 137	I 133

Values are for single bonds except where otherwise stated (in parentheses).

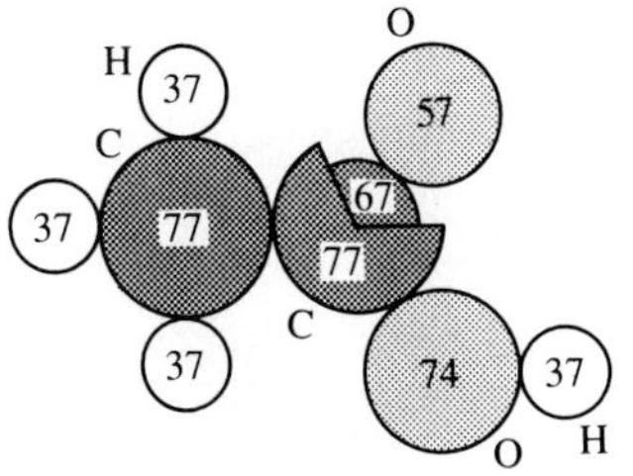

Fig. C.20 An indication of how covalent radii are used to predict bond lengths (in acetic acid). Note that the covalent radius corresponding to the bond order must be used, so the carboxylic C atom appears with two values (one for C—H and C—O and another for C=O).

Further information

Useful discussions of the reliability and significance of the concept of covalent radius have been given by Coulson (1979), Huheey (1991), and Wells (1984).

Crystal-field theory

See °ligand-field theory.

d orbitals

The ***d* orbitals** of a hydrogenic atom are the °atomic orbitals with $l = 2$. There are $2l + 1 = 5$ *d* orbitals in a given shell, corresponding to the five values of m_l that are permitted when $l = 2$ (namely $m_l = +2, +1, 0, -1, -2$). An electron that occupies a *d* orbital is called a ***d* electron**. All *d* electrons have an angular momentum of magnitude $6^{1/2}\hbar$, and the angular momentum of a *d* electron about the *z* axis is $m_l \hbar$. The non-zero angular momentum ensures that a *d* electron never approaches very close to the nucleus: the repulsive centrifugal force of the motion rises more rapidly than the attractive Coulombic force, and the net force is repulsive as *r* approaches zero.

The boundary surfaces of the five real forms of the wavefunctions of the 3*d* orbitals are shown in Fig. D.1 (the boundary surfaces of other *d* orbitals are similar). All the orbitals have two nodal planes. The peculiar shape of the d_{z^2} orbital is a result of forming linear combinations of the °complex forms of the orbitals in order to depict them more simply. Figure D.2 indicates how the five orbitals look when represented as complex wavefunctions, which is appropriate when we want to identify each orbital with a specific value of m_l: now they all look like members of the same family.

The subscripts on *d* indicate the directions in which (in the real form of the orbitals shown in Fig. D.1) the amplitudes of the orbitals are greatest. Specifically, the algebraic form of the d_{xy} orbital is proportional to the product of *xy* and a function of the radius (that is, all the

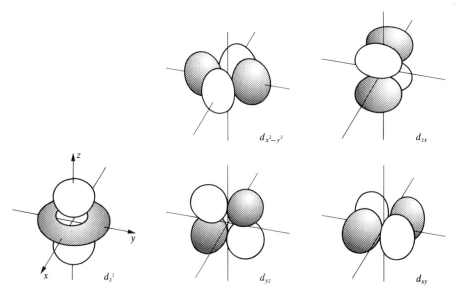

Fig. D.1 The boundary surfaces of the five hydrogenic 3*d* orbitals. The real combinations are shown.

angular dependence of the wavefunction is carried by the factor xy). The orbital is positive where $xy > 0$ and negative where $xy < 0$, and similarly for the other four orbitals. The full algebraic forms of the orbitals are given in Box D.1.

The occupation of the *d* orbitals in the °building-up process is responsible for the occurrence of the **d-block elements** of the periodic table (the 'transition elements'). The occupation of the 3*d* orbitals gives rise to the 'first transition series', the occupation of the 4*d* orbitals to the second series, and that of the 5*d* orbitals to the third series.

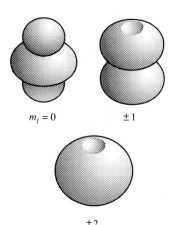

$m_l = 0$ ± 1

± 2

Fig. D.2 A representation of the complex forms of the *d* orbitals (specifically, of the amplitudes of the spherical harmonics with $l = 2$). The real combinations shown in Fig. D.1 are sums and differences of the orbitals with equal and opposite signs of m_l.

Box D.1 *d* orbitals

$\psi = R(r)\, Y(\theta, \phi)$

Complex polar form:

$$m_l = 0, \quad Y = \left(\frac{5}{16\pi}\right)^{1/2} (3\cos^2\theta - 1)$$

$$m_l = \pm 1, \quad Y = \mp \left(\frac{15}{8\pi}\right)^{1/2} \cos\theta \sin\theta\, e^{\pm i\phi}$$

$$m_l = \pm 2, \quad Y = \left(\frac{15}{32\pi}\right)^{1/2} \sin^2\theta\, e^{\pm 2i\phi}$$

Real cartesian form:

$$d_{z^2} \quad r^2 Y = \left(\frac{5}{16\pi}\right)^{1/2} (3z^2 - r^2) = \left(\frac{5}{16\pi}\right)^{1/2} (2z^2 - x^2 - y^2)$$

$$d_{xy} \quad r^2 Y = \left(\frac{5}{4\pi}\right)^{1/2} xy$$

$$d_{yz} \quad r^2 Y = \left(\frac{5}{4\pi}\right)^{1/2} yz$$

$$d_{zx} \quad r^2 Y = \left(\frac{5}{2\pi}\right)^{1/2} zx$$

$$d_{x^2-y^2} \quad r^2 Y = \left(\frac{5}{8\pi}\right)^{1/2} (x^2 - y^2)$$

Radial wavefunctions:

$$n = 3 \quad R = \frac{1}{9\sqrt{30}} \left\{\frac{Z}{a_0}\right\}^{3/2} \rho^2 e^{-\rho/2} \quad \rho = 2Zr/3a_0$$

$$n = 4 \quad R = \frac{1}{96\sqrt{5}} \left\{\frac{Z}{a_0}\right\}^{3/2} (6-\rho) e^{-\rho/2} \quad \rho = Zr/2a_0$$

$$n = 5 \quad R = \frac{1}{150\sqrt{70}} \left\{\frac{Z}{a_0}\right\}^{3/2} (42 - 14\rho + \rho^2) e^{-\rho/2} \quad \rho = 2Zr/5a_0$$

Average values:

$$\langle r \rangle = \frac{3na_0}{Z} \left\{\frac{1}{2} - \frac{1}{n^2}\right\} \quad \langle 1/r \rangle = \frac{Z}{a_0 n^2}$$

See also Box H.7 for general expressions.

Further information

See MQM Chapter 4, the °hydrogen atom, and °atomic orbitals for more information. A very good account will be found in Pauling and Wilson (1935). The depiction of atomic orbitals using their complex forms is described by Breneman (1988). For an account of the role of d orbitals in the transition metals see °ligand field theory and Salahub and Zerner (1989); the energies of the orbitals are discussed by Vanquickenborne, Pierloot, and Devoghel (1989).

δ bond

Two d orbitals on neighbouring atoms can overlap in three distinctive ways (Fig. D.3). Two d_{z^2} orbitals may overlap to give two σ orbitals (one bonding and the other antibonding) with cylindrical symmetry

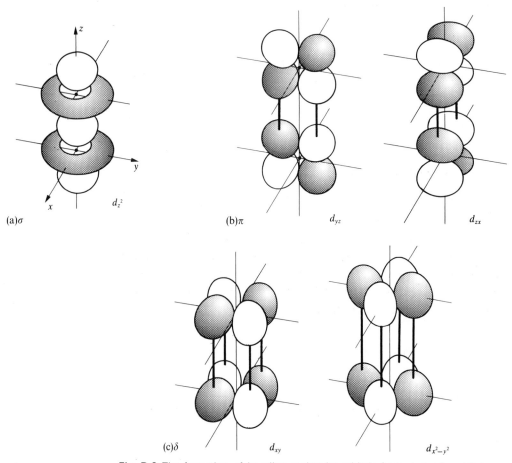

Fig. D.3 The formation of bonding molecular orbitals from d atomic orbitals on neighbouring atoms. (a) a σ orbital, (b) two degenerate π orbitals, and (c) two degenerate δ orbitals. The antibonding orbitals are obtained by reversing the phase of one orbital in each combination.

about the internuclear axis. Two d_{yz} orbitals can overlap to form π orbitals, as can two d_{zx} orbitals. Two $d_{x^2-y^2}$ orbitals have all four of their lobes face on to each other, and overlap to form a distinctively different δ **orbital** with four lobes arrayed around the internuclear axis (Fig. D.3c). Likewise, two d_{xy} orbitals are face-to-face in the same way (but rotated through 45° from the $d_{x^2-y^2}$ lobes), and can also overlap to form bonding and antibonding δ combinations.

One bonding δ orbital occupied by a pair of electrons constitutes a δ **bond**. Such a bond is a contribution to the **quadruple bonding** that is found in some d-metal cluster compounds, as in $[Re_2Cl_8]^{2-}$ in which the Re≡Re bond consists of one σ bond, two π bonds, and one δ bond, and hence may be specified as $\sigma^2\pi^4\delta^2$. Excitation of an electron from a δ orbital to its antibonding δ^* partner is designated a '$\delta^* \leftarrow \delta$ transition', and contributes to the spectroscopic and photochemical properties of such cluster compounds.

Further information

See Shriver, Atkins, and Langford (1990) and Cotton and Walton (1983). I understand that the $\delta^* \leftarrow \delta$ transition is the subject of a detective novel, Wambaugh (1974).

De Broglie relation

The **de Broglie relation** states:

A particle travelling with a linear momentum p has associated with it a wave of wavelength $\lambda = h/p$.

That is, the greater the linear momentum, the shorter the wavelength of the associated wave (Fig. D.4). The relation was proposed by Louis de Broglie (1892–1987) in 1924, in his doctoral thesis. It was a major rung of the intellectual ladder that led shortly after to Schrödinger's formulation of °quantum mechanics in 1926.

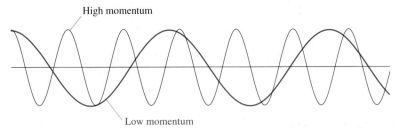

Fig. **D.4** According to the de Broglie relation, a particle with a high momentum has a wavefunction with a short wavelength and a particle with a low momentum has a wavefunction with a long wavelength.

The 'associated wave' of the de Broglie relation is now recognized as the °wavefunction of the particle. The wavefunction of a particle with °linear momentum $p = k\hbar$ is $\cos kx$. Hence, the wavefunction is a wave of wavelength $\lambda = 2\pi/k$, or h/p.

A more qualitative justification of the de Broglie relation is obtained by noting that it can be interpreted as a consequence of the relation between the curvature of a wavefunction and the °kinetic energy that the wavefunction represents. The Schrödinger equation shows that the curvature of a wavefunction increases as the kinetic energy increases. However, since increasing curvature implies that the wave swings from positive to negative displacements more often in a given length, we conclude that the higher the kinetic energy, the shorter the wavelength. Since kinetic energy is proportional to the square of the momentum, it follows that as the momentum increases, so the wavelength decreases.

Further information

Good discussions of the de Broglie relation will be found in Feynman, Leighton, and Sands (1963) and Bohm (1951). For an account of the de Broglie relation by two of his students, which gives an impression of the attempts people once made to interpret it, see Andrade e Silva and Lochak (1969). An excellent account of de Broglie's contribution to quantum theory is given by Jammer (1966).

Degeneracy

Two or more °wavefunctions are **degenerate** if they have the same energy. The three p orbitals of a given shell of a free atom are triply degenerate; the two π orbitals of a diatomic molecule are doubly degenerate.

A criterion for a set of wavefunctions to be degenerate is that they can be interconverted by a °symmetry operation of the system. Thus, the rotation of a free atom through $90°$ about the z axis is a symmetry operation, and as this operation rotates a p_y orbital into a p_x orbital, these two orbitals are degenerate. Either orbital can be rotated into a p_z orbital about other axes, and so the p_z orbital is also degenerate with them both. A $3p$ orbital cannot be generated by rotation of a $2p$ orbital, and so $2p$ and $3p$ orbitals are not degenerate. When a $3d_{z^2}$ orbital is rotated around any axis, the resulting orbital can be expressed as a linear combination of the five $3d$ orbitals; hence the $3d$ orbitals constitute a five-fold degenerate set in a spherical atom.

Two wavefunctions are said to be **accidentally degenerate** if they have the same energy by a numerical coincidence and not (at least, at first sight) as a result of a symmetry operation. For example, it may be the case that at a certain internuclear separation the σ and π orbitals of a diatomic molecule have the same energy to a few significant figures, in which case they would be regarded as accidentally degenerate.

A widely quoted example of accidental degeneracy is that of the °hydrogen atom, in which all the orbitals of a given principal quantum number have the same energy. However, it is probably the case that *exact* accidental symmetries always indicate that there is a *hidden* symmetry operation by which it is possible to transform the wave-functions into each other. Thus, the 'accidental' degeneracy of the two wavefunctions of a rectangular square well shown in Fig. D.5 is exact, and therefore a consequence of a hidden symmetry. In this case it is quite easy to discover the symmetry, since we may rotate the two halves of the well relative to each other and thereby convert one wavefunction into the other.

At first sight it may seem unlikely that there is a symmetry trans-formation that can rotate a 2s orbital into a 2p orbital (and, even more so, that interconverts 3s, 3p, and 3d orbitals). However, an analysis of the Coulomb potential shows that the 2s and 2p orbitals can be treated as projections into three dimensions of a pattern on a four-dimensional sphere, and that all four orbitals may in fact be rotated into each other in four dimensions. The interconversion shown for two dimensions in Fig. D.6 should give a hint about what is involved in four dimensions. All exact 'accidental' degeneracies can be explained in terms of a deeper analysis of the symmetry of the system.

Further information

The most fruitful way to discuss degeneracy is in terms of group theory, so see MQM Chapter 7. For a discussion of the degeneracy of the hydrogen

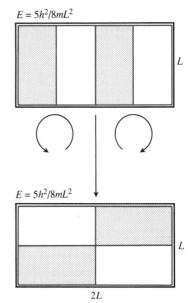

$E = 5h^2/8mL^2$

L

$E = 5h^2/8mL^2$

L

$2L$

Fig. D.5 A rectangular well with sides of length L and $2L$ has a hidden symmetry, because the two halves of the box can be imagined as being rotated separately. When the operation is applied to a square-well wavefunction with three internal nodes, it is converted into another wavefunction with two internal nodes. These two wavefunctions are 'accidentally' degenerate.

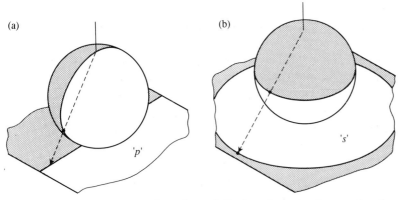

(a)

(b)

'p'

's'

Fig. D.6 A portrayal in three dimensions of the four-dimensional symmetry of a hydrogen atom. (a) A two-dimensional p orbital can be imagined as the projection of the pattern drawn on the surface of a sphere, and (b) a two-dimensional s orbital can be pictured as a projection of the same pattern. Since the sphere can be rotated in three dimensions, there is a hidden symmetry that converts the p orbital into the s orbital. Hence, the two orbitals are degenerate.

atom, see Englefield (1972), McIntosh (1971), and Bander and Itzykson (1966). For the hidden symmetry of the one-dimensional hydrogen atom, see Davytan *et al.* (1987). Accidental degeneracy, both coincidental and hidden, is discussed by McIntosh (1959, 1971). The degeneracies in diatomic molecules are discussed by Judd (1975).

Density matrix

A **density matrix** ρ is an array of (complex) numbers that carries in a compact form all the information needed to compute the observables of a system.

Calculations in elementary quantum mechanics are based on the °wavefunction ψ. However, whenever an actual observable is required we encounter formulae in which the wavefunction appears as its square (specifically, as $\psi^*\psi$; see, for example, °expectation value or °transition probability). We should begin to suspect that there is a formulation of quantum mechanics that deals with the square of ψ directly. Furthermore, the absolute phase of the wavefunction is immaterial, for if ψ is multiplied by an arbitrary phase factor, the product $\psi^*\psi$, and consequently the value of any observable, remains unchanged. Since observables are independent of the absolute phase of ψ, the wavefunction carries around a piece of useless information that would not appear if we dealt with its square directly. Finally, we should also note that the state of a system is only very rarely a single eigenstate of the hamiltonian of the system. More often than not, the observable property of a system is an average over many such states. For instance, if the system were at thermal equilibrium, we would take a Boltzmann average over the available states.

The density matrix is the mathematical object that deals with all these points. Specifically:

- ρ is expressed in terms of $\psi^*\psi$ directly.
- ρ carries only essential phase information.
- ρ carries the information about the composition of mixed states.

The composition of ρ is described in Box D.2. It satisfies an equation that resembles (and is derived from) the Schrödinger equation for the wavefunction. The collection of procedures used to solve many-electron structure calculations in terms of the density matrix directly (without first finding the wavefunction) is called **density-functional theory**.

Further information

For an elementary introduction to the density matrix, see Ziman (1969). Thorough but still introductory treatments of spins in terms of density matrices,

---

Box D.2 The density matrix

For a system described by the wavefunction $\psi = \Sigma_n c_n \psi_n$, where the basis set has N members, the N^2 elements of the density matrix ρ are the numbers $\rho_{nm} = \langle c_n c_m^* \rangle$, the angular brackets signifying an ensemble average. For a many-electron system, the N-particle density matrix is $\psi(r_1', r_2', ..., r_N') \psi^*(r_1, r_2, ..., r_N)$ with the continuously variable position coordinates playing the role of the row and column labels. The elements with $r_i' = r_i$ correspond to the electron density.

The ensemble average expectation value of an observable Ω the system is calculated from

$$\Omega = \text{tr } \hat{\Omega}\rho = \text{tr } \rho\hat{\Omega}$$

where tr signifies the trace of the °matrix. The equation of motion of the density matrix is

$$\frac{\partial \rho}{\partial t} = \frac{1}{i\hbar}[H, \rho]$$

---

which is a very fruitful way of approaching the meaning of a particular class of density matrices (those for a finite space of states), have been given by Slichter (1988) and Munowitz (1988). The application of density matrices to electron densities in molecules is described by McWeeny (1989) and in a more general and far-reaching way by Parr and Yang (1989), who give an excellent introduction to density-functional theory.

Determinantal wavefunction

See °Slater determinant.

Dipole moment

A positive charge q and a negative charge $-q$ separated by a vector r constitute an **electric dipole (1)**. The dipole is represented by a vector μ pointing from the negative to the positive charge, and the **dipole moment** is defined as the product of the charge q and the vector separation r. We normally need to consider only the magnitude μ of the electric dipole moment, which is defined as $\mu = qr$. In elementary work, the direction of the electric dipole moment is often denoted by an arrow $+\!\!\longrightarrow$, with the $+$ located on the positive charge. Note that $+\!\!\longrightarrow$ and μ point in opposite directions.

Molecules act as electric dipoles if the charge distribution in them corresponds to a separation of regions of partial positive and negative charge (as in HCl, for instance). A molecule with a permanent electric dipole moment is called a **polar molecule**. A **nonpolar molecule** is a molecule that, although it might be composed of polar components, has zero net electric dipole moment.

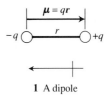

1 A dipole

Typical values of r in molecules are about 100 pm, and partial charges, which arise from differences in °electronegativity, are of the order of $0.1e$. Therefore, a typical molecular electric dipole moment has a magnitude of about $0.1e \times 100$ pm $= 2 \times 10^{-30}$ C m. Electric dipole moments are often reported as multiples of the unit **debye** (D), where 1 D $= 3.3 \times 10^{-30}$ C m, so we expect molecular dipole moments to be of the order of 1 D, as is found in practice (Table D.1). The unit is named after the Dutch physical chemist Peter Debye (1884–1966), a pioneer in the study of polar molecules; its outlandish magnitude is a consequence of its original definition in terms of electrostatic units (specifically, 1 D = 1 esu × 1 angstrom).

The electric dipole moment of a molecule is *approximately* the vectorial sum of the electric dipole moments of its component groups (Fig. D.7). However, modelling accurate electric dipole moments in terms of contributions from the constituent atoms is surprisingly difficult. It is best to use accurate molecular wavefunctions and to evaluate the electric dipole moment directly.

The simplest approach to accounting for the observed magnitudes is the **point-charge model**, in which the electron °charge density on each atom in the molecule and the positive charges of the nuclei are represented by point charges at the centres of the atoms (Fig. D.8a). This model gives poor results for several reasons. One factor it omits is the **asymmetry dipole**, which arises from the distortion in the charge distribution that originates from the °overlap of orbitals of unequal size. For example, if an occupied orbital is formed from the overlap of a

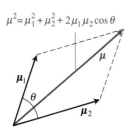

$$\mu^2 = \mu_1^2 + \mu_2^2 + 2\mu_1\mu_2\cos\theta$$

Fig. D.7 The vector addition of two electric dipole moments μ_1 and μ_2 into a resultant μ with a magnitude given by the formula in the diagram.

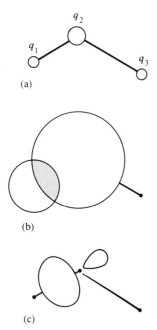

(a)

(b)

(c)

Fig. D.8 (a) The point-charge model used for modelling the electric charge distribution in a molecule and estimating its electric dipole moment. (b) The origin of the asymmetry dipole is the charge distribution arising from the overlap of orbitals of markedly different size. (c) The atomic dipole arises from the charge asymmetry of a hybrid orbital.

Table D.1 Dipole moments

	$\mu/(10^{-30}$ C m)	μ/D
C_2H_5OH	5.64	1.69
$C_6H_5CH_3$	1.20	0.36
CH_2Cl_2	5.24	1.57
CH_3Cl	6.24	1.87
CH_3OH	5.70	1.71
$CHCl_3$	3.37	1.01
CO	3.90	0.117
H_2O	6.17	1.85
HBr	2.67	0.80
HCl	3.60	1.08
HF	6.37	1.91
HI	1.40	0.42
NH_3	4.90	1.47
$o\text{–}C_6H_4(CH_3)_2$	2.07	0.62

Data: Böttcher and Bordewijk (1978)

large *p* orbital on one atom and a small *s* orbital on another, the region of maximum overlap lies closer to the small atom, and there is a considerable accumulation of charge in the overlap region close to the smaller atom (Fig. D.8b). The asymmetry dipole moment is the contribution to the overall moment of this extra charge distribution. It is by no means negligible, for it may amount to about 1 D.

Another major contribution is the **atomic dipole** (which is also called the 'hybridization dipole'). The atomic dipole arises when an electron occupies a °hybridized atomic orbital. Then the centroid of electronic charge on an atom no longer necessarily coincides with the centre of nuclear charge (Fig. D.8c). This separation of charge centroids is particularly important when the hybridized orbital is a °lone pair. The atomic asymmetry for which it is responsible is reflected in a contribution to the total dipole moment of the order of 1 D.

Molecular dipole moments may be measured by making use of the °Stark effect in molecular spectroscopy or determining the relative permittivities of solutions (see °polarizability).

Further information

See Coulson (1979) and McGlynn *et al.* (1972) for a thorough discussion and examples of the calculations involved. For the determination of electric dipole moments see Atkins (1990) for the condensed phase determinations and Hollas (1983, 1987) for the use of spectroscopy of gases. A convenient source of a lot of dipole moments of both organic and inorganic compounds is Dean (1987). Other sources are Smyth (1955), McClellan (1963), and Minkin, Osipov, and Zhdanov (1970).

Dirac equation

The **Dirac equation** is the relativistic version of the °Schrödinger equation. It combines the requirements of quantum mechanics with those of special relativity.

The Schrödinger equation is unsymmetrical with respect to space and time, for it contains first derivatives with respect to time and second derivatives with respect to space. One approach to bringing quantum mechanics into line with relativity is to find an equation that has second-order time derivatives. Such an equation was proposed by a number of people, including Schrödinger himself, and is called the **Klein–Gordon equation** after O. Klein and W. Gordon, who proposed it in 1926. Unfortunately, a property of the solutions of the Klein–Gordon equation is that the total probability of the particle being anywhere in the universe varies with time. When the equation was proposed it was thought that that was unacceptable.

Paul Dirac (1902–1984) approached the problem by keeping the equation first-order in time but forcing the space derivatives to be first derivatives too. He could not do this arbitrarily, for his equation had to agree with the successful Schrödinger equation when the particles were moving at speeds much less than that of light.

The conditions Dirac had to impose led him to conclude that he had to deal with wavefunctions with *four* components. Two of these components correspond to a positive energy for a particle at rest, which is required if the only source of energy is the rest mass mc^2. However, two of the components correspond to an embarrassing *negative* energy, which suggests that they correspond to particles with a negative mass. Instead of forgetting the whole thing, Dirac boldly proposed that all the negative energy states were filled up throughout the universe, and that we experience only the extra particles added to the overlying positive energy solutions. He was therefore able to predict that if a particle is excited out of a negative energy state it leaves a hole with a relative positive charge and (because a particle is *missing* from a negative mass state) a net positive mass. Thus, Dirac predicted the particle we now call the **positron** (e^+), although at the time he thought he was predicting the existence of the proton.

Dirac showed that the two positive energy solutions have the same energy unless a magnetic field is present. When a field is present, the energy of one level rises and the other falls as the field strength is increased. Dirac interpreted this behaviour as indicating that the electron has a magnetic moment that can take two orientations. In other words, he was able to conclude that an electron is a particle with °spin $\frac{1}{2}$. The magnitude of the splitting caused by a given magnetic field is characteristic of a particle of charge $-e$ but with a moment twice as large as would be anticipated on the basis of a classical model of a spinning charge. Thus the anomalous °g-factor of the electron emerges as a natural consequence of the Dirac equation.

Further information

See Dirac (1958), of course. Straightforward accounts of the construction of the Dirac equation are given by Moss (1973) and by Bjorken and Drell (1964). For an account of its solutions, see the latter and also Berestetskii, Lifshitz, and Pitaevskii (1971). See the Dirac *Festschrift* edited by Salam and Wigner (1972) for an account of the activity stimulated by the equation. A very helpful survey of relativistic effects in structural chemistry has been given by Pyykkö (1988).

Direct and indirect gap materials

The distinction between direct and indirect gap materials is the key to why some semiconductors simply get hot when operating but others discard the excess energy as radiation and hence can be used as light-emitting diodes (LED).

The crucial difference between the two types of material lies in the orbitals at the top and bottom of the valence and conduction °bands respectively and whether the electrons in these states have the same linear momentum. In a **direct gap material**, the linear momentum of the electron at the top of the valence band is the same as when it is in the lowest level of the conduction band (Fig. D.9). As a result, it can make a direct transition from one to the other, and it can emit the energy difference as light. In an **indirect gap material**, the linear momentum of the electron is different in the uppermost level of the valence band from what it is in the lowest level of the conduction band. Therefore, the electron can make a transition from the conduction to the valence band only if it can change its linear momentum. This it can do only by giving an impulse to the atoms of the lattice, and hence generating a lattice vibration. Such materials therefore become hot as the conduction electrons return to the valence band.

Direct gap materials include gallium arsenide and indirect gap materials include silicon itself. Gallium arsenide (doped with phosphorus to adjust the band gap so that the emission occurs in the visible region of the spectrum) is used for LEDs. If the direct gap material can be made part of a resonant cavity, the radiation can cause stimulated emission, and hence act as a laser. The resonant cavity is engineered by modifying the refractive index of the material at the junction between two semiconductors to ensure that enough of the radiation is trapped to cause stimulated emission.

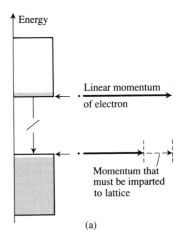

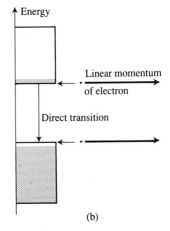

Fig. D.9 The horizontal arrows represent the linear momenta of electrons in states at the top and bottom of valence and conduction bands respectively. (a) In an indirect gap material, a conduction electron must give up linear momentum to the lattice if it is to return to the valence band. (b) In a direct gap material, the linear momenta are the same in both states, and the electron can make a radiative transition between them.

Further information

For inorganic aspects of direct and indirect gap materials see Shriver, Atkins, and Langford (1990). See also Newnham (1974), Harrison (1980), and Kittel (1986, 1987).

Dispersion

The **dispersion** of a property is its variation with frequency. The name derives from the variation of refractive index with frequency that accounts for the separation (the spatial dispersion) of the colours of white light. It applies not only to that variation but also to the frequency-dependence of the response of any system to a disturbing perturbation. For instance, the electric °polarizability, and hence the refractive index, of a molecule varies with the frequency with which the distorting electric field is oscillating (Fig. D.10).

The origin of dispersion lies in the ability of a perturbation to distort

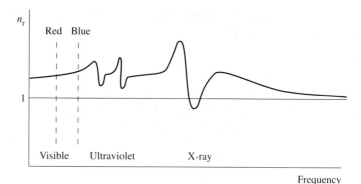

Fig. D.10 A schematic dispersion curve, showing how the refractive index of a substance varies with the frequency of the incident radiation. The sharp oscillations occur when the incident frequency matches an absorption frequency of the substance. Note how the refractive index increases with frequency in the visible region. At very high frequencies the refractive index approaches 1 because (in classical terms) even electrons are too massive to follow the rapidly changing direction of the electric field of the radiation.

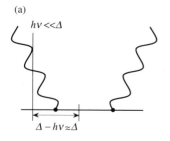

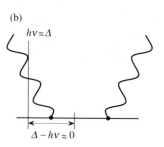

Fig. D.11 The calculation of the refractive index depends on the properties of a Feynman diagram like that shown here. (a) When the energy of a photon of incident radiation is much less than an excitation energy, the energy difference between the states mixed by the perturbation is large. (b) When the energy of the photon matches the excitation energy, the overall energy difference between the mixed states is close to zero and the field interacts strongly with the molecule.

a molecule and the variation of that ability with the perturbation's frequency. This variation is best understood in terms of °Feynman diagrams and a description in which the oscillating field of frequency v supplies photons of energy hv. When such a photon is absorbed by the molecule, the overall change in energy is the excitation of the molecule Δ less the energy of the photon supplied by the perturbation (Fig. D.11). Therefore, the net change in energy is $\Delta - hv$. As the frequency of the perturbation approaches the excitation energy of the molecule, the net excitation energy decreases, and it is zero when the incident frequency matches the molecular excitation energy exactly. At this point the molecule is highly responsive to the perturbation, since overall no energy is required to bring about a change in its electron distribution. At still higher frequencies, the overall change in energy diverges from zero again as the mismatch increases between the photon energy and the molecular excitation energy, and the molecule becomes progressively less easy to distort.

Another way of picturing the origin of dispersion is to draw the comparison between the response of a molecule to a perturbation and the response of a pendulum of natural frequency v_0 to an oscillating driving force. When the frequency of the driving force (the frequency with which the axle of the pendulum is oscillated) is much less than v_0, the motion of the pendulum is only slightly perturbed. The perturbation is greatest when the frequency of the driving force matches the natural frequency of the pendulum. At frequencies very much higher than the natural frequency of the pendulum, the pendulum barely responds at all because the driving force changes direction very rapidly and its effect averages to zero.

In solid state physics, the **phonon dispersion curve** shows the dependence of the frequency of the °phonon on its wavenumber.

Further information

See MQM Chapter 13 for more detail about the frequency dependence of polarizabilities and related properties. A thorough treatment will be found in Böttcher and Bordewijk (1978) and, from a classical viewpoint, Landau and Lifshitz (1960). For the classical approach see Appendix B.3 of Wangsness (1979).

Dispersion forces

The **dispersion force** is the induced-dipole–induced-dipole contribution to °intermolecular forces. The interaction arises from the correlation of fluctuations in the electron distribution of neighbouring molecules. A typical magnitude of the potential energy of the dispersion interaction is about 2 kJ mol^{-1}; it is the second strongest intermolecular force between neutral molecules after °hydrogen bonding, and hence exerts a dominant influence on the cohesive energy of condensed phases and deviations of gases from perfect behaviour. The interaction exists whenever two atoms or molecules are in proximity, whether or not other interactions are also present.

A fluctuation in the electron density on molecule A causes it to possess a transient electric °dipole. The dipole polarizes the electron distribution on a neighbouring molecule (Fig. D.12), the two dipoles attract each other, and the interaction lowers the energy of the pair of molecules. It is essential for there to be a correlation between the directions of the charge fluctuations, for otherwise the effect would disappear (because attractions would occur as often as repulsions). Correlation occurs because one of the dipoles drives the fluctuation that leads to formation of the second dipole, and so the two fluctuating dipoles are in step.

The potential energy of the dispersion interaction depends on the separation R of the molecules as $1/R^6$. The *force*, the first derivative of the potential energy, is proportional to $1/R^7$. The origin of the $1/R^6$ dependence is easy to identify, for the strength of the electric field arising from the initial dipole on A is proportional to $1/R^3$. Therefore, the magnitude of the dipole on B that it induces is also proportional to $1/R^3$ (because the induced dipole is proportional to the field strength at the molecule). Since the potential energy of interaction of two dipoles is proportional to $1/R^3$, the net strength of the interaction varies as $1/R^3 \times 1/R^3 = 1/R^6$.

The full expression for the potential energy of the interaction of two molecules is given by the **London formula** (Box D.3). The αs in the

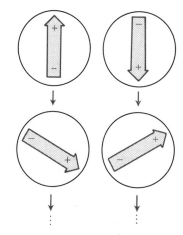

Fig. D.12 The two fluctuating dipole moments of neighbouring molecules are correlated. That is, although they fluctuate in direction, they stay in step in the relative orientation of lowest energy.

formula are the °polarizabilities of the molecules and the Is are their ionization energies. The strength of the interaction is proportional to the product of the polarizabilities because α is a measure both of the ease with which an electric field may distort the molecule and of the size of its charge fluctuations. The °ionization energy appears in the formula because the extent of fluctuation depends on the ease with which a molecule can be excited.

The London formula fails at very large separations on account of the **retardation effect**, which arises from the time it takes the information that a fluctuation has occurred to travel between the atoms. When R is large (in the sense $R \gg hc/\Delta$, where Δ is a typical molecular excitation energy) the $1/R^6$ dependence is replaced by $1/R^7$ and the interaction falls off more rapidly than the London formula predicts.

The retardation effect is best understood in terms of the °Feynman diagrams that occur in the theoretical description of the dispersion interaction (Fig. D.13). A charge fluctuation on A generates a photon that travels to B and distorts it. Molecule B then sends back a photon to A. These photons represent the interaction between two dipoles. When the separation between A and B is large, the photon from A takes so long to travel to B and then back to A that, by the time it has arrived at A, the original fluctuation on A may have changed the charge distribution slightly. As a result, the interaction will be weaker than when there is nearly instantaneous return of the photon and the charge has its original distribution.

Further information

See MQM Chapter 13 for a derivation of the London formula and a further discussion of retardation. Dispersion forces and their influence on bulk properties are treated by Rigby *et al.* (1986). The quantum electrodynamic description of the interaction, which must be used when it is necessary to include retardation, is described by Power (1964) and by Craig and Thirunamachandran (1984).

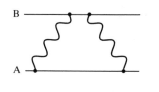

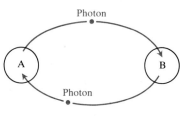

Fig. D.13 In the Feynman diagram description of the dispersion interaction, molecule A transmits a photon to molecule B, which responds by sending a photon back to molecule A.

Dissociation energy

The **dissociation energy**, D_0, is the energy required to separate the fragments joined by a chemical bond. For a diatomic molecule, the dissociation energy is the energy difference between the two infinitely separated atoms and the molecule in the ground vibrational state of a specified electronic state. If not otherwise stated, the initial electronic state is the ground electronic state of the molecule. The dissociation energy must be distinguished from the energy of the lowest point of the molecular potential energy curve, D_e: they differ by the size of the zero point vibrational energy of the bond (Fig. D.14). Since the °zero point energy depends on the masses of the atoms but the potential energy curve might not, isotopically substituted molecules (particularly H—X and D—X molecules) may have significantly different dissociation energies even though their potential energy curves are nearly identical.

The dissociation energy also differs from the **bond dissociation enthalpy**, $B(X—Y)$, the standard enthalpy for the conversion of X—Y to X and Y in the gas phase at a specified temperature. Since 2 mol of gaseous fragments (X and Y) are produced for 1 mol of X—Y molecules that dissociate, B is greater than D_0 by an amount RT. Strictly, D_0 should be replaced by the internal energy of dissociation (the thermal average over the occupied vibrational states at the specified temperature) in this relation, but the internal energy of dissociation is usually negligibly different from D_0 because vibrationally excited states are only negligibly occupied at normal temperatures.

Dissociation energies are calculated by determining the minimum of the molecular potential energy curve that has been calculated by °*ab initio* or °semi-empirical methods and identifying the value of the zero point vibrational energy. The difficulty of estimating the °correlation energy is a major source of error in this procedure. Experimental values are obtained by identifying the dissociation limit of molecular electronic absorption spectra either directly or by an extrapolation procedure originally introduced by R.T. Birge and H. Sponer in 1926. In a **Birge–Sponer extrapolation**, the energy separations of successive vibrational levels are plotted against the vibrational quantum number (Fig. D.15), and the resulting graph is extrapolated linearly to intersect the horizontal axis. The area under the extrapolated curve is equal to the dissociation energy of the bond.

The dissociation energy is a measure of the strength of a chemical bond: the greater the dissociation energy, the stronger the bond. Dissociation energy correlates with °bond order, with multiple bonds stronger than single bonds between the same pair of atoms.

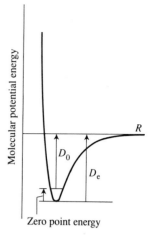

Fig. D.14 The dissociation energy D_0 is the minimum energy needed to dissociate a bond when it is in its lowest vibrational state (and a specified electronic state). It differs from the energy minimum of the potential energy curve, D_e, by an amount equal to the vibrational zero point energy.

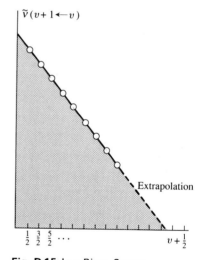

Fig. D.15 In a Birge–Sponer extrapolation, the differences between neighbouring vibrational energy levels are plotted against the vibrational quantum number, and then the curve is extended to cut the horizontal axis. The area under the extrapolated curve is an approximation to the dissociation energy of the bond.

Further information

For an introduction to the Birge–Sponer extrapolation see Atkins (1990). A thorough survey of dissociation energies for diatomic molecules has been given by Gaydon (1968); see Herzberg (1945, 1950, 1966) for more wide-ranging information. More recent introductions to spectroscopy that describe the determination of dissociation energies include the books by Steinfeld (1985) and Struve (1989). See Murrell *et al.* (1984) and Hirst (1985) for introductions to molecular potential energy curves.

Doubling

The expression **doubling** refers to the removal of the °degeneracy of a level by a hitherto neglected interaction with the result that a two-fold degenerate level splits into two distinct nondegenerate levels. Such doubling may appear in a spectrum as an actual doubling of the number of lines, either because transitions occur to either of the two levels (Fig. D.16a) or because some transitions occur to one component of the doublet and others occur to the other component (Fig. D.16b). An example of doubling in atomic spectroscopy is the °fine structure in the spectra of alkali metal atoms. For instance, consider the transition $[Ne]3p^1\ {}^2P \rightarrow [Ne]3s^1\ {}^2S$ of sodium atoms, which generates the characteristic yellow light of sodium (at 589 nm). Close inspection of the spectrum shows that the yellow light consists of two components, one at 588.99 nm and the other at 589.59 nm. The ^2P term has two °levels ($^2P_{1/2}$ and $^2P_{3/2}$) which are degenerate in the absence of spin-

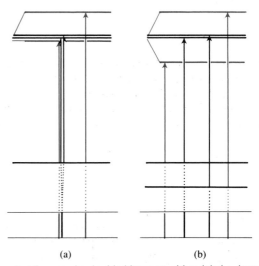

(a)   (b)

Fig. D.16 A spectral line may be doubled because either (a) the degeneracy of one of the states is removed by a perturbation or (b) the degeneracy is removed and the selection rules allow only certain transitions.

orbit coupling but which differ in energy when the coupling is taken into account.

An example of doubling in molecular spectroscopy is **Λ-doubling** (lambda doubling) in which the rotation of a linear molecule removes the degeneracy of the two components of a Π term. It is possible to identify the interaction responsible by considering a diatomic molecule in which a single valence electron occupies a π orbital. Figure D.17 illustrates the case in which the electron occupies either the orbital π or the orbital π'. If the electrons can follow the rotation of the nuclei exactly, with no lag or slip, the two π orbitals would have precisely the same energy. However, since electrons do not follow the motion of the nuclei exactly (see °electron slip and the °Born–Oppenheimer approximation), there is a tendency for them to lag behind the nuclei. The lag has virtually no effect on a π' electron, for to the lowest approximation the nuclei slide a little forward in the orbital's nodal plane and so remain surrounded by the same electron density as in the static case. For a π electron, though, the nuclei slide from the vertical nodal plane into a region of non-zero electron density, and the energy of the electron is modified by the rotation. As a result, the two orbitals move apart in energy and the energy separation, or doubling, increases with the rate of rotation.

Two other examples of doubling are important. The first is **λ-type doubling**, in which the degeneracy of a bending vibrational level of a linear triatomic molecule is removed by molecular rotation (see °Coriolis interaction). The second is °inversion doubling, where the ability of a molecule to invert from one conformation to another doubles its vibrational spectrum.

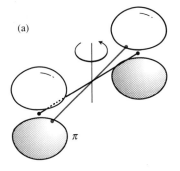

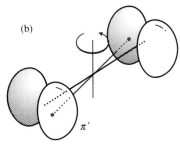

Fig. D.17 The origin of Λ-doubling. If there is imperfect following of the nuclei by the electron distribution, the nuclei (a) remain in the nodal plane of the π orbital but (b) move out of the plane of the π' orbital. As a result, the degeneracy of the π and π' orbitals is removed by the rotation of the molecule.

Further information

See MQM Chapter 11 for more information about the various types of doubling that can occur in molecular spectra. More thorough treatments have been given by Whiffen (1972) and Hollas (1983, 1987). Detailed treatments of Λ-doubling will be found in Herzberg (1950), Kovaćs (1969), and Fischer (1984). Good introductory treatments are given by Steinfeld (1985), Struve (1989), and Graybeal (1988).

Duality

Quantum mechanics introduced the view that matter and radiation have a dual character and that both aspects of behaviour may be exhibited by the same entity. **Duality** is this recognition that entities which traditionally had been classified as particles show behaviour that hitherto had been characteristic of waves, and vice versa.

For instance, in 1927 it was discovered by C. J. Davisson and L. Germer in the Bell Telephone Laboratories (then Western Electric) that crystal lattices could diffract beams of electrons. Similarly, also in 1927, G.P. Thomson working in Aberdeen showed that diffraction occurred when an electron beam passed through a thin gold foil. Conversely, both the °Compton effect and the °photoelectric effect require the energy or momentum of the radiation to possess certain properties of particles. The conclusion to which we are forced is that 'waves' and 'particles' have been so classified because, at the level of the experiments done until the beginning of the present century, one type of behaviour dominated and the other was concealed. Looking more closely at each type of entity reveals the duality of its character, and so the nature of matter and radiation is neither just one nor the other, but a composite of the two. Which aspect dominates depends on the experiment; no experiment can exhibit both aspects of the duality simultaneously (see °complementarity).

Further information

See Jammer (1966, 1974) for historical and philosophical accounts of the emergence of the concept of duality as a fundamental attribute of matter and radiation, and Bohm (1951) for an illuminating discussion. See Hughes (1989) and Hey (1987) for additional viewpoints. Some of the original papers that led to the formulation of the concept, and the intellectual anguish that it inspired, will be found in the collection edited by Wheeler and Zurek (1983).

Eigenfunctions, etc.

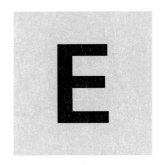

When a mathematical operation (such as multiplication, division, or differentiation) is performed on a function, the result is generally some different function. For instance, differentiation of x^2 yields the different function $2x$. However, for some combinations of operations and functions the same function is regenerated, but perhaps multiplied by a number (Fig. E.1). Thus differentiation of e^{2x} gives $2e^{2x}$, which is the same function multiplied by 2. In such cases, the function is said to be an **eigenfunction** of the operator (in this case the differential operator d/dx), and the numerical factor (2 in the example) is called the **eigenvalue** of the operator. (*Eigen* is the German word meaning 'own' or 'characteristic'.) If an operator Ω operates on a function f and produces ωf, where ω is a number, f is an eigenfunction of Ω and ω is the corresponding eigenvalue. An expression of the form $\Omega f = \omega f$ is called an **eigenvalue equation**.

The importance of eigenfunctions, eigenvalues, and eigenvalue equations in quantum mechanics can be recognized by noting that the Schrödinger equation $H\psi = E\psi$, where H is a differential operator (the °hamiltonian), is an eigenvalue equation, with the energy E as the eigenvalue and the °wavefunction ψ as the eigenfunction. The wavefunction represents a state of the system, and so ψ is often termed the **eigenstate** of the system.

According to the quantum theory of measurement, the determination of the value of an observable always results in an eigenvalue of the operator corresponding to the observable. If the system is in an eigenstate of the observation, every observation will give a single outcome equal to the eigenvalue corresponding to that eigenstate. If the system is not in an eigenstate of the observation, its state may be expressed as a linear superposition of those eigenstates; a single observation then gives an eigenvalue corresponding to any one of the eigenstates in the

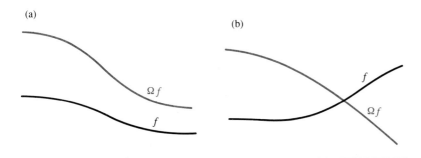

(a)

(b)

Fig. E.1 (a) If when an operator Ω acts on a function f it generates the same function multiplied by a constant, then f is an eigenfunction of the operator. (b) If Ωf is not a constant multiple of f, then f is not an eigenfunction of Ω (but may be an eigenfunction of another operator). In (a), the eigenvalue of Ω is 0.5.

superposition. If the eigenstate occurs in the sum with a coefficient c, the probability that the corresponding eigenvalue will be obtained is proportional to $|c|^2$. The mean of many observations is equal to the °expectation value of the operator for the state in question.

Further information

See MQM Chapter 5 for a discussion of operators and the topics treated here. See Davydov (1976) and Bransden and Joachain (1989) for helpful introductory accounts. For a deep view of eigenfunctions, etc. see Dirac (1958), von Neumann (1955), and Jauch (1968).

Einstein coefficients

The rate of *absorption* of radiation is proportional to the density of radiation present at the frequency of the transition. The coefficient of proportionality between the rate and the radiation density is the **Einstein coefficient of stimulated absorption**, B.

The rate of *emission* is determined by two factors. One contribution to the rate is proportional to the density of radiation at the transition frequency, because the presence of radiation stimulates an excited state to radiate its energy. The coefficient of proportionality between the rate and the radiation density is **Einstein's coefficient of stimulated emission**. This coefficient is identical to the coefficient of stimulated absorption, and is also denoted B. It follows from the equality of the two coefficients that if the populations of the upper and lower states are equal, then the rate of stimulated absorption in a sample is equal to the rate of stimulated emission.

The other contribution to the rate of emission is independent of the radiation already present, for an excited state can discard its energy as radiation even if no radiation is present initially. This 'unstimulated' contribution to the rate is termed **Einstein's coefficient of spontaneous emission**, and denoted A.

Einstein deduced the existence of the spontaneous emission process in 1917 by studying the conditions for thermal equilibrium of a system. He argued that at equilibrium the total rate of absorption must be equal to the total rate of emission, so there is no net exchange of energy between the matter and the electromagnetic field. However, since at thermal equilibrium there are more molecules in the lower state than in the upper state, the two rates would not be equal if the excited state could discard energy only by stimulated emission. To ensure that the rates are equal even though the populations are different there must be an additional mechanism for emission. The spontaneous emission process permits the upper states to return to the lower at a rate independent of the radiation already present. Thus, although the populations may be different, the overall rates of emission and absorption are the same (Fig. E.2).

As shown in Box E.1, $A \propto v^3 B$. That is, the relative importance of the spontaneous emission increases as the cube of the transition frequency. The spontaneous process becomes very important at high frequencies: highly excited states have very short natural lifetimes and rapidly radiate away their energy. A consequence of the increasing importance of spontaneous emission with increasing frequency is the difficulty of producing high-frequency lasers, for laser operation depends on the domination of spontaneous emission by stimulated emission.

Spontaneous emission is not *really* spontaneous (in the sense of not being stimulated), for nothing can happen without a cause. One way of picturing spontaneous emission is to think of the electromagnetic field as a collection of oscillators, with one oscillator for each possible frequency of light. The presence of a °photon of frequency v corresponds to the excitation of the electromagnetic field oscillator of that frequency to its first excited state. The presence of two photons of radiation of frequency v corresponds to the excitation of the same oscillator to a second quantum level, and so on. When we speak of the absence of radiation of frequency v, we mean that the oscillator of

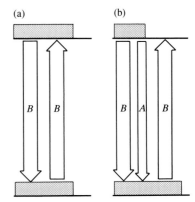

Fig. E.2 (a) If there were no spontaneous emission process, the equilibrium state of a collection of molecules would have equal populations in the upper and lower levels. (b) Since spontaneous emission can occur, the equilibrium populations can be different when the total rates of emission and absorption are the same.

Box E.1 The Einstein coefficients

When a system is in a region where the energy density at the transition frequency is ρ (such as that given by the Planck formula, Box B.1), the **rate of stimulated absorption** is

$$w(\text{absorption}) = B\rho$$

where B is the **Einstein coefficient of stimulated absorption**. The rate of stimulated emission between the same two states is

$$w(\text{emission}) = B'\rho$$

where B' is the **Einstein coefficient of stimulated emission**. The rate of spontaneous emission is

$$w(\text{spontaneous emission}) = A$$

and is independent of the intensity of any radiation that may be present. Einstein established the relations

$$B' = B \qquad A = \frac{8\pi h v^3}{c^2} \times B$$

For °electric dipole transitions,

$$B = \frac{|\mu|^2}{6\varepsilon_0 \hbar^2}$$

where μ is the °transition dipole moment.

that frequency is not excited. However, any °harmonic oscillator has a °zero-point energy. Therefore, even though there is no photon present, the electromagnetic field possesses a zero-point motion that we can picture as a fluctuation of the electric and magnetic fields. These random, stray fields can stimulate the excited state to undergo a transition just like the stronger fluctuating fields that are present when the oscillator is excited (and 'photons are present'). Hence, spontaneous transitions are transitions that are stimulated by the zero-point oscillations of the electromagnetic field.

Further information

See MQM Chapter 8. An accessible account of the Einstein coefficients is given by Bransden and Joachain (1989).

Electric dipole transition

The most intense transitions of atoms and molecules are those caused by the interaction of the electric component of an electromagnetic field with a °dipolar distribution of charge in the system.

A simple example of an electric dipole transition occurs when a polar molecule interacts with an oscillating electric field. The interaction accelerates the rotational motion of the molecule, which makes a transition between its °rotational energy levels. Conversely, a rotating electric dipole can generate an electromagnetic wave that carries away energy and leaves the molecule in a state with lower rotational energy.

Electric dipole transitions can also occur if a vibration is accompanied by the oscillation of an electric dipole. Thus, electric dipole transitions may occur between the °vibrational energy levels of a polar diatomic molecule because a hypothetical observer to the side of the molecule will see an oscillating dipole as the molecule vibrates. The same observer viewing a homonuclear diatomic molecule will detect no such oscillation, and so the vibrations of the molecule cannot interact with the electromagnetic field by an electric dipole mechanism. A nonpolar polyatomic molecule has some vibrational modes that correspond to the transient formation of a dipole (such as a bending motion of CO_2), and these modes can generate and absorb photons by an electric dipole mechanism.

Electric dipole transitions may occur between the electronic states of atoms and molecules. To decide whether or not a given transition can give rise to electromagnetic radiation by an electric dipole mechanism it is necessary to assess whether it is accompanied by a *transient* electric dipole (one that arises from the changing electron distribution and is absent before and after the transition occurs). The transition from a spherically symmetrical $1s$ orbital to a spherically symmetrical $2s$ orbital is accompanied by a symmetrical redistribution of charge (Fig. E.3a), and there is no net dipolar redistribution of charge in the transition. It follows that this transition cannot occur by an electric dipole mechanism. However, there is a transient dipolar redistribution in a transition from an s orbital to a p orbital (Fig. E.3b). If the transition takes place to p_z, the transient dipole moment lies along the z axis and the radiation is emitted with its electric vector in the z direction. If the transition is to p_x (or to p_y) the radiation is x (or y) polarized.

Similar considerations apply in molecules. Thus, a $\pi^* \leftarrow \pi$ transition in a carbonyl group may be accompanied by a migration of an electron from the O atom to the C atom (Fig. E.4a), and hence give an impact to the electromagnetic field. On the other hand, a $\pi^* \leftarrow n$ transition is essentially a rotational displacement of charge (Fig. E.4b), and does not give a dipolar impact to the field.

The general °selection rules for electric dipole transitions are described in that entry. Here we consider the three basic rules for transitions in atoms:

- The **Laporte selection rule**: the °parity must change.

- The **angular momentum selection rules**: $\Delta l = \pm 1$ and $\Delta m_l = 0, \pm 1$.

- The **spin selection rule**: the spin quantum number does not change.

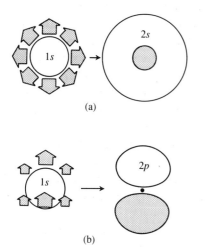

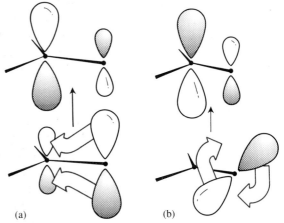

Fig. E.3 (a) The migration of electric charge (shown by the arrows) when a $1s$ electron makes a transition into a $2s$ orbital is spherically symmetrical and is not accompanied by an electric dipole transition moment. (b) In contrast, when the transition is to a $2p$ orbital, the charge migration has a directional (dipolar) character.

Fig. E.4 (a) The migration of charge in a $\pi^* \leftarrow \pi$ transition has a dipolar character and is allowed. (b) The migration in an $\pi^* \leftarrow n$ transition is largely rotational in character, and does not have electric dipole character.

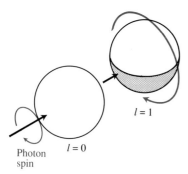

Fig. E.5 When a photon is absorbed by an electron in an *s* orbital (left), the electron acquires orbital angular momentum and makes a transition to a *p* orbital (right, $l=1$ is represented by the nodal structure of the wavefunction).

According to the Laporte rule, g→u and u→g transitions (see °parity) are allowed, but g→g and u→u transitions are not. This rule stems from the fact that there is a net dipolar shift of charge only for transitions in which there is a change in parity. Thus, *s*→*p* and *p*→*d* transitions are Laporte allowed in atoms, for they are accompanied by a dipolar migration of charge; however, *s*→*s* and *p*→*p* transitions are Laporte forbidden, for they are not accompanied by a dipolar migration of charge.

The orbital angular momentum selection rules follow from the possession by the photon of an intrinsic angular momentum (its spin) and the principle of conservation of angular momentum. Since atomic orbitals with different values of *l* correspond to different °angular momenta, when a photon is emitted or absorbed the angular momentum of the atom must change to conserve the total angular momentum. When a photon is absorbed by an *s* electron, that electron acquires the photon's angular momentum and is converted into a *p* electron (Fig. E.5). An *s* electron cannot be converted into a *d* electron by a single photon because the photon does not supply enough angular momentum. When a photon is absorbed by a *p* electron it may be converted into either a *d* electron or an *s* electron depending on the relative orientation of the electron's and photon's angular momenta (Fig. E.6).

Transitions between orbitals with definite values of the °quantum number m_l result in circularly polarized radiation. It follows from the conservation of angular momentum that when $\Delta m_l = +1$ the emitted radiation is right circularly polarized (Fig. E.7). Similarly, when $\Delta m_l = -1$, the radiation is left circularly polarized. The radiation is plane polarized when m_l is unchanged in a transition.

The spin selection rule follows from the inability of the electric field of the radiation to interact directly with the electron spin.

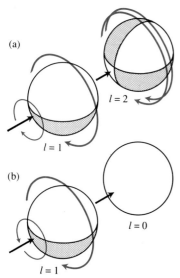

Fig. E.6 Whether *l* increases or decreases by 1 when a photon is absorbed depends on the relative orientations of the photon's and the atom's angular momenta. In (a) the orbital angular momentum is increased when the photon is absorbed (*d*←*p*) but in (b) the two momenta are initially in opposition and the orbital angular momentum is reduced (*s*←*p*).

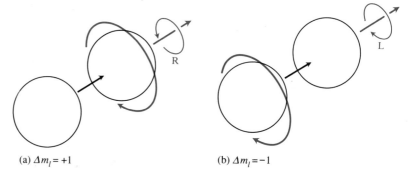

(a) $\Delta m_l = +1$ (b) $\Delta m_l = -1$

Fig. E.7 (a) The emission of a right-circularly polarized photon occurs when $\Delta m_l = +1$. (b) The emission of a left-circularly polarized photon occurs when $\Delta m_l = -1$. In each case, the projection of the total angular momentum on the propagation direction is conserved.

Further information

See MQM Chapter 12 for a discussion of electronic transitions and Chapter 8 for their background in terms of time-dependent perturbation theory. See Murrell (1971) and Sandorfy (1964) for more detailed accounts with special reference to molecular transitions.

Electromagnetic spectrum

The **electromagnetic spectrum** is the entire span of frequencies of electromagnetic radiation (loosely speaking, of 'light'). Each region is commonly identified according to the manner in which it is generated or detected (Fig. E.8). The lowest frequencies, corresponding to wavelengths that may span several kilometres, for instance, fall in the **radiofrequency region** of the spectrum. The highest frequencies and shortest wavelengths are those associated with cosmic phenomena and fall in the **cosmic ray region**. Our eyes are sensitive to frequencies that lie midway between these extremes, and our appreciation of beauty largely depends on the ability of our retinas to respond to a very tiny range of frequencies, the **visible region**, in the middle of the spectrum.

Further information

See §6 of Gray (1972) for information about the sources of different regions of the spectrum. Hollas (1983) gives a survey of the techniques used for generation, dispersal, and detection in different region of the spectrum.

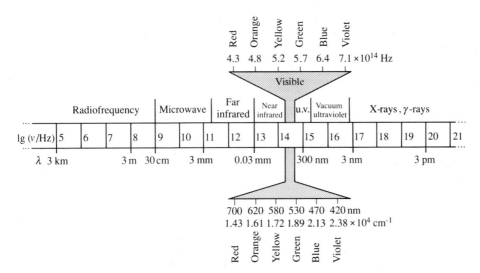

Fig. E.8 The electromagnetic spectrum and the classification of its regions.

Electron affinity

The **electron affinity** E_{ea} of an element is the energy released when an electron attaches to an atom in the gas phase. Strictly speaking, an electron affinity is the negative of the standard enthalpy change that accompanies electron attachment:

$$A(g) + e^-(g) \rightarrow A^-(g) \quad E_{ea} = -\Delta H^{\ominus}$$

However, since the enthalpy of electron attachment differs by only a few kilojoules per mole from the energy change (and the two are identical at $T = 0$), the distinction between the enthalpy and energy change is widely ignored. In elementary work electron affinities are usually treated as energy changes. The electron affinity of an atom A is equal to the ionization energy of the anion A^-.

Electron affinities may be either positive (exothermic electron attachment) or negative (endothermic electron attachment), Table E.1. A positive electron affinity indicates that the anion has a lower energy than the neutral atom and a stationary electron at infinity. Positive electron affinities are found when the incoming electron enters an orbital in which it can interact strongly with the nuclear charge. Negative electron affinities (which occur if A^- has a higher energy than A) are encountered when the repulsion from the electrons already present dominates the nuclear attraction. All *second* electron affinities (the energy released when a second electron is attached to the atom,

Table E.1 Electron affinities of the main group elements, E_{ea}/eV

H							He
0.754							−0.5
Li	Be	B	C	N	O	F	Ne
0.618	−0.5	0.277	1.263	−0.07	1.461	3.399	−1.2
					−8.75		
Na	Mg	Al	Si	P	S	Cl	Ar
0.548	−0.4	0.441	1.385	0.747	2.077	3.617	−1.0
					−5.51		
K	Ca	Ga	Ge	As	Se	Br	Kr
0.502	−0.3	0.30	1.2	0.81	2.021	3.365	−1.0
Rb	Sr	In	Sn	Sb	Te	I	Xe
0.486	−0.3	0.3	1.2	1.07	1.971	3.059	−0.8

To convert to $kJ\,mol^{-1}$, multiply by 96.485. The first values refer to the formation of the ion X^- from the neutral atom X; the second value to the formation of X^{2-} from X^-.
Data: Hotop and Lineberger (1985).

as in the formation of O^{2-} from O^-) are negative, since the incoming electron is strongly repelled by the negative charge already present.

Methods for the determination of electron affinities include:

- Calculation. The electron affinity is essentially the energy of the lowest unoccupied atomic (or molecular) orbital (Fig. E.9); however, Hartree–Fock calculations of electron affinities are notoriously unreliable.

- Measuring the ionization energy of the anion.

- Electron impact studies, and the determination of the appearance potential of the anion in the reaction $XY + e^- \rightarrow X^- + Y$

- Electron attachment measurements (electron affinity spectroscopy).

- Polarography.

- Using the Born–Haber cycle for the lattice energy of an ionic crystal.

Electron affinities are used in the assessment of the energetics of ion formation, in the interpretation of thermodynamic properties — particularly lattice enthalpies and standard reduction potentials — and in the formulation of scales of °electronegativity and °hardness.

Further information

The methods of measuring electron affinities have been surveyed by McDowell (1969) and extensive tables have been published by Hotop and Lineberger (1975) and extracted by Emsley (1989). See Chen and Wentworth (1975) for a survey of trends through the Periodic Table.

Electron correlation

See °correlation energy.

Electron slip

Electrons 'slip' because in a rotating molecule they may be unable to follow the rapid motion of the nuclei. **Electron slip** is a manifestation of the breakdown of the °Born–Oppenheimer approximation as a result of the rotation of a molecule.

There are several consequences of electron slip. One is that it blurs the distinction between s and p orbitals (and between Σ and Π states) and results in Λ-type °doubling. Another consequence is that all molecules possess a magnetic moment, their **molecular magnetic moment**, as a result of their rotation. This moment can be traced to the different

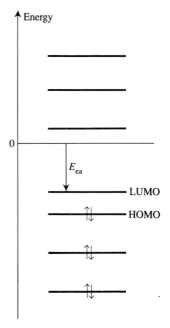

Fig. E.9 The energy released when an electron adds to a molecule is approximately equal to the energy of the LUMO of the molecule. Hence, the electron affinity is determined by the energy of that orbital.

rates of rotation of the positively charged nuclear framework and the negatively charged electron cloud of the molecule. As a result of the different rotation rates there is a net current which gives rise to a magnetic moment. All molecules acquire such a moment when they rotate, and the magnitude of the moment increases with the °rotational quantum number J. Typical molecular magnetic moments are of the order of the °nuclear magneton, and hence are about a thousand times weaker than magnetic moments arising from unpaired electrons.

Further information

See MQM Chapter 12 for a more detailed account of how electron slip mixes terms in rotating linear molecules, and Zare (1987), Struve (1989), Herzberg (1950), and Kovaćs (1969) for quantitative treatments of increasing sophistication. For molecular magnetic moments, see §11.6 of Townes and Schawlow (1955) and Herzberg (1950).

Electron spin resonance

The **electron spin resonance** (ESR) or **electron paramagnetic resonance** (EPR) technique is the observation of the magnetic field needed to bring the separation of electron spin energy levels into resonance with a microwave electromagnetic field.

An electron possesses a spin °magnetic moment, and in the presence of an applied field its two permitted orientations (corresponding to the spin states α and β) have different energies. The α state lies $0.3\ cm^{-1}$ above the β state in a field of 0.3 T. An electron spin makes the transition $\alpha \leftarrow \beta$ most efficiently when the energy separation of the two states matches the energy of the photon, for then the sample and the electromagnetic field are in °resonance (Fig. E.10). Resonance occurs in a 0.3 T field with microwave radiation of wavenumber $0.3\ cm^{-1}$ (wavelength 3 cm, frequency 9 GHz), which is in the microwave X-band region of the electromagnetic spectrum. The apparatus therefore consists of a magnet capable of providing a homogeneous field of 0.3 T, a source of X-band microwaves (a klystron), and a detector to monitor the absorption.

The observation is performed by monitoring the absorption as the applied field is changed while the sample is exposed to a constant microwave radiation. The sample, which must be paramagnetic, may be a solid, a liquid, or (more rarely) a gas. The peculiar appearance of the spectrum (Fig. E.11) arises because the detection technique, which is called **phase-sensitive detection**, monitors the first derivative of the absorption with respect to the field, so the point of zero slope (where the plotted line passes through the horizontal axis) marks the peak of the absorption.

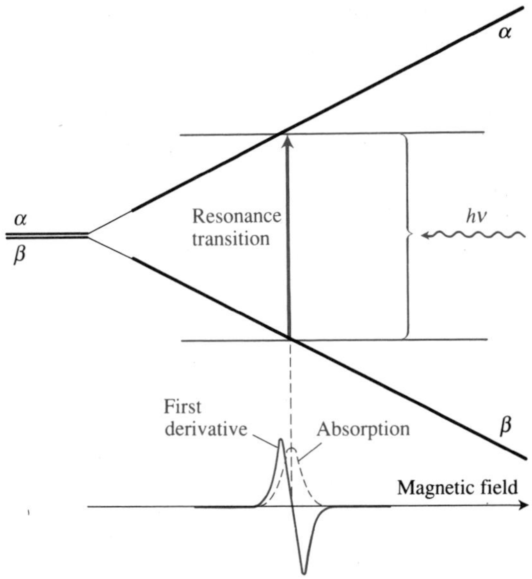

Fig. E.10 The fundamental basis of ESR. The energies of the α and β states of an electron diverge in a magnetic field. When the separation of the states matches the energy of the incident (microwave) photons, there is a strong resonant absorption. The signal (bottom) is the first derivative of the absorption intensity.

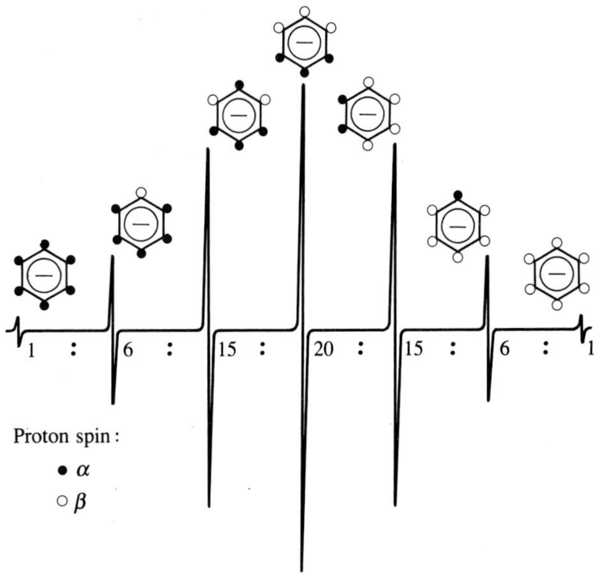

Fig. E.11 The hyperfine structure of the ESR spectrum of the benzene radical anion. There are seven lines in the binomial intensity ratio 1:6:15:20:15:6:1. Each line arises from a molecule with different numbers of α and β *proton* spins, as shown by the colouring at the positions of the H atoms on the ring. For example, the line second on the left can occur with any one of six protons in a β state.

Three principal pieces of information are obtained:

- The position of the centre of the spectrum, as expressed by the °g-factor.
- The structure of the spectrum, its hyperfine structure.
- The widths of the lines.

The magnetic field experienced by an electron in a paramagnetic species differs from the applied field because the latter is able to induce local fields. For a given microwave frequency, the resonance condition is attained at the same *local* field, and therefore at slightly different applied fields in different species. If the microwave frequency is v and the applied field is B, the resonance condition is $g\mu_B B = hv$, where the °g-factor is a parameter that takes into account the possibility that the local field is not exactly equal to B. Measuring the position of the spectrum enables g to be determined and then interpreted in terms of the electronic structure of the paramagnetic species.

The °hyperfine structure of an ESR spectrum arises from the interaction of the electron with nuclei with a spin magnetic moment (such as ^{1}H and ^{14}N but not ^{12}C and ^{16}O). Such magnetic nuclei give rise to a local magnetic field, which, depending on the relative orientation of the nuclear spin and the applied field, can increase or decrease the local field experienced by the electron. As a result, electrons near nuclei with different orientations resonate at different applied magnetic fields. The separation of the hyperfine lines may be interpreted in terms of the distribution of the electrons near the magnetic nucleus in question, and the electron distribution can be mapped over the magnetic nuclei in the molecule.

The widths of the lines depends on the motion of the paramagnetic species, for it is determined by °relaxation processes. An analysis of the shapes and widths is used to infer the rates and anisotropies of the rotational motion of species in fluid solutions.

Further information

For an introduction to ESR see Atkins (1990). For a more detailed introduction see Carrington and McLachlan (1967), Wertz and Bolton (1972), Slichter (1988), and Abragam and Bleaney (1970). Chemical applications are described straight-forwardly in Symons (1978) and Ebsworth, Rankin, and Cradock (1987). For data on organic radicals see Fischer and Hellwege (1977); for inorganic radicals see Atkins and Symons (1967), and for d-metal complexes see Abragam and Bleaney (1970). The calculation of magnetic resonance parameters is introduced in MQM Chapter 14 and described in detail by Memory (1968).

Electronegativity

The **electronegativity** χ of an element is a measure of the power of an atom of the element to attract electrons when it is part of a compound: the greater its electronegativity the greater its drawing power. An element with a low electronegativity (like the alkali metals) is said to be **electropositive**.

The **Pauling scale** of electronegativities, χ_P, is based on the lowering of the calculated energy of a diatomic molecule when ionic structures are included to form a °resonance hybrid. Suppose the energy of the molecule AB is E, but that a calculation using purely covalent structures yielded E_{cov}; then the ionic resonance energy is $E_{res} = E - E_{cov}$. Pauling found that the square root of E_{res} could be set proportional to the difference of two numbers, one characteristic of the element A and the other of B, and that the expression $E_{res}^{1/2} = |\chi_A - \chi_B|$ is valid for a wide range of combinations. To set up the scale it is necessary to estimate E_{cov}. Pauling proposed that a reasonable approximation is the mean of the energies of the molecules A_2 and B_2. This was justified by the view that the energies of A_2, AB, and B_2 should form a simple sequence if polar structures in AB are omitted. Pauling initially used the arithmetic mean, but later (in the third edition of his book) adopted a geometric mean. Some values are listed in Table E.2. Pauling electro-

Table E.2a Electronegativities of the main group elements

Atom	χ_P	χ_{AR}	χ_M	χ_S	Atom	χ_P	χ_{AR}	χ_M	χ_S
H	2.20	2.20	3.06	2.30	Ga	1.81	1.82	1.34	1.76
Li	0.98	0.97	1.28	0.91	Ge	2.01	2.02	1.95	1.99
Be	1.57	1.47	1.99	1.58	As	2.18	2.20	2.26	2.21
B	2.04	2.01	1.83	2.05	Se	2.55	2.48	2.51	2.42
C	2.55	2.50	2.67	2.54	Br	2.96	2.74	3.24	2.69
N	3.04	3.07	3.08	3.07	Kr			2.98	2.97
O	3.44	3.50	3.21	3.61	Rb	0.82	0.89	0.99	0.71
F	3.98	4.10	4.42	4.19	Sr	0.95	0.99	1.21	0.96
Ne			4.60	4.79	In	1.78	1.49	1.30	1.66
Na	0.93	1.01	1.21	0.87	Sn	1.96	1.72	1.83	1.82
Mg	1.31	1.23	1.63	1.29	Sb	2.05	1.82	2.06	1.98
Al	1.61	1.47	1.37	1.61	Te	2.10	2.01	2.34	2.16
Si	1.90	1.74	2.03	1.92	I	2.66	2.21	2.88	2.36
P	2.19	2.06	2.39	2.25	Xe			2.59	2.58
S	2.58	2.44	2.65	2.59	Cs	0.79	0.86		
Cl	3.16	2.83	3.54	2.87	Ba	0.89	0.97		
Ar			3.36	3.24	Tl	2.04			
K	0.82	0.91	1.03	0.73	Pb	2.33			
Ca	1.00	1.04	1.30	1.03	Bi	2.02			

χ_P: Pauling; χ_{AR}: Allred–Rochow; χ_M: Mulliken; χ_S: spectroscopic. Data: Shriver, Atkins, and Langford (1990).

Table E.2b Pauling electronegativities of some *d*- and *f*-block elements.

d block (oxidation numbers as indicated)

+3	+2	+2	+2	+2	+2	+2	+2	+1	+2
Sc	Ti	V	Cr	Mn	Fe	Co	Ni	Cu	Zn
1.36	1.54	1.63	1.66	1.55	1.83	1.88	1.91	1.90	1.65
Y	Zr	Nb	Mo	Tc	Ru	Rh	Pd	Ag	Cd
1.22	1.33		2.16			2.28	2.20	1.93	1.69
La	Hf	Ta	W	Re	Os	Ir	Pt	Au	Hg
1.10			2.36			2.20	2.28	2.54	2.00

f block (oxidation number +3)

Ce	Pr	Nd	Pm	Sm	Eu	Gd	Tb	Dy	Ho	Er	Tm	Yb	Lu
1.12	1.13	1.14		1.17		1.20		1.22	1.23	1.24			1.25

			U	Np	Pu
			1.38	1.36	1.28

Data: Shriver, Atkins, and Langford (1990).

negativities depend on the °oxidation number of the atom and increase with increasing oxidation number. The values in Table E.2 are for the maximum oxidation number of each element.

The **Mulliken scale** of electronegativities, χ_M, gives an *absolute* scale.

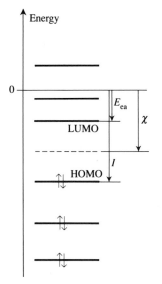

Fig. E.12 The absolute electronegativity is the mean of the energies of the HOMO and the LUMO. Since the HOMO energy determines the ionization energy of the species and the LUMO energy determines its electron affinity (compare Fig. E.9), the electronegativity is also equal to the mean of the ionization energy and the electron affinity.

Mulliken defined the electronegativity of an element as the mean of the °ionization energy and the °electron affinity:

$$\chi = \tfrac{1}{2}(I + E_{ea})$$

Mulliken electronegativities can be expressed in Pauling units and are labelled χ_M in Table E.2.

An advantage of the Mulliken definition is that it is possible to take into account the differences in electronegativities of different orbitals on the same atom and of the dependence of the electronegativity on the atom's state of °hybridization. As illustrated in Fig. E.12, the Mulliken electronegativity is essentially the mean energy of the frontier orbitals of the atom. However, as a typical Mulliken electronegativity refers to the atom in its °valence state (the mixture of spectroscopic states that characterize its actual state when it is part of a molecule), some calculation needs to be done to convert a series of ionization energies and electron affinities for different spectroscopic states into the combination appropriate to the valence state. The Mulliken definition may be applied to establish the absolute electronegativity of a *molecule*, in which case χ is also the mean of the HOMO and LUMO energies.

The **Allred–Rochow scale** of electronegativities, χ_{AR}, is based on the view that χ is an atomic property. It relates the electronegativity to the force exerted by the effective nuclear charge at the periphery of an atom. A spherical ion of radius r and charge number z behaves like a point charge located at the ion's centre. The Coulomb potential at its surface, the **ionic potential** of the ion, is proportional to z/r, and the force it exerts on charges in its vicinity is proportional to z/r^2. The Allred–Rochow definition sets χ proportional to Z_{eff}/r^2, using for r the °covalent radius of the atom:

$$\chi_{AR} = \frac{3590 Z_{eff}}{(r/pm)^2} + 0.744$$

Elements with the highest electronegativity on the Allred–Rochow scale are those with the highest effective nuclear charge and the smallest radii: these lie close to fluorine. Allred–Rochow electronegativities are proportional to the Pauling values and a linear fit permits them to be expressed in 'Pauling units'.

A **spectroscopic electronegativity scale** proposed by L.C. Allen is defined in terms of the average energy of the electrons in the s and p orbitals of an atom:

$$\chi_S = \frac{n_s \varepsilon_s + n_p \varepsilon_p}{n_s + n_p}$$

where n_s and n_p are the numbers of electrons in the s and p orbitals respectively. The orbital energies ε are obtained from averaged spectroscopic data. Since spectroscopic data are available for many elements, it is possible to calculate electronegativities for the noble gases for which the thermochemical data needed for the determination of χ_P are largely unavailable.

The most fundamental definition of electronegativity is that developed extensively by Parr, who identifies the electronegativity with the negative of the chemical potential of the atom (or molecule). The chemical potential, μ, is defined, by analogy with conventional thermodynamics, as $\mu = (\partial E/\partial N)_V$, where E is the electronic energy, N the number of electrons, and the constraint V is the potential arising from the nuclei. To a good approximation, the slope of the electronic energy with respect to the addition of electrons is equal to the mean of the ionization energy and the electron affinity (Fig. E.13), in accord with Mulliken's definition of electronegativity.

The interpretation of electronegativity as the negative of a chemical potential is plausible on a variety of grounds. Most compelling is the fact that electrons tend to flow towards regions of higher electronegativity (lower chemical potential). Moreover, when a molecule forms, the electronegativities of the atoms tend to equalize and acquire the same, uniform value (Sanderson's 'principle of electronegativity equalization').

Electronegativities are used to estimate a number of molecular properties (Box E.2). Thus, they can be used to estimate A—B bond energies if the A—A and B—B bond energies are known (this calculation simply requires a reversal of the Pauling definition). Electronegativities

Box E.2 Electronegativity and molecular properties

Electric dipole moment:

$$\mu/D = \chi_A - \chi_B$$

Percentage ionic character:

$$16|\chi_A - \chi_B| + 3.5|\chi_A - \chi_B|^2$$

Ionic–covalent resonance energy:

$$E/eV = (\chi_A - \chi_B)^2$$

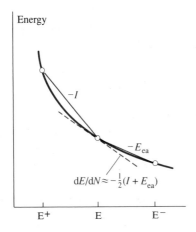

Fig. E.13 The energy of a species plotted against the number of electrons. The difference between the energies of E and E^+, the ionization energy of E, is equal to the slope of the graph to the left of E. The difference between the energies of E^- and E, the electron affinity of E, is equal to the slope on the right of E. The average of the two slopes is the slope of the graph at E itself, and is equal to $\frac{1}{2}(I + E_{ea})$.

may also be used to predict the **ionicity** of a bond (its percentage ionic character) and hence its dipole moment. Pauling electronegativities are best for bond polarities and strengths; the absolute electronegativities (such as Mulliken's) are more reliable for the discussion of properties relating to reactions. The consequence of the principle of electronegativity equalization, that electron distributions adjust until the electronegativities of all the atoms in a molecule are the same, helps to rationalize a large amount of inorganic chemistry (see *Further information*).

Further information

For the original accounts by Pauling and Mulliken see Pauling (1960) and Mulliken (1934). The other scales are described by Allred and Rochow (1958) and by Allen (1989). A brief but helpful synoptic view of the different definitions has been given by Pearson (1990). The account given by Huheey (1983) is a useful survey of the field, as is the critical account by Bratsch (1988a, 1988b). For applications see Atkins (1989), Shriver, Atkins, and Langford (1990), Bratsch (1988c), and Sanderson (1976, 1988, 1989). The *systematic* discussion of electronegativity in terms of density-functional theory, by which the concept and its interpretations (including the principle of electronegativity equalization) can be put on a firm foundation, is given very clearly in Parr and Yang (1989).

Electronic spectra of molecules

The electronic spectrum of a diatomic molecule consists of a number of **bands**, each resulting from a transition from the ground electronic state (denoted X) to an excited electronic state (which is denoted A, B,... or by its °term symbol). The structure of each band is a result of the simultaneous excitation of vibrational modes, with the intensities of the individual vibrational transitions determined by the °Franck–Condon principle. The vibrational transitions have a further °branch structure that arises from the change in the rotational state of the molecule. In a condensed phase the rotational structure of the spectrum is lost, and the vibrational structure is usually so diffuse that the absorption spectrum is often just a series of broad, almost structureless bands.

The vibrational lines converge at high energies as the dissociation limit is approached. The °dissociation energy may be determined by observing the dissociation limit or by a careful extrapolation from lower frequencies. In some cases the rotational and vibrational structure disappears but then reappears before the dissociation limit is reached: this is a sign that °predissociation has occurred. From the spectrum may be determined the °force constants and dissociation energies of the electronic states of the molecule. The electronic spectra of polyatomic molecules are more complicated, but the same principles apply and analogous information is obtained. The return to the ground state is a process of interest, particularly when the molecule loses its energy radiatively by °fluorescence or °phosphorescence.

Other effects may be observed when a molecule is a part of a crystal lattice; one especially important phenomenon is °exciton formation.

Further information

See MQM Chapter 12 for an introduction to the electronic spectra of molecules. Introductory texts include those by Hollas (1987) and Banwell (1983). The surveys by Rao (1975), King (1964), and Lever (1984) should prove helpful, as should Steinfeld (1985) and Struve (1989). Murrell (1971) deals with electronic transitions in organic molecules and Gaydon (1968) shows how to extract information about the potential energy curves of diatomic molecules. Herzberg (1950, 1966) is a mine of detailed information, and is a brilliant example of the application of theoretical concepts to the detailed examination of molecular properties.

Electronvolt

An **electronvolt** (eV) is the energy acquired by an electron when it is accelerated by a potential difference of 1 V. Since the fundamental charge is 1.602×10^{-19} C, and $1\,J = 1\,V\,C$, 1 eV is equivalent to 1.602×10^{-19} J (which corresponds to 96.49 kJmol^{-1}). Another useful conversion is $1\,eV = 8023$ cm^{-1}.

Further information

See Mills *et al.* (1988) for the IUPAC stance on the electronvolt (and for further information on energy units in general).

Equipartition theorem

The **equipartition theorem** states that:

> The mean energy of each quadratic term in the expression for the energy of a classical system in thermal equilibrium at a temperature T is the same, and equal to $\frac{1}{2}kT$.

As an example, the kinetic energy of an atom free to move in one dimension is $\frac{1}{2}mv^2$, corresponding to one quadratic term. The average value of the kinetic energy is therefore $\frac{1}{2}kT$. If the atom is free to move in three dimensions, its total energy is the sum of three quadratic terms (each of the form $\frac{1}{2}mv^2$, one for each coordinate); hence its total average kinetic energy is $\frac{3}{2}kT$ and the total molar kinetic energy is $\frac{3}{2}RT$. A molecule that can °rotate around three axes has a kinetic energy $\frac{1}{2}I\omega^2$ around each axis, and therefore an average energy of $\frac{1}{2}kT$ for each axis, giving a total of $\frac{3}{2}kT$. A one-dimensional °harmonic oscillator has total energy $\frac{1}{2}mv^2 + \frac{1}{2}k_f x^2$, where x is the displacement from equilibrium. Hence, its average energy is kT since there are *two* quadratic terms (v^2 and x^2) in the expression for its total energy.

The equipartition theorem is a consequence of statistical mechanical principles for systems that obey equilibrium classical mechanics. The theorem fails when either the energy separation of quantized levels is large or the temperature is low, and so it is not applicable to the vibration of molecules nor to small rotating molecules at low temperatures. The condition of validity is that $T \gg \theta$, where θ is the **characteristic temperature** of the mode of motion, the typical separation between quantized levels expressed as a temperature (by division by Boltzmann's constant k). Thus, for a harmonic oscillator, for which the separation between neighbouring levels is hv, the characteristic temperature is $\theta_V = hv/k$, or $hc\tilde{v}/k$, where $\tilde{v}$ is the wavenumber of the vibration. For a C—C bond, $\tilde{v} \approx 1000 \text{ cm}^{-1}$, and hence $\theta_V \approx 1500$ K. For a rotating molecule, the typical energy separation is of the order of hcB, where B is the °rotational constant; hence the characteristic rotational temperature is $\theta_R = hcB/k$ and the equipartition theorem applies if the temperature exceeds θ_R. For N_2, $B \approx 2 \text{ cm}^{-1}$, and so $\theta_R \approx 3$ K.

Further information

See Atkins (1990) for a straightforward discussion with applications. The derivation of the equipartition theorem will be found in Hecht (1990) and Davidson (1962). See also §332 of Fowler and Guggenheim (1965), §44 of Landau and Lifshitz (1958b), and Tolman (1938) for erudite discussions.

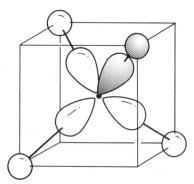

Fig. E.14 One of the four equivalent orbitals of a tetrahedral molecule (such as CH_4). All four equivalent orbitals can be transformed into each other by a symmetry operation of the molecule.

Equivalent orbital

A set of **equivalent orbitals** consists of two or more °localized orbitals that transform into each other under a °symmetry operation of a molecule. As an example, consider one of the C—H bonds in CH_4 formed from (Csp^3,H1s) overlap (Fig. E.14). If the tetrahedral molecule is rotated into another equivalent position, another of the same type of bonds is rotated into the original position. The four σ orbitals constitute a set of equivalent orbitals.

Further information

See Coulson (1979). Further information can be traced through the entry on °localized orbitals.

Exchange energy

The **exchange energy** is a correction that should be made to the °Coulomb integral J to take into account the effect of °spin correlation. The latter quantum mechanical effect gives electrons of opposite spin orientation an enhanced probability of being found near each other but tends to keep apart those of the same spin. Thus if the two electrons have the same spin, the true average repulsion energy will be less than J on account of the intrinsic tendency of such electrons to avoid each other. The average repulsion energy is reduced from J to $J - K$, where K is the **exchange integral** (Box E.3). The exchange integral is always positive and is taken into account automatically in Hartree–Fock theory (see °self-consistent fields).

Box E.3 Coulomb and exchange integrals

The **Coulomb integral**, J, for the interaction between electrons 1 and 2 in orbitals A and B is

$$J = \kappa \int A(1)^2 \times \frac{1}{r_{12}} \times B(2)^2 \, d\tau_1 \, d\tau_2$$

where $\kappa = e^2/4\pi\varepsilon_0$. The corresponding **exchange integral**, K, is

$$K = \kappa \int A(1) B(1) \times \frac{1}{r_{12}} \times A(2) B(2) \, d\tau_1 \, d\tau_2$$

Further information

See MQM Chapters 9 and 10 for a more detailed account and examples of the application of the concept in atoms and molecules. Good discussions will be found in Davydov (1976), McWeeny (1989), Szabo and Ostlund (1982), and Wilson (1984, 1987). For applications to molecules see Coulson (1979), Richards and Cooper (1983), and McGlynn et al. (1972). The last reference (in Appendix E) includes a helpful guide to the computation of exchange integrals. See the entry on °magnetism for further information about the role of exchange in the determination of properties.

Exciton

Consider an electronic excitation of a molecule (or atom or ion) in a crystal. If the excitation corresponds to the removal of an electron from one orbital of a molecule and its elevation to an orbital of higher energy, the excited state of the molecule can be envisaged as the coexistence of an electron and a hole. The particle-like hopping of this **electron–hole pair** from molecule to molecule is the migration of the exciton through the crystal (Fig. E.15). Exciton formation causes spectral lines to shift, split, and change in intensity.

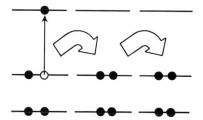

Fig. E.15 The electron-hole pair shown on the left can migrate through a solid lattice as the excitation hops from molecule to molecule. The mobile excitation is called an exciton.

In the 'tight-binding case', the electron and the hole jump together from molecule to molecule as they migrate. A migrating excitation of this kind is called a **Frenkel exciton**. In the 'weak-binding case', the electron and hole are on different molecules, but in each other's vicinity. A migrating excitation of this kind, which is now spread over several molecules (more usually ions), is a **Wannier exciton**. In molecular solids the Frenkel exciton is more common, and we consider it alone.

The migration of an exciton implies that there is an interaction between the species that constitute the crystal, for otherwise the excitation on one unit could not move to another. This interaction affects the energy levels of the system. The strength of the interaction governs the rate at which an exciton moves through the crystal: a strong interaction results in fast migration, and a vanishingly small interaction leaves the exciton localized on its original molecule.

The specific mechanism of interaction that leads to exciton migration is the interaction between the °transition dipole moments of the excitation. Thus, an °electric dipole transition in a molecule is accompanied by a shift of charge, and the transient dipole exerts a force on an adjacent molecule. The latter responds by shifting its charge. This process continues and the excitation migrates through the crystal.

The energy shift arising from the interaction between transition dipoles can be understood in terms of their electrostatic interaction. Because the wavelength of the radiation is long in comparison with the molecular spacing, its electric field has the same phase over a large number of molecules. As a result, it can excite only the all-parallel arrangement of transition dipoles. This arrangement of dipoles, with the orientation shown in Fig. E.16a, is energetically unfavourable, so the absorption occurs at higher frequency than in the isolated molecule. Conversely, if the polarization of the transition dipole moments is along the line of the molecules (Fig. E.16b), the transition occurs at a lower frequency than in the isolated molecules. In this case the transient dipoles lie head-to-tail, which is an energetically favourable orientation.

If there are N molecules per unit cell, there are N **exciton bands** in the spectrum (if all of them are electric dipole allowed). The splitting between the bands is the **Davydov splitting**. To understand the origin of the splitting, consider the case $N = 2$ with the molecules arranged as in Fig. E.17. Let the transition dipoles be along the length of the molecules. The radiation stimulates the collective excitation of the transition dipoles that are in-phase between neighbouring unit cells. *Within* each unit cell the transition dipoles may be arrayed as (a) or (b) in Fig. E.17. Since the two orientations correspond to different interaction energies, the two transitions appear in the spectrum as two bands at different frequencies. The Davydov splitting is determined by the energy of interaction between the transition dipoles within the unit cell.

(a)

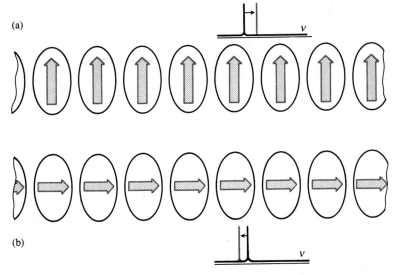

(b)

Fig. E.16 (a) The alignment of transition dipoles (the blue arrows) is energetically unfavourable, and the exciton absorption is shifted to higher energy (higher frequency). (b) The alignment is energetically favourable for a transition in this orientation, and the exciton band occurs at lower frequency than in the isolated molecules.

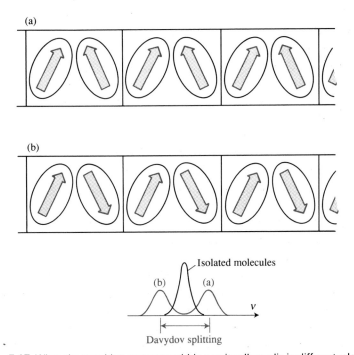

Fig. E.17 When the transition moments within a unit cell may lie in different relative directions, as depicted by (a) and (b), the energies of the transitions are shifted and give rise to the two bands labelled (a) and (b) in the spectrum. The separation of the bands is the Davydov splitting.

Further information

A simple account of the formation of excitons will be found in Murrell (1971), who also discusses 'hypochromism' (the reduction in intensity of absorption) and 'hyperchromism' (the increase in intensity). An introduction to the mathematical theory of excitons is provided by Craig and Walmsley (1968), Kittel (1986, 1987), and Davydov (1976).

Exclusion principle

See °Pauli exclusion principle.

Expansion of the octet

See °hypervalent compounds.

Expectation value

The **expectation value** of an observable is the mean outcome of a set of measurements. Thus, the expectation value of the energy is the mean energy that would be measured in a series of experiments on a collection of identical systems (so long as there is no systematic error in the observations). Similarly, the expectation value of the position is the mean of the positions that would be measured in a series of experiments. If the (°normalized) state of a system is ψ, the expectation value $<\Omega>$ of an observable Ω is

$$<\Omega> = \int \psi^* \Omega \psi \, \mathrm{d}\tau$$

If the system is in a pure state with respect to one of the observables (in other words the state is an °eigenstate of the °operator corresponding to that observable), all identical experiments give identical results, and the expectation value is the result that would be measured in all the experiments. Such results are said to be **dispersion free**. An example is the determination of the energy of an atom or a molecule when it is in a definite energy state. It is then in an eigenstate of the energy operator for the system, and the expectation value of this operator is the energy that would be measured in each and every experiment.

In general, a system is described by a °superposition of pure states. In that case, a particular measurement always gives a result that can be identified as *one* of the eigenvalues of the operator corresponding to the property. The *average* of all the outcomes of identical observations performed on a set of identical systems is the expectation value for the system in that state. An example is the determination of the linear momentum of a system described by a °superposition of states of different momenta (for example, if the particle is described by a °wave packet). Then, if the state with momentum $k\hbar$ occurs in the superposition with coefficient c_k, the experiment will yield the answer $k\hbar$ with a probability proportional to $|c_k|^2$. The average of all such measurements is the expectation value of the linear momentum, written $<p>$.

Further information

See MQM Chapter 5 for an introduction, and any standard book on quantum mechanics, such as those by Rae (1986), Davydov (1976), and Bransden and Joachain (1989). For deeper discussions of the measurement process, see Hughes (1989), Redhead (1987), and the collection edited by Wheeler and Zurek (1983).

Extinction coefficient

See °absorption intensity.

F

f orbital

An **f orbital** is an atomic orbital with $l = 3$; they occur in shells for which n is equal to or greater than 4. There are seven f orbitals in a subshell, corresponding to the seven values of the quantum number m_l, namely $+3, +2, \ldots, -3$. Mathematical expressions for the real forms of the orbitals are given in Box F.1 and their boundary surfaces are shown in Fig. 7.1.

The $4f$ orbitals are occupied starting with lanthanum and are responsible for the existence of the first row of the f block (the 'rare earths' or 'lanthanides'). The $5f$ orbitals are occupied from actinium, and give rise to the second row of the f block (the 'actinides'). One chemically important characteristic of the f orbitals is that electrons that occupy them shield the nuclear charge only very poorly (partly on account of the large number of angular nodes, which denote regions of zero electron density). One consequence is the **lanthanide contraction**, the observation that the atomic radii of elements following the lanthanides are significantly smaller than would be expected in comparison with the previous period of the Periodic Table. The valence shell electrons of these elements, which include gold, mercury, and lead, are strongly attracted by the poorly shielded, highly charged nucleus. Furthermore, the small atomic radii of these heavy (high atomic number) elements accounts for their high densities: osmium and iridium are the densest elements of all. The compact character of the atoms

---

Box F.1 f orbitals

$\psi = R(r)\, Y(\theta, \phi)$

Complex polar form:

$m_l = 0, \qquad Y = \left(\dfrac{63}{16\pi}\right)^{1/2} (\tfrac{5}{3}\cos^3\theta - 1)$

$m_l = \pm 1, \qquad Y = \mp\left(\dfrac{21}{64\pi}\right)^{1/2} \sin\theta\,(5\cos^2\theta - 1)\, e^{\pm i\phi}$

$m_l = \pm 2, \qquad Y = \left(\dfrac{105}{32\pi}\right)^{1/2} \sin^2\theta\,\cos\theta\, e^{\pm 2i\phi}$

$m_l = \pm 3, \qquad Y = \mp\left(\dfrac{35}{64\pi}\right)^{1/2} \sin^3\theta\, e^{\pm 3i\phi}$

Radial wavefunctions:

$n = 4 \qquad R = \dfrac{1}{96\sqrt{35}}\left\{\dfrac{Z}{a_0}\right\}^{3/2} \rho^3\, e^{-\rho/2} \qquad \rho = Zr/2a_0$

$n = 5 \qquad R = \dfrac{1}{300\sqrt{70}}\left\{\dfrac{Z}{a_0}\right\}^{3/2} (8 - \rho)\rho^3\, e^{-\rho/2} \qquad \rho = 2Zr/5a_0$

See also Box H.7 for general expressions.

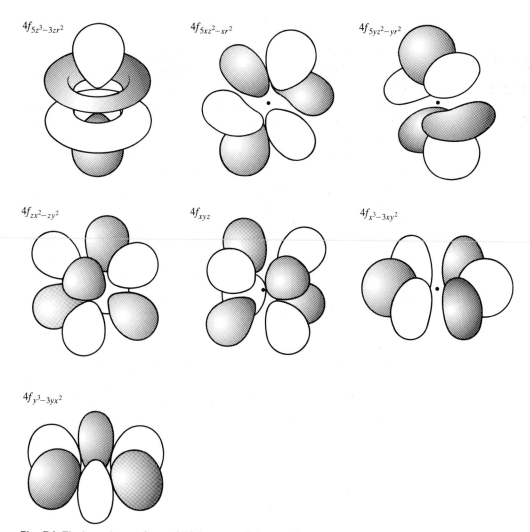

$4f_{5z^3-3zr^2}$ $4f_{5xz^2-xr^2}$ $4f_{5yz^2-yr^2}$

$4f_{zx^2-zy^2}$ $4f_{xyz}$ $4f_{x^3-3xy^2}$

$4f_{y^3-3yx^2}$

Fig. F.1 The boundary surfaces of 4*f* electrons. (Other real linear combinations may also be formed which result in orbitals that look different from these.)

that follow the lanthanides also accounts for the low chemical reactivity of these elements, for their ionization energies are high and the valence electrons are not easy to remove.

Further information

For a modern survey of the consequences of electrons occupying *f* orbitals see Salahub and Zerner (1989). Illustrations of the complex form of *f* orbitals (in the style of Fig. D.2 for *d* orbitals), in a manner that shows most clearly how the seven orbitals form a family, will be found in Breneman (1988). The precise starting point of the lanthanides (at lanthanum or cerium) and of the actinides is questioned by Jensen (1980). The evidence for the lanthanide contraction, and a corresponding 'scandinide' contraction in the d block, has been reviewed by Lloyd (1986).

Fermi contact interaction

The **Fermi contact interaction** is a magnetic °hyperfine interaction between a nucleus and an electron that approach each other so closely that they are effectively in contact. Since only °s electrons have a non-zero probability of being found at their own nucleus, only s electrons can have a Fermi contact interaction.

The energy of the Fermi contact interaction depends on the relative orientations of the spins of the electron and the nucleus. Thus, an electron in a hydrogen atom in which the proton has α spin has slightly different energies when its spin is α from when its spin is β. It follows from this orientation dependence that the contact interaction can be expressed as a magnetic field generated by the nucleus, the strength of the field determining the difference in energy between the α and β states of the electron. The contact interaction for the $1s$ orbital in the H atom is equivalent to a magnetic field of 50 mT acting on the electron; for an electron in a $2s$ orbital of an F atom the field is a massive 1.7 T. The inner electrons of heavier elements can possess very large interactions, amounting to hundreds of tesla. Some representative values are listed in Table F.1.

A pictorial description of the interaction is as follows. First, we regard the magnetic moment of the nucleus as arising from the circulation of a current in a tiny loop with a radius similar to that of the nucleus (Fig. F.2). Far from the nucleus the field generated by this loop is indistinguishable from the field generated by a point magnetic dipole. Close to the loop, however, the field differs from that of a point dipole. The magnetic interaction between this non-dipolar field and the electron's magnetic moment is the contact interaction.

The Fermi contact interaction plays an important role in magnetic resonance. In °nuclear magnetic resonance it contributes to the spin–spin coupling and hence to the fine structure of the spectrum. In °electron spin resonance spectra it contributes to the isotropic component of the °hyperfine structure.

Further information

See MQM Chapter 14 for a further discussion of the interaction. An account of the role of the Fermi contact interaction in magnetic resonance can be found in the books by Harris (1986), Carrington and McLachlan (1967), Homans (1989), Wertz and Bolton (1972), and Abragam and Bleaney (1970). Calculations that depend on the contact interaction are described by Memory (1968) and Freeman and Frankel (1967). The derivation of the form of the

Table F.1 Valence s electron contact hyperfine fields

Nuclide	B/mT
^{1}H	50.8
^{2}H	7.8
^{13}C	113.0
^{14}N	55.2
^{19}F	1720
^{31}P	364
^{35}Cl	168
^{37}Cl	140

Data: Atkins and Symons (1967)

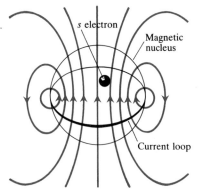

Fig. F.2 A diagrammatic representation of the origin of the Fermi contact interaction. When an electron is very close to the nucleus, it senses a distribution of magnetic field that is typical of a loop of current. The spherical average of the field in the zone occupied by the nucleus is non-zero. Only an s electron can penetrate close enough to the nucleus to detect this contribution.

interaction is described in a simple way in Appendix 22 of MQM and in Slichter (1988). A derivation from the °Dirac equation is given by Griffith (1964) and Bethe and Salpetre (1957).

Fermi–Dirac distribution

The **Fermi–Dirac distribution** describes the occupation of electron °bands in solids and, in general, the populations of °fermions over their available states as a function of temperature. The difference between the Fermi–Dirac and Boltzmann distributions can be traced to the Pauli exclusion principle. The latter effectively inflates the distribution (in the sense of increasing the number of high-energy states that are occupied at a given temperature) by excluding electrons from the low-energy orbitals which they would otherwise be likely to occupy. The Fermi–Dirac distribution governs the populations of states occupied by a system of °fermions, as we shall now explain. The **Bose–Einstein distribution** is the corresponding distribution for populations of °bosons.

Suppose we have 1000 identical particles that must be distributed over a large number of states subject to the constraints that

- the total energy must have a specified value,
- no more than one particle can occupy any state.

The 'state' of a particle includes its spin in this context, so up to two particles can occupy a spatial state so long as they have different spins (the requirement of the °Pauli exclusion principle). Then the probability P of finding a particle in a state of energy E is given by an expression named after Enrico Fermi (1901–1954) and Paul °Dirac:

$$P = \frac{1}{1 + e^{(E - E_F)/kT}}$$

where T is the temperature, k is Boltzmann's constant, and E_F is the **Fermi energy**, the energy of the state in which there is a probability of exactly 0.5 of finding a particle. A graph of the distribution is shown in Fig. F.3.

At $T = 0$ (absolute zero) in a system of N particles, only the lowest $\frac{1}{2}N$ of the levels are occupied. The energy of the uppermost full level at $T = 0$ defines the **Fermi surface** of the distribution. As the temperature is raised, some particles are promoted above the initially sharply defined Fermi surface, and the surface becomes blurred.

At energies well above the Fermi energy, when $E \gg E_F$, the distribution takes the form

$$P = e^{-E/kT}$$

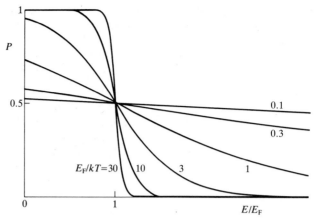

Fig. F.3 The Fermi–Dirac distribution at a series of temperatures. Note the exponentially decaying tail of the function at high energies. The Fermi energy, E_F, is the energy at which the population of a state is 0.5. In this diagram, the energy scale has been expressed as a multiple of E_F.

At such high energies the probability of finding a particle decays exponentially with increasing energy. This behaviour is characteristic of the Boltzmann distribution, for which there is no limitation on the number of particles that may occupy a state. At high energies the populations of the states are so small that it is very unlikely that several particles are available to occupy the same state, so the Pauli constraint is unimportant and the Fermi–Dirac distribution becomes essentially the same as the Boltzmann distribution.

One consequence of the Fermi–Dirac distribution is that metals have low heat capacities. Even though a metal has a large number of mobile electrons, which according to classical physics should all contribute to the heat capacity, only those near the Fermi surface can contribute since only they can be promoted to nearby higher energy levels. Hence, the electronic contribution to the heat capacity arises only from the electrons close to the Fermi surface.

Further information

For a derivation of the Fermi–Dirac distribution, see Davidson (1962). For more recent accounts with applications see Chandler (1987) and Hecht (1990). For application to solid state physics see Anderson (1984), Duffy (1990), and Kittel (1987).

Fermi hole

Because of the quantum mechanical effect of °spin correlation, two electrons with the same spin cannot be found at the same point. Thus, a plot of the probability of finding a second electron relative to the location of the first falls to zero at zero separation (Fig. F.4a). There is a corresponding decrease in amplitude of the wavefunction for the location of the second electron, and that wavefunction is close to zero in a small region surrounding the location of the first electron. This region of almost zero amplitude is the **Fermi hole** in the wavefunction of the second electron.

If the two electrons have opposite spins, there is an enhanced probability of finding the second electron close to the first. That is, instead of a Fermi hole, there is a corresponding **Fermi heap** (although I have never seen it so called), which is an enhanced amplitude in the wavefunction of the second electron wherever the first electron happens to be at any instant (Fig. F.4b).

The existence of Fermi holes and heaps is indirectly responsible for °Hund's rule of maximum multiplicity.

Further information

See §9.6 of MQM for some of the algebra involved in the discussion of Fermi heaps and holes and their consequences. Electron correlation is a major field of current study: see, for example, Wilson (1987) and Szabo and Ostlund (1982). An analysis of the structure of holes in a two-electron model atom has been given by Makerewitz (1988). For more information and further references see °spin correlation and the °exchange integral.

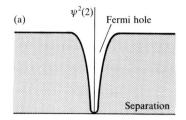

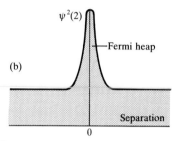

Fig. F.4 (a) A Fermi hole in a wavefunction is a region of the total many-electron wavefunction at which one electron will not be found if another is present with the same spin. (b) A Fermi heap is a region of enhanced probability which shows that if one spin is at a particular location, another of opposite spin is likely to be found there.

Fermion

A **fermion** is a particle with half-integral spin. Examples include the electron ($s=\frac{1}{2}$), the proton ($I=\frac{1}{2}$), the neutron ($I=\frac{1}{2}$), and the neutrino ($s=\frac{1}{2}$). Many nuclei are also fermions; they include the ^{35}Cl nucleus ($I=\frac{3}{2}$), the ^{13}C nucleus ($I=\frac{1}{2}$), and the ^{17}O nucleus ($I=\frac{5}{2}$). The °Pauli principle requires a wavefunction to change sign under the interchange of any pair of identical fermions. In practice, the interchange involves the exchange of labels or the exchange of Cartesian coordinates of pairs of fermions, so that the state $\psi(1,2)$ becomes the state $\psi(2,1)$. A consequence is that *no more than one fermion can occupy a given quantum state*. This constraint has a profound influence on the properties of systems composed of fermions, and distinguishes them sharply from systems composed of °bosons, any number of which may occupy a given state.

Further information

See °spin, the °Pauli principle, and the °Fermi–Dirac distribution for more information.

Feynman diagrams

In the relative antiquity of the 1930s and 1940s, books on quantum theory looked difficult because their pages were covered with integral signs and complicated sums over states. The introduction by Dirac of his °bracket notation rid pages of much of their inhospitality. However, another simplification took place in the 1950s, when physicists began to draw little squiggles like those shown in the margin, and were still able to claim that they were doing calculations.

The person most responsible for the second cultural revolution was the American physicist Richard Feynman (1918–1988), a man of profound physical insight and love of life. He showed how squiggles could be used to portray mathematical expressions systematically, and how large diagrams that represented complex expressions could be built up by combining smaller diagrams that represented the components of the formulae. Thus, each of the diagrams in the margin represents an actual mathematical expression, and the more complicated the diagram, the more complex the expression it represents.

We shall consider only three types of diagram here, simply to give a taste of their interpretation, but others will be found scattered through the book. (We shall not consider a second species of diagrams widely but contentiously used to represent the interactions in many-electron systems.) Each of the wavy lines in the diagrams represents a photon, and each of the straight lines represents a molecular state, technically a **propagator** (since the line represents the mathematical expression for the propagation of the particle through time). Time runs from left to right across the page. The point of intersection of a wavy line and a straight line is called a **vertex**, and represents the interaction of the molecule with the electromagnetic field.

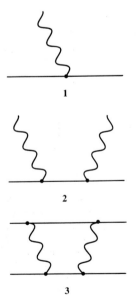

1

2

3

Diagram **1** shows the molecule initially in its ground state, a photon approaching, interacting with it, being absorbed (the vertex), and the molecule being excited to a state of higher energy. This diagram represents the absorption of radiation. We can calculate the probability that absorption will occur by associating the appropriate mathematical expressions with the vertex and the propagators. Diagram **2** shows an event in which the absorption of one photon is followed by the emission of a photon of lower energy, leaving the molecule in an excited state. This diagram contributes to the description of the °Raman effect. Two molecules participate in the process depicted in **3**. One molecule generates a photon, which travels (at the speed of light) across to the second molecule, where it is absorbed. The latter excited molecule then generates a photon, which travels back to the first molecule and is absorbed. Such an exchange of photons is responsible for the cohesion of molecules by the °dispersion interaction.

Once we know the rules for translating the arrangement of vertices into a mathematical expression, we can interpret any of these processes mathematically. We can use them to predict, for instance, the intensities of spectral transitions and the strengths of intermolecular forces.

Further information

Probably the most straightforward introduction to Feynman diagrams in many-particle systems is the book by Mattuck (1976). For a more quantum electrodynamic account see Power (1964) and Craig and Thirunamachandran (1984), who deal with a number of chemically significant processes including intermolecular forces. For the current widespread deployment of Feynman diagrams, open almost any recent book on many-particle systems, solid state physics, and calculations related to quantum electrodynamics. The diagrams were introduced in Feynman (1949), and this paper may be found in the collection of reprints compiled by Schwinger (1958).

Fine structure

The **fine structure** in a spectrum is the detail that arises from the interaction of particles of a similar kind. Thus, in °nuclear magnetic resonance the fine structure arises from the interaction of magnetic nuclei. In °electron spin resonance (of triplet molecules) the fine structure arises from the interaction of unpaired electrons. In atomic spectroscopy the fine structure arises from the interaction of electrons, and specifically from the interaction of the spin of an electron with its own orbital angular momentum.

The fine structure of atomic spectra (which we consider here) arises from the splitting of different levels (different values of J) of a °term by spin–orbit coupling. For example, in atomic sodium, the energies of the terms $^2P_{1/2}$ and $^2P_{3/2}$ are slightly different as a result of spin–orbit coupling, and consequently the transitions $^2P_{1/2} \rightarrow {}^2S_{1/2}$ and $^2P_{3/2} \rightarrow {}^2S_{1/2}$ occur at slightly different frequencies. They give rise to the two closely spaced yellow **D-lines** of alkali metal spectra.

As explained in the entry on °spin–orbit coupling, the orbital motion of an electron gives rise to a magnetic field which interacts with the spin magnetic moment of the electron. Since the interaction energy depends on the relative orientation of the spin and orbital magnetic moments, it also depends on the relative orientations of the two angular momenta (Fig. F.5). The low energy orientation occurs when the two magnetic moments are antiparallel, which is the case when the two angular momenta are antiparallel. Consequently, the lower value of the total angular momentum j corresponds to the lower energy, in accord with °Hund's third rule.

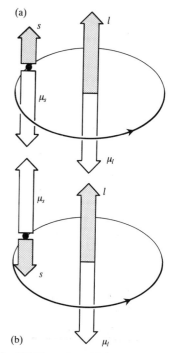

Fig. F.5 The spin–orbit interaction depends on the relative orientations of the spin and orbital magnetic moments, and hence on the relative orientations of the two angular momenta. (a) The high energy (and high j) orientation and (b) the low energy (and low j) orientation of the momenta.

°Terms for which low j levels lie lowest in energy are classified as **regular**. Terms for which low j levels lie highest in energy are **inverted**. Within a given multiplet the spacing of the levels obeys the **Landé interval rule**:

> The interval between pairs of adjacent levels is in the ratio of the j values of the upper level of each pair.

Thus the ratio of the $^3P_2 \rightarrow {}^3P_1$ interval to the $^3P_1 \rightarrow {}^3P_0$ interval is 2:1.

Further information

Atomic fine structure is discussed in more detail in Chapter 9 of MQM. For detailed accounts of its role in atomic spectra, see Condon and Odabaşi (1980). A somewhat simpler account will be found in Woodgate (1980), Shore and Menzel (1968), Corney (1977), and Cowan and Duane (1981).

Fine-structure constant

The **fine-structure constant** α is the dimensionless parameter that determines the strength of interaction between a charged particle and the electromagnetic field. Specifically:

$$\alpha = \frac{e^2}{4\pi\varepsilon_0 hc} = \frac{1}{137.03599}$$

The smallness of α is of great importance, for α determines the size of atoms and the stability of matter. If α were much larger than 1/137, the distinction between matter and radiation would be much less clear. If α were much smaller than 1/137, matter would have virtually no electromagnetic interactions.

These remarks can be illustrated by considering how α determines the sizes of atoms and some of their properties. The °Bohr radius of a hydrogenic atom, which is a measure of an atom's characteristic size, can be expressed in terms of α as $\lambda_C/4\pi\alpha$, where λ_C is the °Compton wavelength of the electron (about 2.4 pm). Hence, the size of an atom is roughly 137 times the characteristic 'size' of an electron. If the interaction strength were much smaller than in fact it is, the atom would be much larger. If α were much larger, and comparable to the analogous coupling constant for the interaction between nucleons (the 'strong interaction' as opposed to the electromagnetic interaction), then atoms would be of roughly the same size as their nuclei.

The minimum energy of a hydrogenic atom (the Rydberg constant) expressed in terms of α is $-\frac{1}{2}Z^2\alpha^2 \times m_e c^2$, where $m_e c^2$ is the rest energy of an electron (the energy equivalent to its rest mass). Thus, the total energy of an atom is of the order of $Z^2/137^2$ of the rest energy of the electron. This is a small proportion when $Z \ll 137$, but not in heavy atoms. As a result, ordinary non-relativistic quantum mechanics is good for light atoms, but requires corrections as Z increases.

The order of magnitude of the velocity of an electron in an atom or molecule is about $Z\alpha c$. Since α is so small, electrons move at non-relativistic velocities in light atoms. Another role for α is in the magnitude of the °spin–orbit coupling, which determines the °fine structure of spectra. The strength of the spin–orbit interaction in hydrogenic atoms is proportional to $Z^4\alpha^2 R$, where R is the Rydberg constant; it is from this relation that α gets its name.

The stability of atoms is a consequence (at several levels) of the smallness of α. Thus, from an analysis of the rate of °electric dipole transitions it is possible to conclude that an electron must oscillate $1/\alpha^3 Z^2$ times before it is likely to emit a photon. (For the H atom, this is about 3 million times.) Hence, the excited states of atoms and molecules are moderately long-lived. The probability that two photons are thrown off by an excited atom is of the order of the square of the probability that one will be emitted; hence this process is only rarely observed.

Matter can turn into radiation. The probability that an electron will disappear in this fashion is proportional to the strength of its coupling to the electromagnetic field, and hence varies as α. Put another way, an electron spends α of its time as electromagnetic radiation. Fortunately α is small; if it were closer to unity, matter and radiation would be indistinguishable.

Further information

A good account of how α determines atomic properties is given in Chapter 1 of Thirring (1958). For the cosmological implications of the size of α, a good starting point is Barrow and Tipler (1986) and the extensive references therein.

Fluorescence

Electromagnetic radiation absorbed by a molecule may be degraded into thermal motion or re-emitted. Light emitted from an excited molecule is called **fluorescence** if the emission mechanism does not require the molecule to pass through a state of different spin multiplicity from that of the initial state. (See °phosphorescence for an alternative mechanism of radiative decay.) Fluorescence generally ceases almost immediately after the exciting radiation is removed.

The mechanism of fluorescence is illustrated in Fig. F.6. The incident radiation excites a °singlet ground state S_0 (all spins paired) molecule into an excited singlet state S_1. The molecule is also excited vibra-

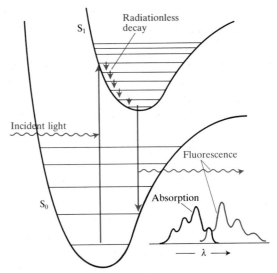

Fig. F.6 The mechanism of fluorescence. The initial excitation takes place between states of the same multiplicity and in accord with the Franck–Condon principle. The vibrationally excited upper singlet state discards energy into the environment, and then undergoes a radiative transition back to the ground electronic state. The insert shows the nearly mirror-image absorption and fluorescence spectra.

tionally during the transition (see °Franck–Condon principle). Collisions with the surrounding medium (which may be a gas, a solvent, or a solid lattice) induce non-radiative vibrational transitions; these occur because the surrounding molecules are able to carry away the vibrational quanta of the molecule, and hence cause it to step down the ladder of vibrational states in the upper electronic state.

Either of two processes can occur when the molecule has reached its lowest vibrational state. In one, the solvent carries away the electronic excitation energy. The solvent may achieve this de-excitation if its molecules have energy levels that match the energy of the excited molecule, for there may then be a °resonant transfer of radiation to the solvent. An alternative mode of decay is the radiative decay of the excited electronic state. In this process, the molecule loses its electronic excitation energy by emitting a photon and collapsing back into the lower electronic state. This emitted light is fluorescence.

Whether or not fluorescence occurs depends on a competition between radiative emission and the radiationless de-excitation by energy transfer to the surrounding medium. If the interaction between the excited molecule and its surroundings is strong, radiationless decay will dominate. If the interaction is ineffective at achieving the large energy transfer needed to take the molecule to a lower electronic state, it may still be able to lower the electronically excited molecule down its ladder of vibrational states until it is vibrationally silent. The molecule can discard its remaining electronic excitation energy only as radiation. In this case the radiative decay dominates and the molecule fluoresces.

Two characteristics of the fluorescence should be noted. The first is that the fluorescence should appear at lower frequency than the incident light. Why this is so may be seen immediately from Fig. F.6, which shows that the energy of the emitted photon differs from the energy absorbed by the amount of vibrational energy lost to the surroundings. Thus, fluorescent substances irradiated with ultraviolet light radiate light shifted towards the red.

The second point is that there may be vibrational structure in the **fluorescence spectrum**, the spectrum of radiation emitted on the decay of the ground vibrational state of the upper electronic state into different vibrational levels of the lower electronic state. Examination of this vibrational structure can provide information about the °force constant of the molecule in its ground state (this is in contrast to electronic absorption spectra, which provide information about force constants of bonds in the *upper* electronic state). It follows that the absorption spectrum and the fluorescence spectrum of a molecule should resemble each other: this is often expressed by saying that one is the 'mirror image' of the other (see Fig. F.6); but that description should not be taken literally, because the vibrational splittings and intensities are not quite the same.

A number of details may be added to this basic description. The first is that the initial absorption might not be to the lowest excited singlet state of the molecule. In this case an **internal conversion** occurs in which collisions cause the higher singlets S_2, S_3, etc. to make a radiationless transition into the lowest excited singlet S_1, which then fluoresces. **Kasha's law** summarizes this process:

The fluorescent level is the lowest level of the specified multiplicity.

Thus, the fluorescent level is often the lowest excited singlet state (S_1, not S_2 etc.).

The intensity of fluorescence depends strongly on the physical state of the sample. Pure, undiluted liquids generally have a very low fluorescent efficiency because an excitation may hop from one molecule to an identical neighbour by a resonant process. (See °exciton for the analogous effect in solids.) Conversely, it is possible to enhance fluorescence by **sensitized fluorescence**. In this process one species of molecule absorbs the incident light and subsequently transfers its excess energy (by a matching of energy levels, and a collision) to a molecule that then fluoresces.

In **resonance fluorescence** the emitted radiation has the same frequency as the incident radiation. Resonance fluorescence is more intense than ordinary fluorescence because the emission process is stimulated by the incident radiation. In normal fluorescence the emission is a spontaneous process (see °Einstein coefficients). Radiation of exactly the same frequency is rare in fluorescing molecules because interactions between the molecule and the solvent slightly shifts the former's energy levels. For example, the solvent may solvate the molecule in one way in its ground state and in a different way in its excited state (the excited state persists long enough for the solvent molecules to organize themselves into a different solvation arrangement). As a result, the absorption and fluorescence frequencies may be different even though the transitions are between the same two states.

Fluorescence is generally extinguished as soon as the incident illumination ceases because all the transitions of interest are allowed and therefore occur very quickly. Nevertheless, there is also the phenomenon of **delayed fluorescence** (which should not be confused with phosphorescence) which may persist for several milliseconds. One mechanism of delayed fluorescence involves the excitation $S_1 \leftarrow S_0$ and the subsequent migration of the excited molecule to another molecule. When the two molecules are in contact they form an excited dimer which is called an **excimer** $(S_0 S_1)^*$ if the molecules are identical and an **exciplex** if they are different. The excimer/exciplex rapidly falls apart with the emission of fluorescence. Another mechanism of delayed fluorescence is by **triplet-triplet annihilation**. In this process the excitation of the molecules leaves them in an excited triplet state (denoted T; see °phosphorescence for the mechanism). Two such molecules migrate together, and their joint excitation energy is enough to raise one of them into an excited singlet state:

$$T + T \rightarrow S_1 + S_0$$

The excited singlet then decays radiatively. The delay in the fluorescence stems from the time it takes the triplet molecules to diffuse together and pool their energy.

Further information

See Chapter 12 of MQM. For chemical aspects of fluorescence, including chemiluminescence, see Wayne (1988), Calvert and Pitts (1966), and Barltrop and Coyle (1975). For a detailed discussion of the quantum mechanics of non-radiative decay, see Englman (1976) and Lin (1980).

Force constant

In a simple °harmonic oscillator, the restoring force is proportional to the displacement from equilibrium. The **force constant** k is the constant of proportionality, and we write $F = -kx$, where x is the displacement.

High force constants imply stiff systems (strong restoring forces for small displacements). In a molecule, the force constant is equal to the second derivative of the molecular potential energy V with respect to the displacement evaluated at the equilibrium position, $k = (d^2V/dx^2)_0$. That is, the greater the curvature of the potential at the equilibrium position, the greater the force constant (Fig. F.7).

The frequency of a harmonic oscillator is related to k by $\omega = (k/\mu)^{1/2}$, where μ is the °reduced mass of the oscillator. The formula shows that the stiffer the oscillator (the larger the value of k), the higher its oscillation frequency. In the quantum mechanical description of the harmonic oscillator, the separation of the quantized energy levels is $\hbar\omega$. Hence, for constant mass, the separation between neighbouring levels increases as k increases.

The force constants of some chemical bonds are given in Table F.2 and are of the order of $100\ \text{N m}^{-1}$. For a bond between two specified atoms, the force constant invariably increases with °bond order. It is generally found that the stronger the bond (the deeper the minimum, D_e, of the molecular energy curve), the greater the force constant (the stiffer the bond), although there is no *direct* connection between k and D_e.

Further information

For more information, including the specific relation between the force constant and the potential energy curve, see Chapter 11 of MQM. The entry on the °harmonic oscillator should also be consulted. A good, simple, but reasonably detailed account is given by Woodward (1972), who shows how to determine k from vibrational data. In that connection see Hollas (1983, 1987), Bunker (1979), Gribov and Orville-Thomas (1988), and Graybeal (1988). The standard monograph is that by Wilson, Decius, and Cross (1955).

Formal charge

The **formal charge** ϕ is a parameter that (like °oxidation number) assesses the relocation of electrons that occurs when atoms link to form molecules. It is calculated by treating each atom as having exactly a half-share in any electron pair that binds it to a neighbouring atom, but as owning all its own °lone pairs. The difference between the number of electrons it 'owns' according to this 'perfect covalence' rule and the number in the unbound, neutral atom, is then the formal charge:

$$\phi = \text{Number of valence electrons} - \{\text{Number of lone pair electrons} + \tfrac{1}{2} \times \text{Number of bonding pair electrons}\}$$

Some formal charges of atoms in characteristic bonding contexts are

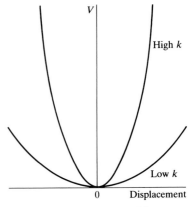

Fig. F.7 The force constant k is a measure of the curvature of the potential at the equilibrium position (zero displacement). The greater the curvature, the larger the value of k.

Table F.2 Typical force constants

	$k/(\text{N m}^{-1})$
H_2^+	160.0
H_2	574.9
H_2	577.0
HF	965.7
HCl	516.3
HBr	411.5
HI	313.8
N_2	2293.8
O_2	1176.8
F_2	445.1
Cl_2	322.7

Data: Gray (1972)

Table F.3 Formal charges on atoms in different bonding arrangements

		IV/14	V/15	VI/16	VII/17	VIII/18
				Group		
—X—		0	+1	+2	+3	+4
:X—	=X—	−1	0	+1	+2	+3
:X:	=X=	−2	−1	0	+1	+2
:X:	=X=	−3	−2	−1	0	+1
:X:	=X=	−4	−3	−2	−1	0

Group III/13: X— 0, —X— −1

given in Table F.3. The sum of the formal charges on all the atoms in a species is equal to the overall charge of the species.

The formal charge on an atom depends on the °Lewis structure used to calculate it, since lone pairs and bonding pairs contribute differently. (In this respect, the formal charge differs from the °oxidation number, which is independent of the bonding model.) This structure-dependence is utilized for the assessment of the relative energies of different Lewis structures in a °resonance hybrid, for it is *broadly* true that a Lewis structure with low formal charges has a lower energy (and hence contributes more strongly) than one in which the formal charges are high. Thus, of the two Lewis structures for BF_3 shown as **4a** and **4b**, the one with zero formal charges (**4a**) has a lower energy than the other (in which the electronegative F atoms are positively charged), and makes the greater contribution to the overall structure in spite of its having a B atom with an incomplete octet. The dominance of structures with formal charges close to zero is the essential content of Pauling's **electroneutrality principle**.

The formal charge can also be used (cautiously) to select the most likely atomic arrangement from several alternatives (for example, to decide whether dinitrogen oxide is NON or NNO). Thus, we evaluate the formal charges on the atoms in each arrangement, and then select the structure that has the lowest formal charges.

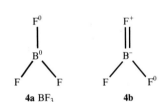

4a BF_3 4b

Further information

Formal charge is a concept more often encountered in elementary chemistry than elsewhere; therefore see Atkins (1989) for an account of how it is used in that context. Pauling's electroneutrality principle is discussed by him in Pauling (1960).

Franck–Condon principle

The **Franck–Condon principle** states that:

> Because nuclei are much more massive than electrons, an electronic transition takes place while the nuclei in a molecule are effectively stationary.

The principle governs the probabilities of transitions between the vibrational levels of different molecular electronic states.

Consider a transition from a state in which the nuclei are stationary at their equilibrium positions. As a result of the electronic transition, electron density is rapidly built up in some regions of the molecule and removed from others. As a result, the initially stationary nuclei suddenly experience a new force field. They respond to the new force by bursting into vibration, and swing backwards and forwards from their original separation (which was maintained during the rapid electronic excitation). The stationary equilibrium separation of the nuclei in the initial electronic state therefore becomes a stationary **turning point**, the point of a vibration at which the nuclei are stationary at the end points of their swing, in the final electronic state (Fig. F.8). Since the nuclear framework remains constant during this excitation, we may imagine the transition as being up the vertical lines in Fig. F.8. The vertical line is the origin of the expression **vertical transition**, which is used to denote an electronic transition that occurs without a change of nuclear geometry.

The quantum mechanical version of the Franck–Condon principle expresses everything in terms of the wavefunctions of the molecule. The following interpretation is often offered.

Consider the case in which before the electronic transition takes place the molecule is in the ground vibrational state of the lower electronic state (Fig. F.9). The form of the vibrational wavefunction shows that the most probable location of the nuclei is at their equilibrium separation R_e. Consequently the electronic transition is most likely to take place when the nuclei have this separation. When the vertical transition occurs, it cuts through several vibrational levels of the upper electronic state. The level marked * is the one in which nuclei are most probably at the same initial separation R_e (because the vibrational wavefunction has maximum amplitude there), and so this is the most probable level for the termination of the transition. However, several nearby levels also have an appreciable probability of the nuclei being at the separation R_e. Therefore, transitions occur to all

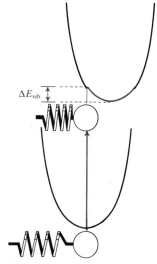

Fig. F.8 The classical version of the Franck–Condon principle. The electronic transition from the lower parabola to the upper parabola occurs without change of nuclear position and without accelerating from rest. As a result, the nuclei are found in a stationary, high potential energy condition immediately after the transition. They respond by breaking into vibration. The vibrational energy of the upper state can be predicted by identifying the point at which the vertical line intersects the potential energy curve.

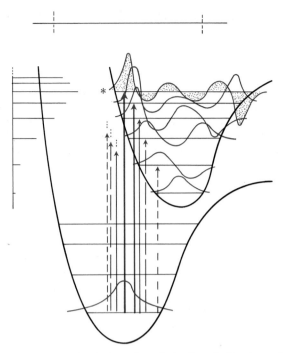

Fig. F.9 The quantum mechanical version of the Franck–Condon principle. The most probable transition occurs from the initial ground state vibrational wavefunction to the vibrational state (*) that it most resembles in the upper electronic state.

the vibrational levels in this region, but most intensely to the level with a vibrational wavefunction that peaks most strongly near R_e.

A more quantum mechanical interpretation is as follows. The quantum version of the statement that the *dynamical condition of the nuclei is unchanged* during an electronic transition is expressed by the statement that the wavefunction of the final vibrational state is the one that *most closely resembles* the initial vibrational state. Therefore, given the initial vibrational wavefunction, we should look for the vibrational wavefunction of the upper electronic state that most closely resembles it, for in that state the nuclei are in a state that most closely resembles their initial state. Since two wavefunctions have the greatest resemblance to each other the closer their °overlap integral is to 1, we should look for the upper vibrational wavefunction that has a peak at similar displacements to the initial vibrational state. As can be seen from Fig. F.9, that is equivalent to selecting a vertical transition.

The vibrational structure of the spectrum depends on the relative displacement of the two potential energy curves. A long **progression** of vibrations (a lot of vibrational structure arising from the transitions $1\leftarrow0$, $2\leftarrow0$,...) is observed if the two states are appreciably displaced. The upper curve is usually displaced to longer equilibrium bond lengths because excited states usually have less bonding character than ground states.

Further information

See MQM Chapter 11 for more information and a derivation of the quantitative form of the principle. See also Hollas (1983, 1987). A thorough discussion of the principle is given in §IV.4 of Herzberg (1950) and developed in more detail in Herzberg (1966). The original formulation of the principle was by Franck (1925), and it was developed mathematically by Condon (1928). A useful summary of modern work on the principle is given by Englman (1976), who works through a lot of useful calculations.

Franck–Hertz experiment

In the experiment carried out in 1914 by James Franck (1882–1964) and Gustav Hertz (1887–1975, the brother of Heinrich Hertz who discovered radio communication and after whom the unit of frequency is named), a beam of electrons was passed through mercury vapour at low pressure and the current arriving at a detector was monitored. In later experiments the energy of the arriving electrons was also monitored. As the energy of the incident electrons was increased it was found that the current dipped sharply when the incident energy was equal to an excitation energy of the atom or molecule (Fig. F.10). It was observed that the sample also emitted light that had a frequency corresponding to the energy of the incident beam.

One of the important features of the Franck–Hertz experiment for the early development of quantum theory was that it showed that energy is quantized even when it is imparted from mechanical motion (as distinct from electromagnetic radiation). The observations can be

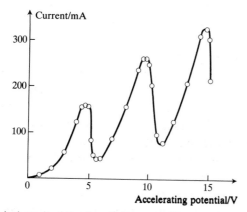

Fig. F.10 A typical result of the Franck–Hertz experiment. The transmitted current dips whenever the kinetic energy of the accelerated electrons matches an excitation energy of the sample.

understood in terms of the quantization of energy, for the electrons are able to transfer their kinetic energy to an atom only if the atom can be excited by that amount. Therefore, the observed, transmitted current of the electron beam will dip each time the energy can be transferred to the quantized system. This is confirmed by the observation of the emitted radiation at the corresponding frequency.

Further information

The original paper is Franck and Hertz (1914). For the historical context of the experiment consult Jammer (1966). For an analysis of the data see Cagnac and Pebay-Peyroula (1975).

Frontier orbitals

The **frontier orbitals** of a molecule are the **highest occupied molecular orbital** (the HOMO) and the **lowest unoccupied molecular orbital** (the LUMO), see Fig. F.11.

The importance of frontier orbitals, and the current considerable interest that there is in them, lies in the roles they play in molecular properties, particularly the reactions that molecules can undergo and the definitions of °electronegativity, °hardness, and °aromaticity. Their importance stems in part from their role in governing the energy needed to rearrange the electrons in a molecule: the LUMO–HOMO gap is the minimum electronic excitation energy. It also stems in part from their role in bond formation, particularly when a molecule is treated as formed by the combination of two fragments. In that case the fragment that acts as a Lewis base (an electron pair donor) supplies the electrons from its HOMO, and the fragment that acts as a Lewis acid (an electron pair acceptor) accommodates the incoming electrons in its LUMO.

As an illustration, we can picture the formation of HCl from a proton and a Cl^- ion in the archetype of a Brønsted acid–base proton transfer reaction as occurring by the overlap of the HOMO of the Cl^- ion (an occupied Cl3p orbital) and the LUMO of the H^+ ion (an unoccupied H1s orbital). The overlap of these two frontier orbitals results in the formation of a σ bonding orbital using the two electrons supplied by the Cl^- ion.

If we relax the definition of frontier orbital to include *any* orbital that is close to the uppermost filled orbital, then we can broaden the scope of the concept to include the molecular orbitals composed primarily of metal d orbitals that are responsible for the properties of d-block metal complexes. Thus, we can take the view that the molecular orbitals of a complex are formed from the overlap of the incompletely

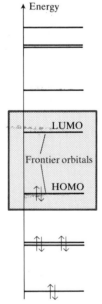

Fig. F.11 The frontier orbitals of an atom or molecule consist of the uppermost filled orbital (the HOMO) and the lowest unfilled orbital (the LUMO).

filled *d* orbitals of the central metal ion, the frontier orbitals of the metal atom, and the appropriate °symmetry adapted combinations of the orbitals of the surrounding ligands. The ligand field transitions of the complex (its $d \leftarrow d$ transitions) can likewise be regarded as transitions between frontier orbitals of the complex.

The theoretical importance of frontier orbitals stems from the general rule that the strongest bonding effects arise from the overlap of orbitals that have a similar energy. Typically, the HOMO of one fragment has an energy quite similar to that of the LUMO of another fragment (at least in the case of fragments for which the combinations are of chemical interest). The separation of the frontier orbitals is used in the definition of °hardness, and the average energy of the two frontier orbitals is a definition of °electronegativity. More recently it has been proposed that the separation of the frontier orbitals in a molecule may be used as a definition of °aromaticity that is independent of the choice of reference state.

Further information

See Shriver, Atkins, and Langford (1990) for an account of the role of frontier orbitals in inorganic chemistry and, in more detail, Albright, Burdett, and Whangbo (1985). Frontier orbitals play an important role in the discussion of chemical reactions in terms of the °Woodward–Hoffmann rules, and that entry should be consulted for further information. The fundamental role of frontier orbitals in the parametrization of molecular properties in terms of electronegativity and hardness is described in those entries and is treated in detail by Parr and Yang (1989). Their role in the definition and assessment of aromaticity is described by Zhou and Parr (1989).

g-value

We shall deal with five types of g-value (or 'g-factor') in this section; each of them is involved in the relation of an angular momentum to a magnetic moment. They are:

- The g-value of the free electron.
- The nuclear g-factor.
- The Landé g-factor.
- The g-factor of electron spin resonance.
- The molecular g-value.

The °magnetic moment $\boldsymbol{\mu}$ of an electron that arises from its °orbital angular momentum $\boldsymbol{l}$ is $\boldsymbol{\mu} = \gamma \boldsymbol{l}$, where γ is the magnetogyric ratio ($\gamma = -e/2m_e$). The magnetic moment of an electron due to its *spin* angular momentum is proportional to the spin angular momentum $\boldsymbol{s}$; however, the actual value is not $\boldsymbol{\mu} = \gamma \boldsymbol{s}$ but $\boldsymbol{\mu} = g_e \gamma \boldsymbol{s}$. In this expression g_e, the **free-electron g-value**, is an 'anomalous' factor found experimentally to be 2.0023 but often approximated as 2. That is, the magnetic moment arising from electron spin is *twice* the value expected from the angular momentum.

It is not surprising to find a factor of 2 appearing in connection with °spin, which involves half-integral quantum numbers, but the full explanation can be found by digging progressively deeper into theory. At the simplest level, we should note that the theory of electron spin reveals that an electron must rotate through 4π (two revolutions) to return to its original orientation, not the 2π (one revolution) of orbital momenta (and familiar objects). Since an electron rotates at twice the rate that its perceived angular momentum suggests, its magnetic moment is twice as large as expected.

More satisfactorily, we can turn to °Dirac's relativistic theory of the electron, which predicts in a straightforward (but non-pictorial way) that $g = 2$ exactly. However, the fact that Dirac's theory leads to the integer 2 for the value of g rather than $g = 2.0023$ shows that his theory is incomplete. The additional 0.0023 required to account for the observed value is explained by °quantum electrodynamics. In this theory, the electron is continuously buffeted by the zero-point fluctuations of the electromagnetic vacuum. In pictorial terms, the buffeting of the electron rocks the electron on its axis as it spins and does so in such a way that the magnetic moment is increased from its Dirac value of $2\gamma \boldsymbol{s}$ to the observed value of $g_e \gamma \boldsymbol{s}$. The calculation of the g-value of the free electron, which has been achieved to an accuracy of 1 part in 10^{11}, is one of the triumphs of quantum electrodynamics.

The **nuclear g-factor**, g_I, is used to relate the magnetic moment of a –

Table G.1 Nuclear g factors

Nuclide	g
^{1}n*	-3.8260
^{1}H	5.5857
^{2}H	0.857 45
^{3}H*	-4.2553
^{13}C	1.4046
^{14}N	0.403 56
^{17}O	-0.7572
^{19}F	5.2567
^{31}P	2.2634
^{33}S	0.4289
^{35}Cl	0.5479
^{37}Cl	0.4561

*Radioactive.

nucleus to its spin angular momentum. Thus, if the spin of a nucleus is I, its magnetic moment is written $\mu = g_I \gamma_N I$, where γ_N is the nuclear °magneton ($\gamma_N = e/2m_p$). Nuclear *g*-factors are empirical parameters of the order of 1 (Table G.1). Their calculation requires a greater knowledge of nuclear structure than is currently available.

The **Landé *g*-factor** is used to express the magnetic moment of an atom that has both spin and orbital angular momenta S and L combined together to give a total angular momentum J. Alfred Landé explored whether, given that the total spin magnetic moment is $2\gamma S$ and the total orbital magnetic moment is γL, it is possible to write the total magnetic momentum as proportional to γJ. Landé found that the overall magnetic moment can in fact be written

$$\mu = g_J \gamma J$$

where

$$g_J = 1 + \frac{J(J+1) + S(S+1) - L(L+1)}{2J(J+1)}$$

When $L = 0$ (so that $J = S$, and there is only spin angular momentum), $g_J = 2$, the normal value for a collection of spins. When $S = 0$ (so $J = L$, and there is only orbital angular momentum in the system), $g_J = 1$, the normal value for a spinless system. The expression for g_J stems from the anomalous *g*-value. The vector-coupling picture of an atom with spin momentum S and orbital momentum L coupled to give a resultant J is shown in Fig. G.1. Both L and S °precess around J. Collinear to L and S we may draw vectors representing the corresponding magnetic moments, but the spin magnetic moment must be drawn twice as long in proportion on account of the factor 2. It follows that the resultant of the orbital and spin magnetic moments does not lie along the direction of J (it would if $g_e = 1$) but at an angle to it. The resultant magnetic moment vector precesses about the direction of J, for it follows the precession of L and S about J. Only the component along J does not average to zero as a result of this precession, and so the effective magnetic moment is parallel to J, and we can write $\mu \propto \gamma J$. After some straightforward vector algebra, the constant of proportionality turns out to be the factor g_J in the expression above.

The **g-factor** of °electron spin resonance is a measure of the difference between the local magnetic field and the field B applied to the sample. The resonance condition for a free electron is $h\nu = g_e \mu_B B$, where μ_B is the Bohr °magneton. In a molecular environment (such as a radical or a complex), the electron experiences the *local* magnetic field, not the applied field itself. The difference between the local and applied fields is absorbed into a parameter g, and we write the resonance condition with g in place of g_e in the expression above. The g-factor would be

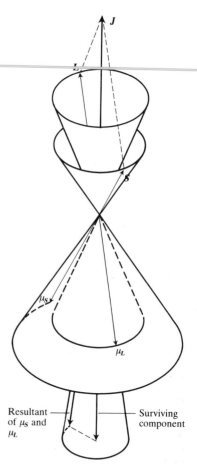

Fig. G.1 The vector addition diagram used to derive the formula for the Landé *g*-factor. Because the spin magnetic moment vector is twice as long (in proportion to the angular momentum) as the orbital magnetic moment vector, the resultant of the magnetic moments is not collinear with the resultant of the angular momenta. Precession, however, averages out all except the collinear component.

equal to g_e if the electron were not in a molecular environment, and it would be equal to the Landé factor g_J if the electron were part of an atom.

The value of g in a molecular environment (a radical or a complex) is related to the ease with which the applied field can stir up currents through the molecular framework and the strength of the magnetic field this electronic (orbital) current generates. The ease of stirring up currents is inversely proportional to the separation of energy levels Δ in the molecule, and the strength of the field generated by these currents is proportional to the °spin–orbit coupling constant ζ. Hence we can expect the g-factor of a radical to differ from g_e by an amount that is proportional to ζ/Δ. This proportionality is widely observed: the g-factors of many inorganic and organic radicals are close to 2.00, and deviate more from this value the smaller the value of Δ and the larger the value of ζ. The g-factors of d-metal complexes often differ considerably from g_e since in them Δ is small (see °ligand field theory) and in some cases the electrons occupy almost degenerate orbitals (so orbital angular momentum is not °quenched).

The **molecular *g*-value** is used to relate the molecular °magnetic moment to the °rotational angular momentum of the molecule, and we write $\boldsymbol{\mu} = g\gamma_N \boldsymbol{J}$. The nuclear magneton has nothing fundamental to do with molecular magnetic moments: it is used in this expression only because molecular magnetic moments are about the same size as nuclear magnetic moments, and expressing the magnetic moment as a multiple of γ_N ensures that molecular g-factors are then close to 1. Thus, for NH_3, $g \approx 0.053$, which implies that when the molecule is in a rotational state with $J \approx 10$, its magnetic moment is approximately $5.3\mu_N$.

Further information

For an account of the origin of the anomalous magnetic moment of the electron see the entry on the °Dirac equation; for its *non-relativistic* origin see Galindo and Sanchez del Rio (1961). An account of the derivation of the Landé g-factor will be found in Chapter 9 of MQM. Further discussions will be found in Woodgate (1980), Condon and Odabaşi (1980), and Cowan and Duane (1981). The factor plays an important role in the explanation of the °Zeeman effect and of the magnetic properties of d-metal complexes (see °ligand field theory). For an account of the role of the g-factor in °electron spin resonance, see that entry and its bibliography, particularly Abragam and Bleaney (1970) and Hoff (1989). The g-factors of radicals are listed by Fischer and Hellwege (1977). Molecular magnetic moments are discussed in Chapter 11 of Townes and Schawlow (1955) and, in a more modern format, by Flygare (1978). The latter book gives a table of values and theoretical expressions. The book by Jelley (1990) describes the modern position in relation to the calculation of nuclear magnetic moments. Alfred Landé's contributions to modern physics are assessed by Yourgrau and van der Merwe (1979).

Gaussian atomic orbitals

One major time-consuming aspect of molecular structure calculations is the number and complexity of the electron–electron interaction integrals that must be evaluated. These integrals are over atomic orbitals on up to as many as four centres. One simplification of these multi-centre integrals stems from the adoption of a suggestion made by the Cambridge theoretical chemist S.F. Boys, that the atomic orbitals should be expressed in terms of **gaussian type orbitals** (GTO), which are functions of the form e^{-ar^2}, instead of as the °Slater atomic orbitals (STO), which are exponential functions of the form e^{-ar}.

The advantage of replacing exponential functions by gaussians stems from the property that the product of two gaussians based on different centres is itself a gaussian based on a point lying between the centres (Fig. G.2). Therefore a complicated three- or four-centred integral can be expressed as a relatively simple equivalent two-centred integral, which can be evaluated speedily. The disadvantage of the method lies in the fact that an atomic orbital is not well represented by a simple gaussian function, and so each atomic orbital has to be expressed as a sum of several gaussians. (About two or three times as many GTOs are needed as STOs.) Therefore, although each integral is simpler, very many more of them must be evaluated. However, this is no longer much of a problem with modern computers, and will probably become less so.

Further information

See Richards and Cooper (1983) for a simple, straightforward introduction, including a description of 'contracted' gaussians, and McGlynn *et al.* (1972), Hinchcliffe (1988), Clark (1985), and Hehre *et al.* (1986) for more detail. The

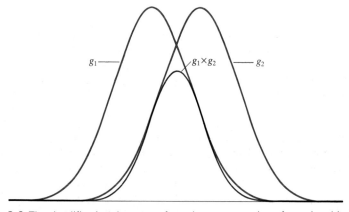

Fig. G.2 The simplification that stems from the representation of atomic orbitals by gaussian functions arises from the fact that the product of two gaussian functions is itself a gaussian function. In this illustration the two gaussian functions are equivalent, and their product is centred on the point half way between them.

method was initiated by Boys (1950), and its subsequent development can be traced by reference to the historical bibliography compiled by Schaefer (1984). A useful general reference is the handbook compiled by Poirier, Kari, and Csizmadia (1985).

gerade and *ungerade* (g and u)

The German words *gerade* (even) and *ungerade* (odd) are added as labels to states to denote the **parity** of their °wavefunctions; that is, to denote whether they do not (g) or do (u) change sign under inversion.

A simple example is the classification of the orbitals of a homonuclear diatomic molecule. In such a molecule there is a centre of inversion midway between the two nuclei. We consider the amplitude of a σ bonding orbital at an arbitrary point of the molecule, project a straight line from this point through the centre of inversion, and travel an equal distance to the other side (Fig. G.3). The σ orbital has the same sign and amplitude at the new point. Hence, the orbital is even (g) under this symmetry operation, and is labelled σ_g. On the other hand, the same journey in an °antibonding σ orbital begins at a point of positive amplitude, passes through a node, and ends at a point of equal magnitude but negative sign. The parity of this orbital is odd (u), and so it is labelled σ_u. The π orbitals of a homonuclear diatomic molecule behave differently under inversion (see Fig. G.3): the bonding π orbital changes sign and hence is u (and denoted π_u); the antibonding π orbital is g (π_g).

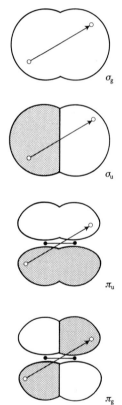

The overall parity of a many-electron atom or molecule is obtained by determining whether the state is even or odd overall. To do this, we form the product of the parities of its occupied orbitals using the rules

$$g \times g = g \quad u \times u = g \quad g \times u = u$$

All closed shell homonuclear diatomic molecules are g because each orbital is doubly occupied and all the g and u multiply out to g. The ground state of the F_2 molecule, for example, is $^1\Sigma_g$. The excited °configuration $\sigma_g^1 \sigma_u^1$ of the H_2 molecule is u because $g \times u = u$; it is denoted Σ_u.

The parity classification is inapplicable when there is no centre of inversion. Thus, parity labels may be added to the states and orbitals of homonuclear diatomics, symmetrical linear molecules (such as CO_2 and HCCH), and centrosymmetric octahedral complexes, but not to heteronuclear diatomics or to tetrahedral complexes, which do not have centres of inversion. The parity is useful when discussing °selection rules because the only °electric dipole transitions allowed are those in which the parity changes; thus $g \rightarrow u$ and $u \rightarrow g$ transitions are allowed, but $g \rightarrow g$ and $u \rightarrow u$ transitions are forbidden.

Fig. G.3 The parities of orbitals in centrosymmetric molecules (here, but not necessarily, diatomic molecules). Relative phases are shown by tinting. An orbital that has the same sign under inversion is *gerade* (σ_g and π_g in these examples), and an orbital that changes sign is *ungerade* (σ_u and π_u).

Further information

Parity is an aspect of molecular symmetry; therefore, see MQM Chapter 7 and Cotton (1990). Banwell (1983), Herzberg (1950), King (1964), Graybeal (1988), Struve (1989), and Hollas (1983, 1987) all discuss the role of parity in the formulation of selection rules.

Grotrian diagram

In a **Grotrian diagram** (Fig. G.4) the °terms of an atom are displayed as a ladder of lines classified into convenient groups (such as all singlets in one column, all triplets in another, and so on). Spectral transitions are represented by lines connecting the terms between which they take

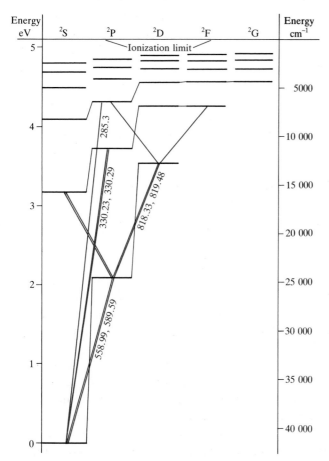

Fig. G.4 A Grotrian diagram shows the energies of atomic terms and the wavenumbers (or wavelengths) of the allowed transitions; the thicker the transition line, the more intense the transition. This is the Grotrian diagram for atomic sodium; the numbers on the transition lines are wavelengths (in nm).

place, and the wavenumber (or wavelength) of the transition is often added to the transition line. The relative intensities of the transitions may be indicated by the thickness of the connecting lines. The diagrams were introduced by W. Grotrian in 1928.

Further information

A good, extensive introduction to Grotrian diagrams, with lots of examples, is Candler (1964). More up to date, and very thorough, is the compilation edited by Bashkin and Stoner (1976 *et seq.*).

Group theory

Group theory is the mathematical theory of symmetry. It puts on a formal basis our intuitive notions about the symmetry of objects, and so enables unambiguous deductions to be drawn about the consequences of their symmetry.

By taking full advantage of the symmetry of a system we can reduce the labour of calculations. Moreover, we can often draw conclusions without the need for elaborate manipulations. In quantum chemistry, group theory is used to find the appropriate °symmetry-adapted linear combinations of atomic orbitals for the °molecular orbitals of a molecule, to classify atomic and molecular states and hence to express the °selection rules that govern the transitions between them, and to set up °correlation diagrams (such as °Walsh diagrams and °Tanabe–Sugano diagrams). It is also used to find the °normal modes of °vibration of molecules and the permitted states of many electron atoms, molecules, and ions. It is particularly useful for systems that have a high degree of symmetry, such as octahedral complexes. °Angular momentum theory can be regarded as a branch of group theory arising from the consideration of the symmetry properties of spherical systems.

Group theory deals with the °symmetry transformations of molecules and solids. In the applications most widely used in chemistry and physics, the effect of a symmetry operation on a basis (a set of orbitals, for instance) is expressed in terms of °matrix multiplication. The most important feature of a matrix used to represent the effect of a symmetry operation is its **character**, χ, the sum of its diagonal elements. Most of the manipulations of group theory used in the description of atoms, molecules and periodic solids make use of the properties of characters, hence the importance of **character tables** like those in Table G.2. A brief summary of the properties of °characters (for those already familiar with their general significance) is given in Box G.1.

Table G.2   Character tables

C_{2v} (2 mm)	E	C_2	$\sigma_v(xz)$	$\sigma_v'(yz)$		
A_1	1	1	1	1	z	$x^2,\ y^2,\ z^2$
A_2	1	1	-1	-1	R_z	xy
B_1	1	-1	1	-1	$x,\ R_y$	xz
B_2	1	-1	-1	1	$y,\ R_x$	yz

C_{3v} (3 m)	E	$2C_3$	$3\sigma_v$		
A_1	1	1	1	z	$x^2+y^2,\ z^2$
A_2	1	1	-1	R_z	
E	2	-1	0	$(x,\ y)\ (R_x,\ R_y)$	$(x^2-y^2,\ xy)\ (xz,\ yz)$

C_{4v} (4 mm)	E	$2C_4$	C_2	$2\sigma_v$	$2\sigma_d$		
A_1	1	1	1	1	1	z	$x^2+y^2,\ z^2$
A_2	1	1	1	-1	-1	R_z	
B_1	1	-1	1	1	-1		x^2-y^2
B_2	1	-1	1	-1	1		xy
E	2	0	-2	0	0	$(x,\ y)\ (R_x,\ R_y)$	$(xz,\ yz)$

C_{5v}	E	$2C_5$	$2C_5^2$	$5\sigma_v$		
A_1	1	1	1	1	z	$x^2+y^2,\ z^2$
A_2	1	2	2	-1	R_z	
E_1	2	$2\cos 72°$	$2\cos 144°$	0	$(x,\ y)\ (R_x,\ R_y)$	$(xz,\ yz)$
E_2	2	$2\cos 144°$	$2\cos 72°$	0		$(x^2-y^2,\ xy)$

C_{6v} (6 mm)	E	$2C_6$	$2C_3$	C_2	$3\sigma_v$	$3\sigma_d$		
A_1	1	1	1	1	1	1	z	$x^2+y^2,\ z^2$
A_2	1	1	1	1	-1	-1	R_z	
B_1	1	-1	1	-1	1	-1		
B_2	1	-1	1	-1	-1	1		
E_1	2	1	-1	-2	0	0	$(x,\ y)\ (R_x,\ R_y)$	$(xz,\ yz)$
E_2	2	-1	-1	2	0	0		$(x^2-y^2,\ xy)$

$C_{\infty v}$	E	$2C_\infty^\phi$	$\ldots$	$\infty\sigma_v$		
$A_2,\ \Sigma^+$	1	1	$\ldots$	1	z	$x^2+y^2,\ z^2$
$A_2,\ \Sigma^-$	1	1	$\ldots$	-1	R_z	
$E_1,\ \Pi$	2	$2\cos \phi$	$\ldots$	0	$(x,\ y)\ (R_x,\ R_y)$	$(xz,\ yz)$
$E_2,\ \Delta$	2	$2\cos 2\phi$	$\ldots$	0		$(x^2-y^2,\ xy)$
$E_3,\ \Phi$	2	$2\cos 3\phi$	$\ldots$	0		
$\ldots$	$\ldots$	$\ldots$	$\ldots$	$\ldots$		

D_{3h} ($\bar{6}m2$)	E	$2C_3$	$3C_2$	σ_h	$2S_3$	$3\sigma_v$		
A_1'	1	1	1	1	1	1		$x^2+y^2,\ z^2$
A_2'	1	1	−1	1	1	−1	R_z	
E'	2	−1	0	2	−1	0	(x, y)	(x^2-y^2, xy)
A_1''	1	1	1	−1	−1	−1		
A_2''	1	1	−1	−1	−1	1	z	
E''	2	−1	0	−2	1	0	(R_x, R_y)	(xz, yz)

D_{6h} ($6/mmm$)	E	$2C_6$	$2C_3$	C_2	$3C_2'$	$3C_2''$	i	$2S_3$	$2S_6$	σ_h	$3\sigma_d$	$3\sigma_v$		
A_{1g}	1	1	1	1	1	1	1	1	1	1	1	1		$x^2+y^2,\ z^2$
A_{2g}	1	1	1	1	−1	−1	1	1	1	1	−1	−1	R_z	
B_{1g}	1	−1	1	−1	1	−1	1	−1	1	−1	1	−1		
B_{2g}	1	−1	1	−1	−1	1	1	−1	1	−1	−1	1		
E_{1g}	2	1	−1	−2	0	0	2	1	−1	−2	0	0	(R_x, R_y)	(xz, yz)
E_{2g}	2	−1	−1	2	0	0	2	−1	−1	2	0	0		(x^2-y^2, xy)
A_{1u}	1	1	1	1	1	1	−1	−1	−1	−1	−1	−1		
A_{2u}	1	1	1	1	−1	−1	−1	−1	−1	−1	1	1	z	
B_{1u}	1	−1	1	−1	1	−1	−1	1	−1	1	−1	1		
B_{2u}	1	−1	1	−1	−1	1	−1	1	−1	1	1	−1		
E_{1u}	2	1	−1	−2	0	0	−2	−1	1	2	0	0	(x, y)	
E_{2u}	2	−1	−1	2	0	0	−2	1	1	−2	0	0		

T_d ($\bar{4}3m$)	E	$8C_3$	$3C_2$	$6S_4$	$6\sigma_d$		
A_1	1	1	1	1	1		$x^2+y^2+z^2$
A_2	1	1	1	−1	−1		
E	2	−1	2	0	0		$(2z^2-x^2-y^2, x^2-y^2)$
T_1	3	0	−1	1	−1	(R_x, R_y, R_z)	
T_2	3	0	−1	−1	1	(x, y, z)	(xy, xz, yz)

O_h ($m3m$)	E	$8C_3$	$6C_2$	$6C_4$	$3C_2$ ($=C_4^2$)	i	$6S_4$	$8S_6$	$3\sigma_h$	$6\sigma_d$		
A_{1g}	1	1	1	1	1	1	1	1	1	1		$x^2+y^2+z^2$
A_{2g}	1	1	−1	−1	1	1	−1	1	1	−1		
E_g	2	−1	0	0	2	2	0	−1	2	0		$(2x^2-x^2-y^2, x^2-y^2)$
T_{1g}	3	0	−1	1	−1	3	1	0	−1	−1	(R_x, R_y, R_z)	
T_{2g}	3	0	1	−1	−1	3	−1	0	−1	1		(xz, yz, xy)
A_{1u}	1	1	1	1	1	−1	−1	−1	−1	−1		
A_{2u}	1	1	−1	−1	1	−1	1	−1	−1	1		
E_u	2	−1	0	0	2	−2	0	1	−2	0		
T_{1u}	3	0	−1	1	−1	−3	−1	0	1	1	(x, y, z)	
T_{2u}	3	0	1	−1	−1	−3	1	0	1	−1		

Box G.1 Characters

The **definition** of a character, χ, of an operation:

$$\chi = \text{tr } \mathbf{D} = \sum_{\mu} D_{\mu\mu}$$

where $\mathbf{D}$ is the representative matrix of the operation.

Orthogonality of the characters of different irreducible representations:

$$\sum_{R} \chi^{(i)}(R)^{*}\chi^{(j)}(R) = h\delta_{ij}$$

The **reduction** of a representation:

$$\Gamma = \sum_{i} a_i \Gamma^{(i)} \qquad a_i = \frac{1}{h}\sum_{R} \chi^{(i)}(R)^{*}\chi(R)$$

The **projection** of a basis function, $f^{(i)}$, for an irreducible representation $\Gamma^{(i)}$ from a general function f or set of functions:

$$p^{(i)}f = f^{(i)} \quad p^{(i)} = \frac{l^{(i)}}{h}\sum_{R} \chi^{(i)}(R)^{*}R$$

where $l^{(i)}$ is the dimension of $\Gamma^{(i)}$ and h is the order of the group.

Selection rules in terms of characters and irreducible representations: the integral

$$I = \int f^{(i)}f^{(j)}f^{(k)}*\mathrm{d}\tau$$

is zero unless the totally symmetric irreducible representation occurs in the decomposition of the product $\Gamma^{(i)} \times \Gamma^{(j)} \times \Gamma^{(k)}$.

Further information

For a very qualitative, applications-orientated introduction to group theory, see Atkins (1990); for a more rigorous introduction see Chapter 7 of MQM. Two of the most accessible introductions to the chemical applications of group theory are Cotton (1990) and Kettle (1985). See also Douglas and Hollingsworth (1985), Harris and Bertolucci (1978), and Bunker (1979) for books at a similar level. For more advanced treatments (with applications more to physics) see Tinkham (1964), Hamermesh (1962), Wigner (1959), and Wybourne (1974), perhaps in that order. Piepho and Schatz (1983) give a very thorough treatment of the applications of group theory to spectroscopy and magnetic properties of complexes and Altmann (1986) gives a very thorough treatment of symmetries of spherical systems, including those with spin. For the subtleties associated with the symmetry properties of flexible molecules see Ezra (1982). For solids see Wooster (1973) and Altmann (1991b) for applications, Lockwood and Macmillan (1978) for pictures, and Bradley and Cracknell (1972) for exhaustive mathematics. A brief collection of tables for group theory has been compiled by Atkins, Child, and Phillips (1970).

Hamiltonian

Sir William Rowan Hamilton (1805–1865), who was Astronomer Royal of Ireland while still an undergraduate, set up a system of classical mechanics which was to turn out to be ideally suited to reformulation as quantum mechanics. Indeed, it is arguable that he almost discovered quantum mechanics itself, and quite possibly would have done so if experiments had required it during his lifetime. For the systems that concern us, in classical mechanics the **hamiltonian**, H, of a system is the sum of the kinetic energy, T, and potential energy, V, of all the particles present:

$$H = T + V$$

(H can also be defined more generally, more subtly, and consequently more powerfully than as this sum.) In quantum mechanics, H is interpreted as the **hamiltonian operator** and treated as the operator corresponding to the total energy of the system. Some typical forms of the hamiltonian operator are listed in Box H.1.

The time-independent °Schrödinger equation is often written in terms of the hamiltonian as $H\psi = E\psi$. This form of the equation shows that the Schrödinger equation is an °eigenvalue equation: the eigenvalue of the equation is the total energy E of the system, and the eigenfunction is the wavefunction. Since only certain eigenfunctions are acceptable (because they have to satisfy certain requirements), only the corresponding eigenvalues are permitted; hence, the energy is °quantized.

The hamiltonian determines the rate of change of the wavefunction and all the observables of a system. The rate of change of a wavefunction is given by the time-dependent Schrödinger equation (see Box H.1). For a stationary state (that is, an eigenstate of the hamiltonian, one for which $\Psi^*\Psi$ does not change with time), the larger the eigenvalue of the hamiltonian, the more rapidly the wavefunction oscillates between real and imaginary values (see °complex wavefunctions). For instance, if we think of an H1s wavefunction as frozen in time (and hence as defining the zero of energy), an H2s wavefunction oscillates between real and imaginary values rapidly because it has a higher energy, and an H3s wavefunction oscillates even more rapidly because it has a still higher energy (Fig. H.1).

The rate of change of the °expectation value of an operator is also proportional to the hamiltonian, but in a more subtle way. The expression in Box H.1 states that the rate of change of the expectation value of an operator is proportional to its °commutator with the hamiltonian. If an operator commutes with the hamiltonian of the system, its expectation value does not change with time. If the hamiltonian and the

Box H.1 Hamiltonians

In elementary work, a **hamiltonian operator**, H, is the sum of the operators for kinetic and potential energy: $H = T + V$. The rate of change of a wavefunction is given by the °Schrödinger equation:

$$\frac{\partial \psi}{\partial t} = \frac{1}{i\hbar} H\psi$$

The rate of change of an expectation value is

$$\frac{\mathrm{d}}{\mathrm{d}t}\langle \Omega \rangle = \frac{1}{i\hbar}\langle [H, \Omega] \rangle$$

Particle in constant potential:

$$\text{One dimension: } H = -\frac{\hbar^2}{2m}\frac{\mathrm{d}^2}{\mathrm{d}x^2} + V$$

$$\text{Three dimensions: } H = -\frac{\hbar^2}{2m}\nabla^2 + V$$

where ∇^2 is the °laplacian.

Harmonic oscillator of force constant k:

$$H = -\frac{\hbar^2}{2m}\frac{\mathrm{d}^2}{\mathrm{d}x^2} + \tfrac{1}{2}kx^2$$

Rotation of body of moment of inertia I:

$$\text{On a plane: } H = -\frac{\hbar^2}{2I}\frac{\mathrm{d}^2}{\mathrm{d}\phi^2}$$

$$\text{On a sphere: } H = -\frac{\hbar^2}{2I}\Lambda^2$$

where Λ^2 is the legendrian operator (see °laplacian)

$$\text{In general: } H = -\frac{\hbar^2}{2I_{xx}}J_x^2 - \frac{\hbar^2}{2I_{yy}}J_y^2 - \frac{\hbar^2}{2I_{zz}}J_z^2$$

Hydrogenic atom of atomic number Z and °reduced mass μ:

$$H = -\frac{\hbar^2}{2\mu}\nabla^2 - \frac{Ze^2}{4\pi\varepsilon_0 r}$$

Collection of particles with charges $z_i e$ and masses m_i:

$$H = -\tfrac{1}{2}\hbar^2 \sum_i \frac{1}{m_i}\nabla_i^2 - \sum_{i>j}\frac{z_i z_j e^2}{4\pi\varepsilon_0 r_{ij}}$$

The restriction on the double sum ($i > j$) ensures that interactions are not counted twice.

Electric dipole in an electric field E:

$$H = -\boldsymbol{\mu}.\boldsymbol{E}$$

Magnetic dipole in a magnetic field B:

$$H = -\boldsymbol{\mu}.\boldsymbol{B}$$

operator do not commute, then the more widely separated the eigenvalues of the hamiltonian (the more widely separated the energy levels of the system), the faster the time-evolution of the expectation value of the operator. For example, the expectation value of the position oper-

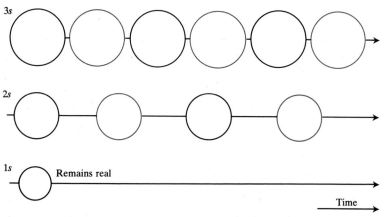

3s

2s

1s   Remains real

Time

Fig. H.1 The real part of the wavefunction is drawn in black and the imaginary part is drawn in blue. If the 1s orbital is taken as defining the zero of energy, it remains real for all times; the 2s wavefunction changes from real to imaginary, and back again; the 3s oscillates more rapidly.

ator of a harmonic oscillator varies more rapidly with time the greater the force constant and the smaller the mass. In the classical world of our perceptions, we see this faster variation as the more rapid motion of the oscillator. More generally and grandiosely, perceived motion is the manifestation of non-commutation of the observable and the hamiltonian: progress in the 'macroworld' is non-commutation in the 'microworld'.

The formulation of the hamiltonian of a system is the first step to setting up a quantized description (in one formulation of quantum theory). Once the hamiltonian is available, and expressed in terms of the positions q and linear momenta p of the system, quantization is introduced by requiring that the qs and ps are to be interpreted as operators that satisfy the fundamental °commutation rule of quantum mechanics (that $[q,p] = i\hbar$).

Further information

See Chapters 2 and 5 of MQM for some more information about the hamiltonian in quantum mechanics. All texts on quantum mechanics describe the properties, formulation, and eigenvalues of hamiltonians, so see Das and Melissinos (1986) and Bransden and Joachain (1989), as well as many others. For the classical background and Hamilton's formulation of classical mechanics see Goldstein (1982); for a very elegant approach to the quantization of arbitrary hamiltonians see Dirac (1964). A biographical note on Hamilton, with some details about his somewhat unconventional personal habits, has been given by Whittaker (1954); for more complete biographies, see Hankins (1980) and O'Donnell (1983). The formulation of quantum mechanics in a manner that Hamilton might have achieved is essentially that described in Feynman and Hibbs (1965) and Rivers (1987).

Hardness

The concept of chemical **hardness**, and of its counterpart **softness**, is an elaboration of the role of °polarizability that traditionally has been used to rationalize certain trends in the Periodic Table. The concept grew out of the recognition that elements generally fall into two classes. In one (specifically 'Class A', a class later to be more evocatively renamed the 'hard' elements), the strength of bonding to halogens decreased in the order $F > Cl > Br > I$. In the second class (Class B; later, the 'soft' elements) the strength of the bonds increased in the order $F < Cl < Br < I$.

If we think of a polarizable atom such as an I atom as being 'soft' and of the small, unpolarizable F atom as being 'hard', then we can identify a hard element as one that forms strong bonds with a hard element and a soft element as one that forms a strong bond with a soft element. Expressed slightly differently, soft–soft combinations tend to be covalently bonded and hard–hard combinations tend to be largely ionic. A list of hard and soft atoms and ions is given in Table H.1.

A quantitative definition of hardness has been proposed so that, in the same spirit that motivates °electronegativity, we can put the elements in order of hardness and hence begin to make broadly reliable predictions about their chemical properties. The definition proposed by the American chemist R.G. Pearson is that the **absolute hardness**, η, is half the separation of the °frontier orbitals (Fig. H.2). Since the energy of the HOMO determines the ionization energy I of the molecule, and the energy of the LUMO determines its electron affinity E_{ea}, we can express the absolute hardness as

$$\eta = \tfrac{1}{2}(I - E_{ea})$$

(The same quantity has been shown to correlate with the °aromaticity of planar, cyclic, conjugated hydrocarbons.) Some values of the abso-

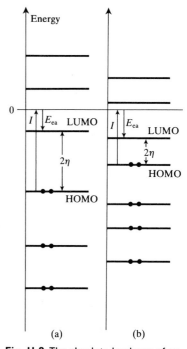

Fig. H.2 The absolute hardness of an atom (or molecule or ion) is defined as half the separation of the frontier orbitals (the HOMO and the LUMO energy gap). (a) A hard system; (b) a soft system.

Table H.1 The classification of Lewis acids and bases

	Hard	Borderline	Soft
Acids	H^+, Li^+, Na^+, K^+	Fe^{2+}, Co^{2+}, Ni^{2+}	Cu^+, Ag^+, Au^+, Tl^+, Hg^+
	Be^{2+}, Mg^{2+}, Ca^{2+} Cr^{2+}	Cu^{2+}, Zn^{2+}, Pb^{2+}	Pd^{2+}, Cd^{2+}, Pt^{2+}, Hg^{2+}
Bases	F^-, OH^-, H_2O, NH_3 CO_3^{2-}, NO_3^-, O^{2-} SO_4^{2-}, PO_4^{3-}, ClO_4^-	NO_2^-, SO_3^{2-}, Br^- N_3^-, N_2	H^-, R^-, CN^-, CO, I^- SCN^-, R_3P, C_6H_6 R_2S

Data: Source: Shriver, Atkins, and Langford (1990)

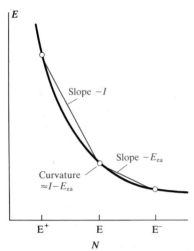

Fig. H.3 According to density functional theory, the hardness of a species can be identified with the curvature of the plot of electronic energy E against the number of electrons N. The identification of the hardness with the difference $I - E_{ea}$ then follows from the approximation that d^2E/dN^2 is equal to the difference in the slopes of the curve on either side of the species.

lute hardness calculated in this way are given in Table H.2. A more fundamental interpretation of the absolute hardness (which leads to the same expression) is obtained from density functional theory (see °density matrix); here the absolute hardness is equated with the curvature, specifically $\eta = d^2E/dN^2$, of the plot of electronic energy E against number of electrons N (Fig. H.3; the *slope* of the graph determines the °electronegativity).

A sense of why the absolute hardness depends on the separation of the frontier orbitals is provided by °perturbation theory. According to that theory, a system is easy to distort if its energy levels are close together. Therefore, an atom (or molecule) is responsive to the presence of another atom if its frontier orbitals are close together, which corresponds to a low value of the absolute hardness (and low curvature of the E against N plot). On the other hand, if the frontier orbitals are widely separated in energy, the species will be 'stiff' and unresponsive to a perturbation. In this case the atom will be unpolarizable and chemically hard. A polarizable, soft atom is one that can lend its orbitals to the formation of covalent bonds (a form of distortion). A hard atom generally has a high ionization energy (in the sense that its LUMO is relatively deep in energy, Fig. H.2) and is likely to form ionic bonds.

Table H.2 Absolute hardness

H						
6.43						

Li	Be	B	C	N	O	F
2.39	4.5	4.01	5.00	7.23	6.08	7.01
35.12	67.84					

Na	Mg	Al	Si	P	S	Cl
2.30	3.90	2.77	3.38	4.88	4.14	4.68
21.08	32.55	45.77				

K	Ca	Ga	Ge	As	Se	Br
1.92	4.0	2.9	3.4	4.5	3.87	4.22
13.64	19.52	17				

Rb	Sr	In	Sn	Sb	Te	I
1.85	3.7	2.8	3.05	3.80	3.52	3.69
11.55	16.3	13				

Cs	Ba	Tl	Pb	Bi		
1.71	2.9	2.9	3.53	3.74		
10.6		10.4				

First entries are for the neutral atoms; second entries are for the representative cation of the group (M^+ for Group 1, M^{2+} for Group 2, and so on).
Data: Pearson (1988). The reference also gives values for members of the d block and for molecules.

Further information

An introduction to the deployment of hardness and softness in chemistry will be found in Shriver, Atkins, and Langford (1990). A very helpful survey is that by Pearson (1987), who originated the terminology. For an excellent account of the analysis of the concepts in terms of density functional theory see Parr and Yang (1989).

Harmonic oscillator

Harmonic oscillations occur in classical mechanics when the restoring force on a body obeys **Hooke's law**:

The restoring force on a body is proportional to its displacement from equilibrium: $F = -kx$.

The constant of proportionality k is the °force constant of the system (for example, the force constant of a chemical bond). Because force is the negative slope of potential ($F = -dV/dx$), a force $F = -kx$ implies the existence of a **parabolic potential energy**, $V = \frac{1}{2}kx^2$ ($y = $ constant $\times x^2$ is the formula of a parabola, Fig. H.4). Hence, harmonic oscillation is characteristic of parabolic potentials.

The properties of a classical harmonic oscillator are summarized in Box H.2. The essential features are that the amplitude of swing increases with the total energy of the oscillator, and the frequency of the motion (the number of cycles of displacement per unit time) is determined only by the mass of the oscillator and its °force constant and is independent of the amplitude of the motion. The lighter the mass the higher the frequency (because a given restoring force can pull back a light mass more readily than a heavy mass). The greater the force constant (the stiffer the spring), the higher the frequency. The harmonic oscillator is stationary at the **turning points** at the end of the swing (when its total energy is entirely potential energy); it moves most rapidly at zero displacement (when its energy is entirely kinetic). The displacement at which a classical oscillator will most probably be found is therefore at either of its turning points, where it is (instantaneously) stationary (Fig. H.5).

The properties of a harmonic oscillator in quantum mechanics are determined by solving the °Schrödinger equation with a parabolic potential energy. The conclusions are summarized below and expressed quantitatively in Box H.2 (Fig. H.6).

- The energy is quantized, with evenly spaced energy levels.
- The separation of adjacent levels is $\hbar\omega$, where $\omega = (k/m)^{1/2}$.

Therefore, the stiffer the restoring force and the lighter the mass, the greater the separation of the levels. Note that the *separation* of the

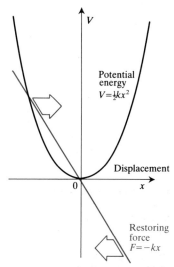

Fig. H.4 A body undergoes harmonic oscillation if the restoring force is proportional to the displacement. Such a force corresponds to a parabolic potential energy. The strength of the force and the narrowness of the parabola increase with increasing force constant k.

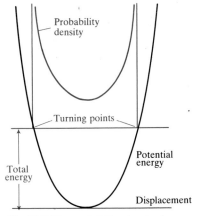

Fig. H.5 The probability density for finding the oscillator with a given displacement. Note that the probability density is greatest at the turning points (when the total energy is equal to the potential energy), for then the oscillator is momentarily stationary.

Box H.2 Properties of a harmonic oscillator

Classical mechanics:

For $x = 0$ at $t = 0$,

$$x = A \sin \omega t \qquad p = m \omega A \cos \omega t$$

with $\omega = (k/m)^{1/2}$ and $E = \frac{1}{2} kA^2$. The turning points are at $x = \pm A$. The probability of finding the harmonic oscillator in the range dx at the displacement x is $P\,dx$, where

$$P = \frac{1}{\pi} \frac{1}{(A^2 - x^2)^{1/2}}$$

Quantum mechanics:

The **energy levels** are

$$E_v = (v + \tfrac{1}{2}) \hbar \omega, \qquad \omega = (k/m)^{1/2} \text{ and } v = 0, 1, 2, \ldots$$

The **wavefunctions** are

$$\psi_v = N_v H_v(y)\, e^{-y^2/2}, \qquad N_v = \left\{ \frac{1}{\alpha \mu^{1/2} 2^v v!} \right\}^{1/2}$$

where the H_v are **Hermite polynomials** in the variable y,

$$y = x/\alpha, \qquad \alpha = \left\{ \frac{\hbar^2}{mk} \right\}^{1/4}$$

The explicit form of some of the hermite polynomials are given in Box H.3. They satisfy the equation

$$H_v'' - 2y H_v' + 2v H_v = 0$$

An important **recursion relation** is

$$H_{v+1} = 2y H_v - 2v H_{v-1}$$

An important **integral** is

$$\int_{-\infty}^{+\infty} H_v H_v\, e^{-y^2}\, dy = \begin{cases} 0 & \text{if } v' \neq v \\ \pi^{1/2} 2^v v! & \text{if } v' = v \end{cases}$$

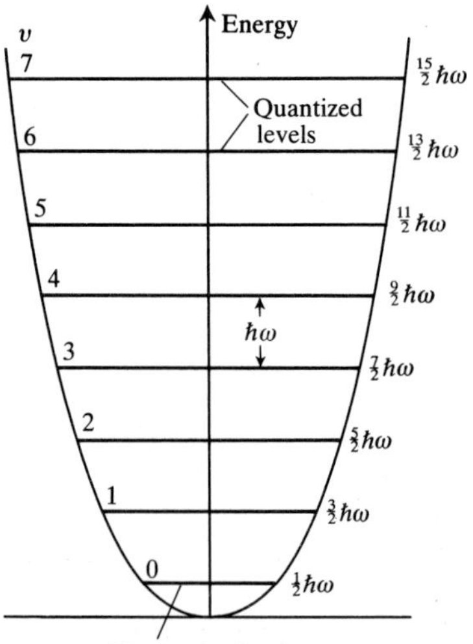

Fig. H.6 The energy levels of a harmonic oscillator superimposed on a drawing of its potential energy. Note the uniform (and infinite) array of levels and the existence of a zero point energy.

quantized levels (expressed as a frequency) is equal to the frequency of the classical oscillator.

● There is a °zero-point energy of magnitude $\frac{1}{2}\hbar \omega$.

Therefore, the stronger the restoring force and the smaller the mass, the greater the zero-point energy. The existence of the zero-point energy is consistent with the °uncertainty principle, for a stationary oscillator would have simultaneously a well defined location (at $x = 0$) and a well defined momentum ($p = 0$), which is forbidden.

● The wavefunctions are polynomials in the displacement (the 'Hermite polynomials') multiplied by a gaussian function.

The explicit form of some of the wavefunctions is given in Box H.3;

Box H.3 The Hermite polynomials up to $v = 6$

v	$H_v(y)$
0	1
1	$2y$
2	$4y^2 - 2$
3	$8y^3 - 12y$
4	$16y^4 - 48y^2 + 12$
5	$32y^5 - 160y^3 + 120y$
6	$64y^6 - 480y^4 + 720y^2 - 120$

some of their more important technical properties are included in Box H.2. The shapes of a few of them are drawn in Fig. H.7.

In the state of lowest energy (the ground state of the oscillator, the state with $v = 0$), the polynomial that occurs in the wavefunction is simply the factor 1. Hence, the ground state wavefunction is a bell-shaped gaussian curve with its maximum at zero displacement. In the ground state the oscillator will be found with displacements close to zero, but not strictly confined to zero. The oscillator has non-zero °kinetic energy because the wavefunction is curved, and it has non-zero potential energy because the wavefunction spreads into regions of non-zero potential. The polynomial for the first excited state wave-function ($v = 1$) is x itself. The product of this factor and the gaussian is a function with a °node at zero displacement (at $x = 0$). The wavefunc-tion has a higher energy because it is more sharply curved (has higher °kinetic energy) and penetrates further into the potential (in classical terms, the oscillating body swings both faster and further). The wave-function with $v = 2$ has a peak at $x = 0$ and also has significant accumu-lations of probability in the regions of higher potential. Note that successive wavefunctions are alternately even (°*gerade*) and odd (*unger-ade*) with respect to inversion.

The gradual migration of probability towards higher displacements shown in Fig. H.7 is a general feature of a harmonic oscillator as it becomes more highly excited. The regions of greatest probability shift towards the turning points of the classical motion (which is where, classically, the oscillator is most likely to be found).

● If the position of a harmonic oscillator is known to lie in a finite range, its wavefunction is a superposition of individual wave-functions of different energies.

A superposition with a gaussian probability distribution—a gaussian °wavepacket—swings from one turning point to the other with a frequency ω, just like the classical motion of the oscillator with the

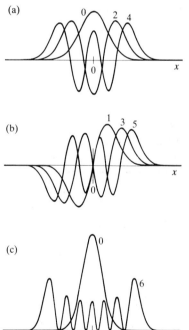

Fig. H.7 The first few harmonic oscillator wavefunctions (a) for even v and (b) for odd v. (c) A comparison of the probability distributions for the ground state and an excited state ($v = 6$); note how the latter is beginning to resemble the classical distribution shown in Fig. H.5.

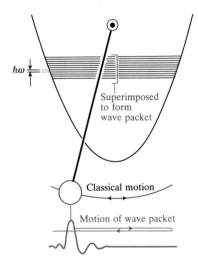

Fig. H.8 A wavepacket formed from a superposition of harmonic oscillator wavefunctions reproduces the motion of a classical harmonic oscillator. It oscillates between the classical turning points with a frequency equal to the separation of the quantum levels.

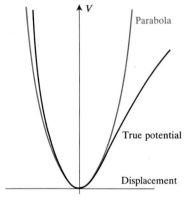

Fig. H.9 A parabola is a good approximation to almost all potential energy curves at small displacements. The formal justification of this remark depends on being able to show that the first non-vanishing term in a Taylor expansion of the potential around its minimum is proportional to the square of the displacement in all except atypical cases.

same mass and force constant (Fig. H.8). That is, the quantum spacing ω becomes the frequency of the classical oscillator. This is an example of how quantum-mechanical principles underlie classical mechanics (see °correspondence principle).

The importance of the harmonic oscillator arises from a number of features. First, oscillations in nature are often harmonic to a very good approximation. This is because almost all naturally occurring potential wells are parabolic close to their minima (Fig. H.9). Consequently, the theory of the harmonic oscillator can be used in the description of the vibration of molecules and of atoms in solids. One such application is to the description of molecular vibrational spectroscopy; another is to the calculation of °heat capacity and the details of thermal motion in solids. For quite different reasons, the harmonic oscillator algebra occurs naturally when the electromagnetic field is quantized, and successive excitations up a ladder of energy levels are interpreted as the generation of an increasing number of °photons of a given frequency.

A further, more technical point, is that (as in classical mechanics) the algebra of harmonic oscillation is closely related to that of °rotational motion. Specifically, harmonic oscillation is the projection of the steady motion of a point on a circular trajectory on to the diameter of the circle (Fig. H.10). It should therefore not be surprising to see harmonic oscillator algebra appearing in discussions of °angular momenta.

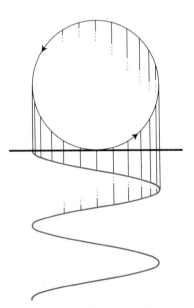

Fig. H.10 The projection of steady motion on a circle on to a diameter (or, as here, a line parallel to a diameter) is equivalent to simple harmonic motion on the line.

Further information

The details of the quantum mechanics of the harmonic oscillator are described in Chapter 3 of MQM. The solution there is in terms of annihilation and creation operators, which are widely used in quantum mechanics (see °second quantization). For a more ponderous solution, in the manner originally used by Schrödinger, see Pauling and Wilson (1935). See Das and Melissinos (1986) and Bransden and Joachain (1989) for other versions. The properties of the Hermite polynomials are set out in Abramowitz and Stegun (1965), who also give numerical values. The manipulation of these polynomials is described by Margenau and Murphy (1956) and Morse and Feshbach (1953). The relation of harmonic oscillator algebra to angular momentum is described by Englefield (1972), in a virtuoso performance by Schwinger (1965), and by Biedenharn and Louck (1981, 1984). Books on the quantized electromagnetic field make full use of harmonic oscillator algebra, so see Louisell (1973).

Heat capacity

The **heat capacities** of a substance at constant pressure C_p and constant volume C_V are defined as

$$C_p = \left(\frac{\partial H}{\partial T} \right)_p \qquad C_V = \left(\frac{\partial U}{\partial T} \right)_V$$

The two properties measure the response of the enthalpy, H, and the internal energy, U, of a sample to an increase in temperature. The energy (or enthalpy) of a substance with a large heat capacity increases sharply as its temperature is increased. Conversely, the higher the heat capacity of a substance, the greater the heat needed to increase its temperature by a given amount. A sample with a high heat capacity has a high 'capacity' for heat, for it undergoes only a small rise in temperature as energy is transferred to it as heat. Heat capacity is an extensive property; **molar heat capacity** (the heat capacity per mole) and **specific heat capacity** (the heat capacity per unit mass) are intensive properties.

The historical starting point for the theoretical explanation of the heat capacities of solids is the law proposed by the French chemist Pierre Dulong and the French physicist Alexis-Thérèse Petit in 1819 on the basis of somewhat meagre experimental evidence. In modern terms, **Dulong and Petit's law** states:

The heat capacity of a solid is about $25 \, \text{J K}^{-1}$ per mole of atoms present in the sample.

In the early days of chemistry the law was used to obtain atomic weights (relative atomic masses) from measurements of the specific heat capacities of elements.

Dulong and Petit's law is easily explained in terms of classical physics and the °equipartition principle. Let the sample contain N atoms (they need not all be atoms of the same element). Each atom can vibrate relative to its neighbours in three perpendicular directions, and so the sample behaves like a collection of $3N$ oscillators. The °equipartition theorem implies that the mean energy of each oscillator is kT; therefore the contribution of the thermal motion of the atoms in the block to its total internal energy at a temperature T is $3NkT$, or $3nRT$. It follows that the heat capacity at constant volume is $3R$ per mole of atoms present. Since $3R = 24 \, \text{J K}^{-1} \, \text{mol}^{-1}$, the calculation accounts for the value suggested by Dulong and Petit's observations.

Unfortunately for Dulong, Petit, and classical physics, more extensive measurements show that in fact few solids accord with their law and that none does at low temperatures. At first there was no explanation for the disagreement, since classical physics leads inexorably to the value $3R$ for all elemental solids. However, the explanation is quite easy to find now that we know that energy is quantized.

In the **Einstein model**, which was proposed by Albert Einstein in 1906, it is supposed that every oscillator in the sample has the same fundamental frequency v. If energy (as heat) is transferred to the metal at a low temperature, it can be used to activate only a very small number of oscillators. This is because no oscillator can take up an energy of less than hv, and if only a little energy is supplied there is enough available to excite only a few oscillators. Since only a few oscillators in the sample can be excited, there is a reduction of the heat capacity below the value expected classically (when an oscillator can be excited by even an infinitesimal amount of energy). At higher temperatures the energy supplied may be distributed over more oscillators, and so more are able to accept energy: the heat capacity is therefore greater than at lower temperatures. At very high temperatures, when the energy of the block greatly exceeds the excitation energy of the oscillators, all oscillators are active and may be stimulated to high quantum levels. At this stage the sample attains its classical heat capacity of $3R$. Box H.4 gives the Einstein expression for the heat capacity, and Fig. H.11 shows the predicted temperature dependence.

The Einstein heat capacity depends on the temperature and a characteristic constant known as the **Einstein temperature**, $\theta_E = hv/k$. The higher the Einstein temperature the lower the heat capacity at a given temperature. Since θ_E is highest in rigid materials composed of light atoms (like diamond), such materials have a lower heat capacity at a given temperature than solids composed of heavy atoms (like lead) bonded weakly together. The classical (Dulong and Petit) heat capacity is approached for temperatures significantly higher than θ_E.

The **Debye model**, which was proposed by Peter °Debye in 1912, is

Box H.4 Molar heat capacities of solids

Dulong and Petit:

$$C_V = 3R$$

Einstein:

$$C_V = 3Rf \quad f = \left\{\frac{\theta_E}{T}\right\}^2 \left\{\frac{e^{\theta_E/2T}}{1 - e^{\theta_E/T}}\right\}^2$$

where $\theta_E = h\nu_E/k$.

Debye:

$$C_V = 3Rf \quad f = 3\left\{\frac{T}{\theta_D}\right\}^3 \int_0^{\theta_D/T} \left\{\frac{x^2}{1 - e^x}\right\}^2 dx$$

where $\theta_D = h\nu_D/k$. The **Debye temperature** is related to the speed of sound, v, in the solid by

$$4\pi v_D^3 = 3\rho v^3$$

where ρ is the number density of atoms. At low temperatures ($T \ll \theta_D$),

$$C_V \approx aT^3 \quad a = \frac{12\pi^4 R}{5\theta_D^3}$$

a modification of the Einstein model. It takes into account the fact that the atoms vibrate not as individuals at a single frequency but *collectively* with a range of frequencies from zero up to a limit ν_{max}. We can see why there is a range of collective modes of vibration by considering a one-dimensional chain of atoms (Fig. H.12). If we

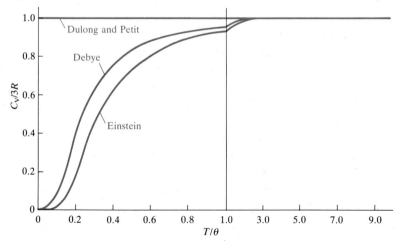

Fig. H.11 The temperature dependence of the heat capacity of an elemental solid as predicted by the laws of Dulong and Petit, Einstein, and Debye. The temperature is expressed as a multiple of the characteristic Einstein or Debye temperatures. Note that the Einstein formula predicts a lower heat capacity than the Debye formula at all temperatures (except in the classical limit). At very low temperatures the Debye curve approaches zero as T^3.

Table H.3 Debye temperatures of solids, θ_D/K

Li	344	C(d)	2230	Cu	343	Al	428
Na	158	C(gr)	420	Ag	225	Fe	467
K	91			Au	165	Pb	105
Rb	56					Hg	72
Cs	38						
NaF	492	KF	336				
NaCl	321	KCl	235				
NaBr	225	HBr	174				
NaI	164	KI	132				

Data: Gray (1972).

consider only transverse vibrations (those perpendicular to the line of atoms) it should be clear that the vibration with the highest restoring force is the one in which neighbouring atoms are displaced in opposite directions and so stretch the bond between them by the greatest amount for a given displacement. The vibration with least restoring force is the one where all the atoms are displaced in the same direction. As there are also intermediate modes of displacement we can expect a range of restoring forces, and so a range of fundamental frequencies up to a maximum. The Debye model takes the distribution of frequencies into account by calculating the number of oscillators at each frequency and then finding their total contribution to the heat capacity (see Box H.4 and Fig. H.13).

Since there are oscillators at lower frequency than in the single-frequency Einstein model, the Debye model predicts higher values of

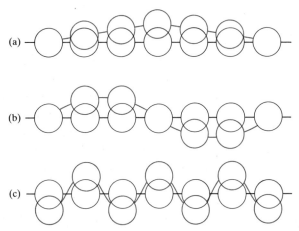

Fig. H.12 (a) The vibrational mode of this form has the lowest frequency because the *relative* motion of the atoms is not very great and the restoring force is weak. (b) A mode corresponding to a higher frequency: the relative motion of the atoms is greater and the restoring force is stronger. (c) The greatest frequency is found for this mode because the atoms cannot undergo a greater relative displacement.

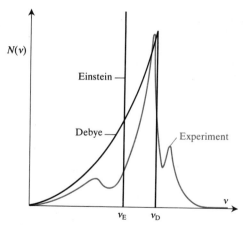

Fig. H.13 The spectrum of vibrational frequencies in a solid according to the Einstein and Debye models, and the experimentally determined spectrum (for copper).

the heat capacity than the Einstein model (Fig. H.11), and hence it is in better accord with observation. The Debye heat capacity depends on the **Debye temperature** $\theta_D = h\nu_{max}/k$, and the higher the Debye temperature the lower the heat capacity at a given temperature. Broadly speaking, the Debye temperature varies with the solid in much the same way as the Einstein temperature, with diamond having a high value (2230 K) and lead a low value (105 K). Some representative values of θ_D are listed in Table H.3. The classical (Dulong and Petit) heat capacity is approached at temperatures significantly higher than θ_D.

The Debye model predicts that at low temperatures the heat capacity of a solid should be proportional to $(T/\theta_D)^3$. This **Debye T^3 law** is obeyed reasonably well by many substances, and is used in the calorimetric determination of entropies.

Individual molecules in the gas phase also have heat capacities that show consequences of quantization. Thus, according to the equipartition theorem, the translational energy of a gas of N particles is $\frac{3}{2}NkT$, which gives a contribution to the molar heat capacity at constant volume of $\frac{3}{2}R$ and at constant pressure of $\frac{5}{2}R$; this full classical value is always achieved because so many translational states are populated. A non-linear polyatomic molecule has a rotational energy of $\frac{3}{2}kT$, and hence there is an additional contribution of $\frac{3}{2}R$ to the molar heat capacity; at ordinary temperatures so many rotational states are occupied that this classical contribution is also fully achieved in most cases. Each vibrational mode of a molecule contributes a further R to the heat capacity, but this value is attained only if many energy levels are populated at the temperature in question, which is rarely the case. Hence, as in solids, the contribution of the vibrations of individual molecules to the heat capacity is much less than expected classically. However, it should not be forgotten that many-atom polyatomic molecules have so many °vibrational modes that the *total* vibrational contribution to the heat capacity is often quite significant and cannot be ignored. For example, the benzene molecule has $3 \times 12 - 6 = 28$ modes of vibration, and their total contribution is significant.

Further information

The calculation of the Einstein formula for the heat capacity is described in MQM Chapter 1 and in Atkins (1990); see the latter for applications to the heat capacities of gas phase molecules. A very good discussion of the thermal properties of solids, including the full details of the Einstein and Debye calculations, will be found in Kittel (1987), Anderson (1984), and Ashcroft and Mermin (1976). A compilation of numerical data, including tables of the Einstein and Debye expressions and the Debye temperatures of many materials, is given in Gray (1972). Abramowitz and Stegun (1965) also tabulate the Debye function. For comments on the electronic contribution to the heat capacities of metals see the entry on the °Fermi–Dirac distribution.

Hellmann–Feynman theorem

The **Hellmann–Feynman theorem**, which is ascribed to the German theoretical chemist H. Hellmann (who used it in 1933) and which was elaborated by Richard °Feynman (in 1939), states:

> The slope dE/dP of the energy E with respect to a parameter P is equal to the °expectation value of the derivative of the °hamiltonian with respect to P:
>
> $$dE/dP = \langle (\partial H/\partial P) \rangle$$

The parameter may be a molecular bond length or angle, a nuclear charge, the strength of an applied field, and so on. An advantage of the theorem is that since $\partial H/\partial P$ is often a very simple operator, dE/dP may be evaluated very simply.

One application of the Hellmann–Feynman theorem is to the calculation of the response of molecules to electric and magnetic fields (see °polarizability); another is to the study of the geometry and °force constants of molecules. The latter application stems from a remarkable consequence of the Hellmann–Feynman theorem which is known as **Feynman's electrostatic theorem**:

> The force on a nucleus in a molecule may be calculated from classical electrostatics if the exact charge distribution is known.

That is, if we know the electron density everywhere, which is known if the °wavefunction is known, then the force on a nucleus can be calculated by considering the Coulombic force that the same *classical* charge distribution exerts on a point nucleus. The °force constant for any distortion of the molecule can then be calculated by working out the restoring force on a nucleus when the geometry is distorted.

Although the electrostatic theorem appears to convert the problem of molecular geometry into a problem in classical electrostatics, we have to remember that the quantum mechanical aspects of the problem, such as the exchange interactions, are hidden in the difficult task of determining the correct, quantum-mechanical electron density for a given nuclear configuration. Nevertheless, the theorem does remove some of the mystery about the shapes of molecules, for the geometry can be understood in terms of a balance of electrostatic forces, even though the distribution responsible for the balance is governed by quantum mechanics.

Further information

A simple derivation of the theorem is given in MQM Chapter 8, which describes its application to electric and magnetic properties of molecules. For more information, see the very thorough reviews by Deb (1972, 1981). An analysis of molecular structure in terms of the theorem from the viewpoint of °density matrices (specifically, density-functional theory) is given in Parr and Yang (1989). The original papers on the theorem are Hellmann (1933) and Feynman (1939). The theorem was in the literature before Hellmann used it: see for instance van Vleck (1932) and Pauli (1933). It seems best to regard Hellmann as the last person to discover the theorem, not the first. For an example of how the theorem may be deployed and extended see Singh and Singh (1989).

Hermitian operators

An °operator Ω is hermitian if

$$\int \psi_f{}^* \Omega \psi_i \, d\tau = \left(\int \psi_i{}^* \Omega \psi_f \, d\tau \right)^*$$

In the Dirac °bracket notation this criterion is written

$$\langle f|\Omega|i \rangle = \langle i|\Omega|f \rangle^*$$

Hermitian operators are important in quantum theory because:

- Their eigenvalues are real.
- States belonging to different eigenvalues are °orthogonal.

It follows from the first property that observables are represented by hermitian operators, for that guarantees that the outcome of observations (the eigenvalues of the operators) are real. It follows from the second property that many °overlap integrals can be set equal to zero without any calculation at all. Thus, we know at once that a hydrogenic $1s$ orbital is orthogonal to all other hydrogenic orbitals of the same atom, since all other orbitals correspond to different eigenvalues of the hermitian energy operator (the °hamiltonian).

Further information

See MQM Chapter 5 for a discussion of hermiticity and proofs of the properties mentioned above. For detailed, fundamental accounts of the hermiticity of operators see Dirac (1958), von Neumann (1955), Jordan (1969), and Jauch (1968).

HOMO and LUMO

See °Frontier orbitals.

Hückel method

By taking into account the symmetry of a molecule in a wise way, and by making drastic simplifications, the German physicist and theoretical chemist Erich Hückel (1896–1980) developed a simple scheme for calculating the contribution of the π electrons to the total energy of conjugated molecules. His approximations, which we list below, are extreme, for they neglect almost every complication. On the other hand, they cut through to the heart of the problem and help us to identify a parameter that can be used to correlate a variety of molecular properties. The parameter β of Hückel theory is the two-dimensional analogue of the parameter Δ in the °ligand field theory of d-metal complexes.

The **Hückel approximations** are as follows:

- The σ electrons are ignored.

The σ electrons are present implicitly because they are largely responsible for the shape of the molecule, but their interactions with the π electrons are neglected. The justification is that the energies of σ and π orbitals are very different, and systems with widely different energies interact only weakly (a short, high frequency pendulum hardly affects a long, slowly swinging pendulum hanging from the same support).

- All overlap is neglected.

The justification of this approximation is that $C2p\pi$ overlap is small (but not very: $S(2p\pi, 2p\pi)$ is about 0.25). In fact, it turns out that overlap can be included quite simply, and that its inclusion does not alter the energy savagely because it affects both the numerators and denominators in the expressions that need to be evaluated.

- All one-atom integrals over the hamiltonian are set equal to the same value α.

The integral α is called the **Coulomb integral** and is approximately equal to the energy of an electron occupying a C2p orbital. The justification of this approximation is the Coulson–Rushbrooke theorem (see °altern-ant hydrocarbons).

- All integrals over the hamiltonian that involve two non-adjacent atoms are set equal to zero.

The justification is the hope (which is reasonably well supported by calculation) that, as these integrals are broadly proportional to the overlap, and the overlap of orbitals on non-adjacent atoms is small, they are small enough to ignore.

- All the integrals over the hamiltonian that involve two adjacent atoms are set equal to the same value β.

The integral β is the **resonance integral**. The justification of this step is the near equality of bond lengths and (in °alternant hydrocarbons) the similarity of all the C atoms.

- All π electron–electron interactions are ignored.

With this considerable simplification, the π-electron energy is the sum of the individual orbital energies calculated using the previous approximations.

With these approximations in hand, the °variation method is applied to determine the best °linear combinations of atomic p orbitals to describe the structure of the molecule. This procedure leads initially to a °secular determinant. Each diagonal element of the determinant is equal to $\alpha - E$ and each off-diagonal element is zero except those corresponding to adjacent atoms, which are equal to β (Box H.5). The roots of the determinant give the energies of the π orbitals (Fig. H.14) and the eigenvectors give the coefficients of the linear combinations (see, for instance, °benzene).

Once the Hückel orbitals have been derived, we can go on to calculate the π electron energy as the sum of the occupied orbital energies, the °delocalization energy, the energy of electronic transitions within the π system, the °charge density on the carbon atoms, the π electron contribution to the °dipole moment, and the °bond order. In calculations of the delocalization energy a reasonable value of β is -0.69 eV (-67 kJ mol^{-1}), but for transition energies a better value is -2.71 eV ($-21\,900$ cm^{-1}). A sample calculation is illustrated in Box H.5.

Box H.5 Hückel calculations

In general,

- Write each MO as $\psi = \sum_i c_i \phi_i$

- Set up the °secular determinant $|H_{ij} - ES_{ij}|$ as follows:

$$S_{ij} = 1 \text{ if } i = j \text{ but } 0 \text{ otherwise}$$

$H_{ij} = \alpha$ if $i = j$, β if i and j are neighbours, and 0 otherwise.

- Solve $|H_{ij} - ES_{ij}| = 0$ for the energies E (in practice, find the eigenvalues of the secular matrix).

- Solve the secular equations for the coefficients c for each value of E (in practice, obtain the eigenvectors for each eigenvalue E).

As an example, for butadiene,

$$\psi = c_1 \phi_1 + c_2 \phi_2 + c_3 \phi_3 + c_4 \phi_4$$

where the ϕ are C2$p\pi$ orbitals. The secular determinant is

$$\begin{vmatrix} \alpha - E & \beta & 0 & 0 \\ \beta & \alpha - E & \beta & 0 \\ 0 & \beta & \alpha - E & \beta \\ 0 & 0 & \beta & \alpha - E \end{vmatrix} = 0$$

The roots are $E = \alpha \pm 1.6\beta$ and $\alpha \pm 0.6\beta$ and the four normalized molecular orbitals are

$$\psi_1 = 0.37\phi_1 + 0.60\phi_2 + 0.60\phi_3 + 0.37\phi_3 \text{ for } E = \alpha + 1.6\beta$$

$$\psi_2 = 0.60\phi_1 + 0.37\phi_2 - 0.37\phi_3 - 0.60\phi_4 \text{ for } E = \alpha + 0.6\beta$$

$$\psi_3 = 0.60\phi_1 - 0.37\phi_2 - 0.37\phi_3 + 0.60\phi_4 \text{ for } E = \alpha - 0.6\beta$$

$$\psi_4 = 0.37\phi_1 - 0.60\phi_2 + 0.60\phi_3 - 0.37\phi_4 \text{ for } E = \alpha - 1.6\beta$$

Since β is negative, ψ_1 lies lowest in energy and is most bonding in character.

In general, for cyclic hydrocarbons of N atoms, the orbital energies are given by

$$\varepsilon_i = \alpha + 2\beta \cos \frac{2\pi i}{N}$$

with $i = 0, 1, \ldots \frac{1}{2}N$ (N even) or $\frac{1}{2}(N-1)$ (N odd). The energy levels can be constructed by inscribing an N sided equilateral polygon inside a circle of radius 2β with one corner at the lowest point. The other corners denote orbital energies. Their energies are given in units of β by the vertical distance from the horizontal diameter of the circle. The method is illustrated for $N = 7$ in diagram (1).

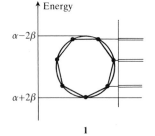

Energy

$\alpha - 2\beta$

$\alpha + 2\beta$

1

The Hückel method is moderately satisfactory as a first approximation because it emphasizes the role of the symmetry of the molecule: the orbitals are essentially classified according to their symmetry, and then the numbers of bonding and antibonding neighbours are counted by the coefficient multiplying β in the expression for the orbital energy. The latter effectively provides an indication of the order in which the orbitals lie (the more nodes, the higher the energy).

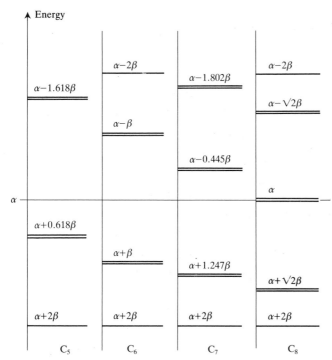

Fig. H.14 The π molecular orbital energy levels of four cyclic hydrocarbons according to the Hückel method.

There is obviously enormous room for improvement in the method, and it should be regarded as no more than a rough sighting-shot for developing a more reliable description. A trivial improvement is the inclusion of overlap between neighbours: this squashes the lower, bonding orbitals together and separates the upper, °antibonding levels without significantly affecting their order.

The next improvement often employed in simple calculations notes that the energy of an electron on a C atom, as measured by α, depends on the charge density on that atom. The Hückel method ignores this dependence and gives the same value of α to all the atoms irrespective of the local accumulation of electron density. The **ω-technique** seeks to overcome this deficiency by setting α proportional to the charge density (with ω as the constant of proportionality; specifically, the Coulomb integral at an atom i is written $\alpha_i = \alpha + (1 - q_i)\omega\beta$, where q_i is the charge on the atom), and is therefore a simple example of a °self-consistent calculation. A Hückel calculation is first done to find the charge density, then each α is modified appropriately, and the calculation is then repeated with the new set of αs. The scheme is repeated until the charge density remains constant through a cycle of the calculation.

The only satisfactory improvement, however, is to adopt one of the °semi-empirical or, better still, °*ab initio* procedures, and to calculate the electronic structure of a molecule properly. The Hückel method has been historically important, and it is still useful pedagogically and for formulating explanations. However, there is now little excuse for not using one of the widely available program packages and doing a reliable calculation.

Further information

See MQM Chapter 10. Many books on quantum chemistry deal with the Hückel method in detail; see, for example, Coulson (1979), Coulson, O'Leary, and Mallion (1978), Murrell, Kettle, and Tedder (1970), Salem (1966), and McGlynn *et al.* (1972). Streitweiser (1961) and Yates (1978) work through many calculations on organic molecules and show how the results may be applied to the prediction of their chemical properties. For the method in the context of modern computational procedures see Cook (1978), Murrell and Harget (1972), and Szabo and Ostlund (1982). For pictorial representations of orbitals calculated within the Hückel scheme see Tedder and Nechvatal (1985), Hout, Pietro, and Hehre (1984), and Jørgensen and Salem (1973). For elaborations see °semiempirical methods. A program for the calculation of Hückel orbitals is available in the collection of software written by Ayscough (1990). See °aromaticity for further information.

Hund coupling cases

There are several sources of angular momentum in a diatomic molecule, notably the °spins of the electrons, their °orbital motion about the internuclear axis, and the °rotation of the nuclear framework. The total angular momentum, the vector sum of all these momenta, can be arrived at in a variety of different ways. The **Hund coupling cases**, which were devised by the German spectroscopist Friedrich Hund are idealized ways of performing the sum. We shall describe two of the schemes to give a sense of what is involved.

In **Hund's case (a)** the strong, axially symmetric field generated by the two nuclei dominates the interactions in the molecule. This field allows only orbital circulation about the axis to survive. The °spin–orbit coupling is incorporated by combining the total spin of the electrons with the surviving component of the orbital momentum. The orbital angular momentum around the axis is $\Lambda\hbar$ and the spin component on that axis is $\Sigma\hbar$, giving a total momentum around the axis of $\Omega\hbar$ with $\Omega = \Lambda + \Sigma$. The nuclear framework also rotates with a momentum that can be represented by a vector $\boldsymbol{O}$ perpendicular to the axis (Fig. H.15a). It follows that the total angular momentum of the molecule is represented by $\boldsymbol{J}$, the resultant of $\boldsymbol{O}$ and $\Omega\boldsymbol{k}$, where $\boldsymbol{k}$ is a unit vector along the internuclear axis.

Hund's case (c) applies to a molecule in which the °spin–orbit coupling is so strong that the spin and orbital momenta couple strongly into a resultant $\mathbf{J'}$. This resultant °precesses around the internuclear axis with a component $\Omega\hbar$ (Fig. H.15b). The angular momentum of the rigid rotation of the nuclear framework couples to $\mathbf{J'}$, and the whole system yields a total momentum $\mathbf{J}$. Hund's case (c) is important for linear molecules containing heavy atoms because the spin–orbit coupling in them is large.

It should be clear that these two cases are only a few of the many possible; for instance, the nuclei themselves may possess °spin and hence contribute to the total angular momentum of the molecule. Furthermore, the cases are extreme, ideal, or pure cases; in any real molecule there are various competitions between different angular momenta, and for none can there be complete victory. The contamination of one of the pure Hund's cases by another is referred to as **decoupling**.

Further information

See MQM Chapter 12 for a closer look at the coupling cases, and Barrow (1962), King (1964), and Herzberg (1950) for descriptions of how they are used. Chapter 5 of Herzberg (1950) is a splendid discussion of all the cases. For more advanced treatments of angular momentum in linear molecules see Kovaćs (1969) and Judd (1975).

Hund rules

The rules devised by Friedrich Hund (see preceding entry) provide a simple way of predicting which °term of a given °configuration of an atom has the lowest energy:

- The term with the highest °multiplicity (equivalently, the highest value of the total spin S) lies lowest in energy.

- For a given multiplicity, the term with the highest °orbital angular momentum L lies lowest in energy.

- For a given multiplicity and orbital angular momentum, the °level with the lowest (highest) value of the total angular momentum J lies lowest in energy if the valence shell is less (more) than half full.

The reversal of the order of J levels when a shell is more than half full is called the **inversion** of the levels. The rules are reasonably reliable for predicting the term of lowest energy, but are not reliable for predicting the relative order of the other terms of the configuration (see °Racah parameters).

As an example, consider the configuration d^2 and the terms ^1G, ^3F, ^1D, ^3P, and ^1S to which it gives rise. Application of the first rule suggests that the ^3F and ^3P terms should lie lower than the singlet

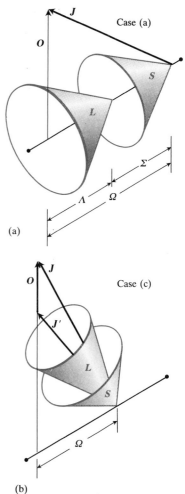

(a)

(b)

Fig. H.15 (a) Hund's coupling case a: the orbital angular momentum is well-defined only about the internuclear axis, and the total spin couples to that component to give a total projection designated by the quantum number Ω. (b) Hund's coupling case c. The spin–orbit coupling is so strong that the orbital and spin angular momenta couple together before their resultant couples to the internuclear axis. Although Ω is still a meaningful quantum number, Λ and Σ are not.

terms. The second rule then selects ^{3}F (with $L = 3$) as the lower of the two terms (since a P term has $L = 1$). The ^{3}F term has three levels, with $J = 4$, 3, and 2 (see the °Clebsch–Gordan series). Since the configuration corresponds to a less than half full subshell, the $J = 2$ level (the level with the lowest value of J) lies lowest in energy. Thus, we predict a ^{3}F$_2$ term for the ground state of a d^2 configuration.

The explanation of the first rule (of maximum multiplicity) can be traced to the effects of the Pauli principle. To achieve the maximum value of S, as many electrons as possible should occupy different orbitals, for then the maximum number of electrons may have parallel spins. Since they occupy as many different orbitals as possible, the electrons of a term with high multiplicity are spatially separated to the greatest extent, and hence repel each other least.

The conventional explanation of why, given that they occupy different orbitals, the electrons spins are in fact parallel is that electrons with parallel spins tend to keep apart (see °spin correlation and °Fermi holes), so their repulsion is less than when they are paired. However, the actual reason is rather more subtle. Detailed calculation on He has shown that the repulsion between the electrons is *greater* in triplet states than in singlets, and that the lowering of, the energy is due to the increase in the *electron–nucleus* attraction. Thus, if the electrons have parallel spins, the electron distribution contracts and is stabilized by the improved nuclear attraction. The electron repulsion indeed rises because the electrons are closer together, but this increase does not defeat the improvement in the nuclear attraction. Presumably the formation of a Fermi hole helps to stop the electron–electron repulsion rising faster than the nuclear attraction. There is no protective Fermi hole when the electrons are paired (in fact, there is an anti-protective Fermi heap), and so the orbitals occupied by the electrons are prevented from shrinking. In fact, the orbitals expand slightly to relieve the effect of Fermi heaping, and so the favourable interaction of the electrons with the nucleus is reduced.

The second (high L) rule reflects the tendency of electrons to stay apart if their orbital angular momentum carries them in the same direction. Electrons circulating in the same direction, with a high total orbital angular momentum, can stay apart (like cars on a traffic circle). Electrons orbiting in opposition will meet frequently, and so have a large repulsive interaction.

For an explanation of the third rule we must consider the °spin–orbit coupling. As explained in the entry on °fine structure, the energy is least when the orbital and spin magnetic moments of an electron in an atom are opposed. However, such an arrangement of magnetic moments implies that the two angular momenta are also opposed, which corresponds to a low total angular momentum. The inversion of the levels when the shell is more than half full reflects the accompanying change in the sign of the spin–orbit coupling constant in such cases.

Further information

See MQM Chapter 9 for a little more information and Candler (1964) for quite a lot more. The validity of the rules, and their reliability for terms other than the ground term, can be assessed by considering the °Racah parameters for the terms of a configuration; see that entry and Griffith (1964). The view that the conventional explanation of the maximum multiplicity rule is erroneous is based on the work of Lemberger and Pauncz (1970), Katriel (1972), and Colpa and Islip (1973).

Hybridization

A **hybrid orbital** is a linear combination of atomic orbitals centred on a single atom. An **sp hybrid**, for example, is an atomic wavefunction composed of equal proportions of s and p orbitals of the same atom, and an electron that occupies it has 50 per cent s character and 50 per cent p character. An **sp^3 hybrid** has 25 per cent s character and 75 per cent p character (that is, contributions in the ratio 1:3). An **sp^3d^2 hybrid** is an orbital composed of s, p, and d orbitals in the ratio 1:3:2. If N atomic orbitals contribute to the hybridization, N orthogonal hybrids may be formed. Thus, there are two orthogonal sp hybrids, four sp^3 hybrids, and six sp^3d^2 hybrids.

The shapes of hybrid orbitals are obtained by noting the constructive and destructive interference between the contributing atomic orbitals. The two sp hybrids formed from an s and a p orbital (Fig. H.16) have their major lobes at 180° to each other; the four sp^3 hybrids that may be formed are arranged tetrahedrally around their central atom, and the six sp^3d^2 hybrids are arranged octahedrally. In each case the hybrid orbitals are constructed so that they are directed along bond directions (or towards regions of space that will be occupied by lone pairs). Since all the hybrids mentioned above have some s character, none of them has a node running precisely through the nucleus: in each case the nodes typical of p and d orbitals are displaced slightly away from the nucleus (Fig. H.17). Common hybridizations and the geometrical arrangement of the lobes of maximum probability are listed in Table H.4. Note that the higher the proportion of p character, the smaller the angle between the hybrids (Fig. H.16).

Hybrid orbitals were originally introduced in the context of °valence bond theory, where each bond is formed by pairwise overlap of one orbital on each atom of a pair. For example, in the tetrahedral CH_4 molecule it is assumed that each bond is formed by the overlap of an

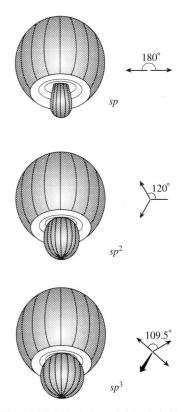

Fig. H.16 A depiction of three hybrid orbitals formed by superimposing hydrogenic $2s$ and $2p$ orbitals on the same atom. Note that as the proportion of p character increases (from top to bottom), the hybrid increasingly resembles a p orbital with two equal lobes.

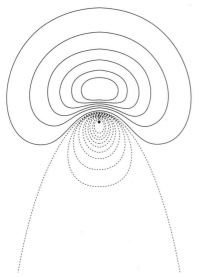

Fig. H.17 Contours of equal amplitude in an *sp* hybrid. Note how the nodal plane of the original $2p$ orbital is displaced slightly from the nucleus as a result of the admixture of $2s$ orbital.

Table H.4 Some hybridization schemes

Coordination number	Arrangement	Composition
2	Linear	sp, pd, sd
	Angular	sd
3	Trigonal planar	sp^2, p^2d
	Unsymmetrical planar	spd
	Trigonal pyramidal	pd^2
4	Tetrahedral	sp^3, sd^3
	Irregular tetrahedral	spd^2, p^3d, pd^3
	Square planar	p^2d^2, sp^2d
5	Trigonal bipyramidal	sp^3d, spd^3
	Tetragonal pyramidal	$sp^2d^2, sd^4, pd^4,$ p^3d^2
	Pentagonal planar	p^2d^3
6	Octahedral	sp^3d^2
	Trigonal prismatic	spd^4, pd^5
	Trigonal antiprismatic	p^3d^3

H1s orbital and one of the four sp^3 tetrahedral hybrid orbitals on the C atom. In the hybridization scheme used to describe the trigonal pyramidal NH_3 molecule one of the hybrids is different from the other three. Since the HNH angle of 107° is less than tetrahedral, it follows that the hybrids used to form the N—H bonds must have a higher proportion of p character than sp^3; the precise proportion can be calculated from the expressions given in Box H.6 and corresponds to $sp^{3.42}$. Since the bond hybrids have a higher proportion of p character, the orbital destined to hold the lone pair has a higher proportion of s character.

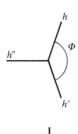

1

Box H.6 Hybridization ratios

The angle between two sp^λ hybrid orbitals on an atom is related to the angle between them, Φ, by

$$\cos \Phi = -1/\lambda$$

The composition of two equivalent hybrid orbitals of the form (1)

$$h = as + bp \qquad h' = as + bp'$$

is given by

$$a^2 = \frac{\cos \Phi}{\cos \Phi - 1} \qquad b^2 = 1 - a^2$$

A third hybrid composed of an s orbital and a p orbital along the bisector of the angle Φ has the composition

$$h'' = a's + b'p' \qquad a'^2 = \frac{1 + \cos \Phi}{1 - \cos \Phi}$$

The precise hybridization needed to account for the shape of a molecule is established by varying the geometry, matching the hybrid orbitals on the central atom to that geometry by varying their composition until they point along the bond directions, and calculating the overall energy of the molecule. The arrangement corresponding to the lowest energy gives the hybridization scheme. Thus, the energy of the NH_3 molecule decreases as the·HNH angle is changed from 120° (corresponding to sp^2 trigonal planar hybridization) through 109° (for sp^3 tetrahedral hybridization), and starts to rise again after the angle reaches 107° of the actual trigonal pyramidal molecule. Note that hybridization does not directly *determine* the shape of the molecule, but *follows* the changing shape.

The concept of hybridization is sometimes used in discussions of molecular orbital theory, and molecular orbitals are pictured as formed from the overlap of hybrid orbitals. However, it is usually better to regard molecular orbitals as formed from the appropriate °symmetry-adapted linear combinations of the entire basis set of atomic orbitals, and to avoid the language of hybridization. Hybridization may be useful when discussing molecules in terms of °localized orbitals, where each bond is pictured as formed by the pairwise overlap of neighbouring orbitals, much as in valence bond theory. Thus, in the description of the structure of °benzene, it is convenient to regard the σ orbitals as formed from sp^2 hybrids on the C atoms and to treat the π orbitals as delocalized molecular orbitals.

Further information

See MQM Chapter 10 for a derivation of the formulae given in Box H.6. The role of hybridization is described by Coulson (1979), DeKock and Gray (1980), Burdett (1980), Williams (1979), and Murrell, Kettle, and Tedder (1970, 1985). McGlynn *et al.* (1972) describe the formation of hybrids in some detail, and evaluate a number of useful integrals.

Hydrogen atom

The hydrogen atom, which consists of an electron and a proton, was one of the fences that classical mechanics failed to take, and one of the remarkable early successes of quantum mechanics and its later developments.

Spectral observations on the hydrogen atom showed that it emits and absorbs radiation at a series of discrete wavenumbers (Fig. H.18). In 1884, the Swiss school teacher Johann Balmer (1825–1898), who was fascinated by numerical relations, identified a relation satisfied by the wavenumbers of the lines that lie in the visible region of the

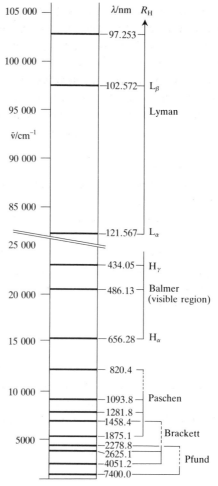

Fig. H.18 The spectrum of atomic hydrogen showing the wavelengths of the lines (in nm) and their classification into series.

spectrum. The formula summarizing this **Balmer series**, as it is called (Fig. H.18), is now written

$$\tilde{v} = R\left\{\frac{1}{4} - \frac{1}{n^2}\right\} \qquad n = 3, 4, \dots$$

As other lines were identified, the Swedish spectroscopist Johannes Rydberg (1854–1919) found (in 1890) that all the known lines fitted the expression

$$\tilde{v} = R\left\{\frac{1}{n_1^2} - \frac{1}{n_2^2}\right\} \qquad n_1 = 1, 2, \dots; n_2 = n_1 + 1, n_1 + 2, \dots$$

R is the °Rydberg constant, with the empirical value 109 677 cm^{-1}. In due course the members of several series of lines were identified, and all of them fitted Rydberg's formula. In 1906 Theodore Lyman discovered the **Lyman series** in the ultraviolet, which fits this formula with $n_1 = 1$, and in 1908 F. Paschen discovered the **Paschen series** in the infrared, which fits the formula with $n_1 = 3$. In 1922 F.S. Brackett identified several members of the **Brackett series** which lies in the infrared (with $n_1 = 4$), and in 1924, A.H. Pfund discovered the **Pfund series** in the far infrared, for which $n_1 = 5$. The **Humphreys series** ($n_1 = 6$) was identified in 1953 by C.J. Humphreys, and lies in the very far infrared.

In 1913 Neils °Bohr proposed a model of the hydrogen atom based on Rutherford's nuclear model and the quantum hypothesis (see °Bohr model). However, that model was only a preliminary version of the structure deduced by °Schrödinger from his equation in 1926. A principal feature of the quantum mechanical treatment of the hydrogen atom, and of **hydrogenic atoms** in general (those with one electron and any nuclear charge), is that the boundary conditions the wavefunctions must satisfy imply that the energy of the bound state of the atom is °quantized and limited to the values

$$E = -\frac{hcR}{n^2} \quad \text{with} \quad n = 1, 2, \ldots \quad \text{and} \quad hcR = \frac{Z^2 e^4 m_e}{8h^2 \varepsilon_0^2}$$

The allowed energy levels are shown in Fig. H.19. The integer n is called the **principal** °**quantum number**. The numerical value of R for $Z = 1$ is almost exactly equal to the empirical Rydberg constant. It follows that a transition from the level n_2 to the level n_1 results in a change in energy

$$\Delta E = hcR \left\{ \frac{1}{n_1^2} - \frac{1}{n_2^2} \right\}$$

Since this energy must be carried away by a photon (see the °Bohr frequency condition), the radiation emitted by the atom has exactly the wavenumbers observed. In practice, not all transitions are observed (see °selection rules), and those that are observed occur with a range of intensities (see °electric dipole transitions).

The wavefunctions of the electron in a hydrogenic atom are called °atomic orbitals, and that entry should be consulted for information beyond what is given below. The main features of atomic orbitals are as follows.

● Each atomic orbital is distinguished by a set of three °quantum numbers:

The **principal quantum number**, n, with $n = 1, 2, \ldots$

The °**angular momentum quantum number**, l, with $l = 0, 1, \ldots, n-1$. (l is still widely called the 'azimuthal quantum number'.)

The **magnetic quantum number**, m_l, with $m_l = 0, \pm 1, \ldots, \pm l$.

Fig. H.19 The energy levels of atomic hydrogen. Note the degeneracy of states of the same n but different values of l. The numbers in brackets are the numbers of orbitals in each subshell. The zero of energy corresponds to an infinitely separated stationary proton and electron. Above the zero of energy lies a continuum of energy states corresponding to the unbound states of the atom in which the electron has too much energy to be attached to the proton. All hydrogenic atoms have a similar array of energy levels but with the vertical scale increased by the factor Z^2.

The energy of a bound state of the atom, which is equal to $-hcR/n^2$, is determined by the value of n and is independent of the values of l and m_l. The magnitude of the orbital angular momentum of the electron is equal to $\{l(l+1)\}^{1/2}\hbar$, and the orientation of the orbital angular momentum is determined by m_l (see °angular momentum).

All orbitals of the same value of n belong to the same **shell** of the atom. All orbitals of a shell with the same value of l belong to the same **subshell**. There are n^2 orbitals in a shell with quantum number n, and there are $2l+1$ orbitals in a subshell with quantum number l. All n^2 orbitals of a given shell have the same energy (because the energy depends only on n).

It is conventional to refer to subshells by letters:

$$l = 0 \quad 1 \quad 2 \quad 3 \quad 4 \quad 5 \quad 6 \quad 7 \quad 8 \quad \ldots$$

Designation of subshells: $s \quad p \quad d \quad f \quad g \quad h \quad j \quad k \quad l \quad \ldots$

Number of orbitals: $\quad 1 \quad 3 \quad 5 \quad 7 \quad 9 \quad 11 \quad 13 \quad 15 \quad 17 \quad \ldots$

(Note that i is not used to designate a subshell because some national alphabets do not distinguish between i and j.) In chemistry it is rare to encounter orbitals with l greater than 3. The first four letters are derived from the names *sharp, principal, diffuse,* and *fundamental* which were once used in spectroscopy and which denoted transitions to the orbitals of shells with $l = 0$ to 3. In any given shell there is an s subshell with a single °**s orbital**, a p subshell (if $n > 1$) with three °**p orbitals**, a d subshell (if $n > 2$) with five °**d orbitals**, an f subshell (if $n > 3$) with seven °**f orbitals**, and so on. The shapes of these orbitals are shown in the respective entries (see also °atomic orbitals).

A striking feature of the structure of the hydrogen atom is the dependence of the energy on only the principal quantum number n and its independence of the orbital angular momentum quantum number l (Fig. H.19). The origin of the n^2-fold degeneracy, which is found only for hydrogenic atoms, is the very high symmetry of the central Coulomb potential (see °degeneracy for a pictorial discussion). The degeneracy is absent when more than one electron is present (see °penetration and shielding).

The shape of the lowest energy orbital of a hydrogenic atom is worth discussing in greater detail. On classical grounds we expect the electron to spend as much time as possible near the nucleus, for that minimizes its potential energy. The ideal position of the electron is at the nucleus. But the ground state of an H atom, an electron in a $1s$ orbital, has an electron distributed in regions close to the nucleus, not wholly confined to it. The classical explanation of the apparent repulsion of the electron is its angular momentum and the centrifugal force to which the orbital motion gives rise. Indeed, this repulsive force was used in °Bohr's

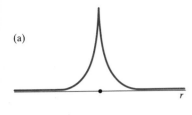

(a)

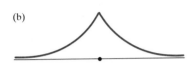

(b)

(c)

Fig. H.20 (a) A hypothetical $1s$ orbital with low potential energy but high kinetic energy. (b) The actual $1s$ orbital. (c) A hypothetical $1s$ orbital with low kinetic energy but high potential energy.

model. However, since $l = 0$ for an s orbital, an s electron has no angular momentum, and so centrifugal force cannot be the explanation of the repulsion from the nucleus.

The quantum mechanical explanation lies in the connection between the shape of the radial wavefunction and the °kinetic energy of the electron. Figure H.20 shows three possible hypothetical $1s$ orbitals. In Fig. H.20a the electron is strongly confined near the nucleus, but at the expense of increased curvature of the wavefunction. This wavefunction therefore corresponds to a state with a very high component of kinetic energy along the radial direction. Figure H.20c shows an orbital with a much lower kinetic energy, but at the expense of a less favourable potential energy. It is probable that an electron in this orbital will be found at large distances from the nucleus, in regions of high potential energy. The actual $1s$ orbital is a compromise (Fig. H.20b) in which the electron achieves a balance between a moderate kinetic energy and a moderate potential energy.

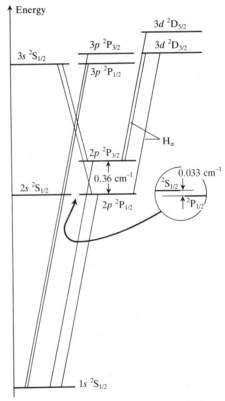

Fig. H.21 The fine structure in the energy levels of a hydrogen atom. The Dirac equation predicts that the $2s\,^2S_{1/2}$ and $2p\,^2P_{1/2}$ levels should be exactly degenerate. They in fact differ by an energy called the Lamb shift, which is explicable only within the context of quantum electrodynamics, for it stems from the influence of the zero-point fluctuations of the electromagnetic vacuum on the atom.

Box H.7 Hydrogenic atoms

The **energy levels** of a hydrogenic atom of atomic number Z and reduced mass μ are

$$E_n = -\left\{\frac{Z^2 \mu e^4}{32\pi^2 \varepsilon_0^2 \hbar^2}\right\} \times \frac{1}{n^2} \qquad n = 1, 2, \dots$$

Each level is n^2-fold degenerate. The **wavefunctions** are

$$\psi_{n,l,m_l} = R_{n,l}(r)\, Y_{l,m_l}(\theta, \phi)$$

with $l = 0, 1, 2, \dots, n-1$ and $m_l = l, l-1, \dots, -l$. The Y are °spherical harmonics. The **radial wavefunctions**, R, are proportional to the **associated Laguerre functions**, L. All the radial wavefunctions have the form

$$R = \left\{\frac{Z}{a_0}\right\}^{3/2} N f\, e^{-\rho/2} \qquad \rho = \frac{2Zr}{na_0}$$

n	l	N	f
1	0	2	1
2	0	$1/2\sqrt{2}$	$2 - \rho$
2	1	$1/2\sqrt{6}$	ρ
3	0	$1/9\sqrt{3}$	$6 - 6\rho + \rho^2$
3	1	$1/9\sqrt{6}$	$(4 - \rho)\rho$
3	2	$1/9\sqrt{30}$	ρ^2
4	0	$1/96$	$24 - 36\rho + 12\rho^2 - \rho^3$
4	1	$1/32\sqrt{15}$	$(20 - 10\rho + \rho^2)\rho$
4	2	$1/96\sqrt{5}$	$(6 - \rho)\rho^2$
4	5	$1/96\sqrt{35}$	ρ^3

Expectation values of powers of the radius:

$$\langle r^2 \rangle = \frac{a_0^2 n^4}{Z^2}\left\{\frac{5}{2} - \frac{3l(l+1) - 1}{2n^2}\right\}$$

$$\langle r \rangle = \frac{a_0 n^2}{Z}\left\{\frac{3}{2} - \frac{l(l+1)}{2n^2}\right\}$$

$$\langle 1/r \rangle = \frac{Z}{a_0 n^2}$$

$$\langle 1/r^2 \rangle = \frac{Z^2}{a_0^2 n^3 (l + \frac{1}{2})}$$

$$\langle 1/r^3 \rangle = \frac{Z^3}{a_0^3 n^3 l(l + \frac{1}{2})(l + 1)}$$

The °**spin–orbit coupling constant** ζ is

$$\zeta = \frac{\alpha^2 Z^4 R}{n^3 l(l + \frac{1}{2})(l + 1)}$$

where α is the °fine-structure constant and R is the °Rydberg constant.

The major features of the structure and spectrum of the H atom are well explained by non-relativistic (Schrödinger) quantum mechanics. However, a closer scrutiny of the spectrum reveals that the lines depicted in Fig. H.18 in fact have a °fine structure. This structure can be explained, as described in that entry, in terms of the coupling of the spin and orbital momenta. The size of the splitting emerges naturally from the relativistic °Dirac equation. Nevertheless, a slight discrepancy remains even after the application of that beautiful theory, and the **Lamb shift** (Fig. H.21), a splitting of only 0.033 cm^{-1} between $^2S_{1/2}$ and $^2P_{1/2}$ (which in the Dirac theory are exactly degenerate) was explained only when the H atom was treated in terms of °quantum electrodynamics.

Further information

See MQM Chapter 4. For the explicit solution of the Schrödinger equation for a hydrogenic atom, with all the steps carefully explained, see Pauling and Wilson (1935). For the solution by the more elegant method of factorization, see Infeld and Hull (1951), Englefield (1972), and Wybourne (1974). For a discussion of why the degeneracies are not accidental see Englefield (1972), Bander and Itzykson (1966), and McIntosh (1959, 1971). The spectrum of atomic hydrogen is described in detail by Kuhn (1962), Herzberg (1944), and Woodgate (1980); see particularly Series (1957) for an ancient but compact survey. The one-dimensional hydrogen atom is described by Davytan et al. (1987). Through its bibliography, the article by McIntosh (1971) will open up a path of boundless fascination: through accidental degeneracy, projections of hydrogen atoms on to hyperspheres and hyperhyperbolas, and the two- and four-dimensional atoms. Theoretical data on hydrogenic atoms are collected in Box H.7.

Hydrogen bond

A **hydrogen bond** is a link of the form A—H$\cdots$B, where A and B are °electronegative atoms (which may form part of a larger molecule) and B has a °lone pair of electrons. The most significant hydrogen bonding occurs where A and B are atoms of N, O, and F, but weak hydrogen bonding can also be discerned between less electronegative atoms.

The simplest picture of a hydrogen bond is as an electrostatic interaction between the proton of A—H and the lone pair of B. The A atom needs to be electronegative in order to polarize the A—H bond and hence leave the charge of the proton partially unshielded. The B atom needs to be electronegative so that it has retained a high electron density in the molecule of which it may be part and hence can interact reasonably strongly with the proton's exposed charge. In molecular orbital terms, a hydrogen bond consists of four electrons in the three molecular orbitals that can be built from the appropriately directed atomic orbitals of A, H, and B (Fig. H.22).

Hydrogen bonds between identical atoms may be symmetrical (with H lying at the mid point of A—H—A) or unsymmetrical (with H lying closer to one A atom than the other). The two cases correspond to different potential energy curves (Fig. H.23), the symmetrical case corresponding to a well with a single minimum and the unsymmetrical case to a well with a double minimum. Strong hydrogen bonds (as in F—H—F) are likely to be symmetrical, and weak hydrogen bonds (as in Cl—H—Cl) are likely to be unsymmetrical. **Deuterium bonds** (A—D⋯B) are generally stronger than hydrogen bonds because the greater mass of the D atom results in its having a smaller °zero-point vibrational energy, and hence the energy of the A—D⋯B is closer to the foot of the potential well than in A—H⋯B.

The formation of hydrogen bonds accounts for a wide variety of physical properties and for anomalies in their trends. A famous example is the low volatility of water compared with hydrogen sulfide. Another example is the lower density of ice than liquid water at 0°C. In the latter case the hydrogen bonds between neighbouring H_2O molecules result in a more open structure than in the partially collapsed structure of the liquid phase. Hydrogen bonds are of immense importance in biomacromolecules, for they account for the rigidity of cellulose and

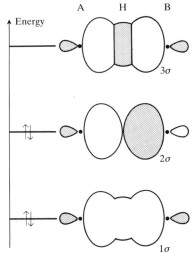

Fig. H.22 The molecular orbital description of hydrogen bonding. The diagrams show the three molecular orbitals of an AHB system. The A–H component supplies two electrons, and the :B component supplies another two (its lone pair). The fully antibonding orbital remains unoccupied, so the net effect is a lowering of energy when AHB forms.

Table H.5 The role of hydrogen bonds

Reduced vapour pressure
Higher boiling points
Higher viscosity
Low density of ice
Hydration by water
Solubility in water
Acid–base strengths
Structural strength
Conformational organization
(shapes of biological macromolecules)

Table H.6 Hydrogen bond enthalpies, $B/(\text{kJ mol}^{-1})$

Hydrogen bond (⋯)		Covalent Bond (—)	
$HS–H⋯SH_2$	−7	S–H	−363
$H_2N–H⋯NH_3$	−17	N–H	−386
$HO–H⋯OH_2$	−22	O–H	−459
$F–H⋯F^-$	−29	F–H	−565
$HO–H⋯Cl^-$	−55	Cl–H	−428
$F⋯H⋯F^-$	−165	F–H	−565

Data: Shriver, Atkins, and Langford (1990).

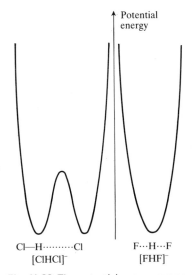

Fig. H.23 The potential energy curves of an unsymmetrical (left) and symmetrical (right) hydrogen bond.

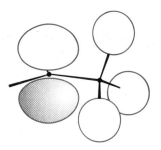

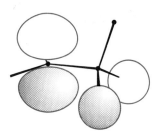

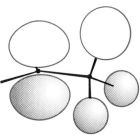

Fig. H.24 The orbital basis of hyperconjugation. The group orbitals that can be formed from the three H1s orbitals of a methyl group attached to a carbon atom that is part of a conjugated ring. One of the group orbitals (bottom) has the same symmetry as the local π orbital of the ring, and hence may overlap with it. This overlap extends the molecular orbital system from the ring to the methyl group, and the latter is able to make available its electrons to the ring.

the secondary structures of proteins. A summary of some consequences of hydrogen bonding is given in Table H.5 and some typical hydrogen bond strengths are given in Table H.6.

Further information

See Shriver, Atkins, and Langford (1990) for a discussion of the role of hydrogen bonding in inorganic chemistry. See Coulson (1979) and Webster (1990) for a good discussion of hydrogen bond formation. For information about the role of hydrogen bonds in the structures of solids see Hamilton and Ibers (1968).

Hyperconjugation

Hyperconjugation (or the 'Baker–Nathan effect') is the °overlap of σ and π orbitals on neighbouring atoms, such as the overlap of the methyl C—H σ bonds with the ring π orbitals in methylbenzene. It provides a mechanism by which the CH_3 group can behave as an electron donor, and its consequences include increased electron density at *ortho* and *para* sites and the resulting effects on aromatic reactivity.

The quantum mechanical description of hyperconjugation considers the three σ orbitals of the CH_3 group as a basis from which three **group orbitals** may be constructed (Fig. H.24). Of the three possible combinations of the σ orbitals, one has the same local symmetry as the π orbital on the ring C atom to which the CH_3 group is attached, and hence may overlap with it and form a molecular orbital spreading round the ring and over the CH_3 group. In this way, the electrons of the CH_3 group are made available for distribution around the ring.

Further information

See Coulson (1979) and Streitwieser (1961) for many references. The evidence for and consequences of hyperconjugation are discussed by Baker (1952, 1958); for a more modern view of hyperconjugation in organic chemistry see Volhardt (1987). Group orbitals are discussed by McGlynn *et al.* (1972).

Hyperfine interactions

A **hyperfine interaction** is an interaction between an electron and a nucleus other than by their point-charge Coulombic interaction. **Magnetic hyperfine interactions** arise if the nucleus has a °magnetic dipole moment. **Electric hyperfine interactions** occur if the nucleus has an electric °quadrupole (or higher multipole) moment.

A nucleus with non-zero °spin has a magnetic dipole moment. This dipole gives rise to a magnetic field with which the electron spin and orbital magnetic moments can interact. Since the energy of the electron depends on the relative orientations of its magnetic moment and the field of the nucleus, the hyperfine interaction results in the α and β spin states of the electron having different energies. The interaction itself may be a classical **dipolar interaction**, in which the magnetic moments of the electron and the nucleus interact like two point dipoles (Box H.8). Alternatively, if the electron can get very close to the nucleus (if it is an *s* electron centred on the nucleus), the two particles may interact by a °Fermi contact interaction.

Dipolar and contact hyperfine fields at the position of an electron may amount to several millitesla; some typical values are given in Table H.7. These values are for an *s* or a *p* electron confined completely to the atom containing the magnetic nucleus.

A nucleus has an electric °quadrupole moment if its spin quantum number *I* is 1 or more. The existence of a quadrupole moment implies an asymmetry in the distribution of charge in the molecule (Fig. H.25). In a **prolate nucleus** there is an excess of positive charge in the polar regions of the nucleus (with respect to the axis of spin), and a compensating slight relative deficiency in the equatorial zone. In an **oblate nucleus** the charge distributions are reversed. An electric °quadrupole moment interacts with the field gradient at the nucleus, not with the field itself (see Fig. Q.3). If, therefore, there is an electric field gradient at the quadrupolar nucleus, the energy of the molecule will depend weakly on the orientation of the nucleus in the molecular framework (Fig. H.26). There is no field gradient at the nucleus if the electrons are spherically distributed around the nucleus (or if their distribution is tetrahedral, octahedral, or icosahedral). In other cases there may be a very strong field gradient (as in Fig. H.26). Therefore, the magnitude of the electric quadrupole interaction in a spectrum gives information about the °hybridization of the orbitals occupied by the electrons surrounding it (that is, the amount of *p*-orbital character in the sur-

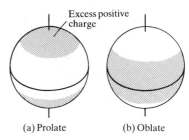

Fig. H.25 The distribution of charge in (a) a prolate nucleus, (b) an oblate nucleus. The density of shading represents the excess relative to a spherical distribution.

Box H.8 Hyperfine interactions

The hamiltonian for the **Fermi contact interaction** between an electron and a nucleus is

$$H = -\tfrac{2}{3}\mu_0\, g_e\, g_I \gamma_e \gamma_N |\psi(0)|^2\, \mathbf{s}.\mathbf{I}$$

where μ_0 is the vacuum permeability. The corresponding expression for the **dipole–dipole interaction** is

$$H = \frac{\mu_0\, g_e\, g_I \gamma_e \gamma_N}{r^3}\left\{ \mathbf{s}.\mathbf{I} - \frac{3\mathbf{s}.\mathbf{r}\mathbf{r}.\mathbf{I}}{r^2}\right\}$$

When the two spins are parallel (**1**), the expression simplifies to

$$H = \frac{\mu_0\, g_e\, g_I \gamma_e \gamma_N}{r^3}(1 - 3\cos^2\theta)\, s_z I_z$$

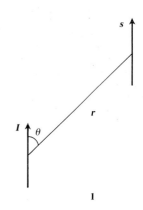

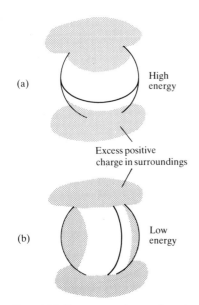

(a) High energy

Excess positive charge in surroundings

(b) Low energy

Fig. H.26 The energy of a quadrupolar nucleus depends on its orientation if the charge density in its vicinity is non-spherical (and, specifically, if there is a field gradient). The shading on the nucleus has the same significance here as in Fig. H.25.

Table H.7 Hyperfine coupling constants for atoms, a/mT

Nuclide	Spin	Isotropic coupling (valence s electron)	Anisotropic coupling (valence p electron)
^{1}H	$\frac{1}{2}$	50.8	
^{2}H	1	7.8	
^{13}C	$\frac{1}{2}$	113.0	6.6
^{14}N	1	55.2	4.8
^{19}F	$\frac{1}{2}$	1720	108.4
^{31}P	$\frac{1}{2}$	364	20.6
^{35}Cl	$\frac{3}{2}$	168	10.0
^{37}Cl	$\frac{3}{2}$	140	8.4

Data: Atkins and Symons (1967).

rounding electrons). The distribution of charge near the nucleus may be estimated by pretending that the molecule consists of an array of point charges (see °dipole moment). The ionicity of bonds has been estimated in this fashion.

One complication is that the valence electrons may distort the underlying closed shells. This distortion gives rise to a field gradient at the nucleus and therefore decreases or increases the strength of the quadrupole coupling. The decrease is called **Sternheimer shielding** and the increase is called **Sternheimer anti-shielding**.

Further information

See MQM Chapter 14 for a more detailed account of magnetic hyperfine interactions. Their role in magnetic resonance is described in the entries on °nuclear magnetic resonance and °electron spin resonance; see those entries for further information. The role of hyperfine interactions in spectra, and the structural information they yield, are described in Hollas (1983) and Townes and Schawlow (1955). For a general survey see Freeman and Frankel (1967).

Hypervalent compounds

In the °Lewis approach to chemical bonding, atoms are presumed to share electron pairs until they have completed an octet. In a **hypervalent compound** there are more atoms around the central atom, or more bonds have been formed to the central atom, than can be explained by octet completion. An example is SF_6, in which six atoms are bound to a single central atom by six electron pairs (**1**).

The bonding in hypervalent compounds can be explained in Lewis terms only if the central atom is allowed to expand its octet until it has ten, twelve, or more electrons in its valence shell. Hence, the earlier

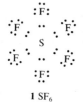

1 SF_6

name for hypervalence is 'octet expansion', and hypervalent compounds were said (and still widely are said) to have an 'expanded octet'. However, the neutral term 'hypervalence' does not ascribe the richer bonding to a particular explanation, but merely implies that more bonds appear to be formed by the central atom than the octet rule allows.

Atoms that can become hypervalent are principally those of Period 3 and later (but even Period 2 elements can have hypervalence forced on them, as in eight-coordinate C). Since Period 3 and subsequent atoms also have low-lying, unfilled d orbitals available (S, for example, has the configuration $[Ar]3s^23p^4$, with its $3d$ orbitals not much higher in energy), the *conventional* explanation of hypervalence is the availability of d orbitals. However, in chemistry we must beware of correlations that are not necessarily causal connections. This appears to be the case with hypervalence, for self-consistent field calculations on SF_6, among other compounds, have shown that $3d$ orbitals play a minimal role in the bonding, and hence it is inappropriate to ascribe octet expansion to the availability of unfilled d orbitals.

The simplest plausible explanation of octet expansion is that it is merely a consequence of size: unless there are special stereochemical constraints in action, Period 2 atoms are too small to have more than four atoms packed around them, but Period 3 and subsequent atoms are larger and can readily accommodate more atoms. The bonding resources needed to bind the stereochemically permitted number of atoms can be explained readily once we can see that the s and p orbitals of the central atom can participate in the formation of enough molecular orbitals to bind the atoms together without it being necessary to invoke d orbitals.

For example, in SF_6 the S atom supplies four orbitals (one $3s$ and three $3p$ orbitals) and each of the six F atoms provides one orbital directed towards the central S atom. From these ten orbitals we can construct ten molecular orbitals (Fig. H.27), about half of which will be largely bonding and about half largely antibonding (two orbitals are in fact nonbonding). There are twelve electrons to accommodate (six from the S atom, six from the F atoms), and so only the lowest six orbitals will be occupied. The overall effect is bonding, and the SF_6 molecule has a lower energy than the separated atoms.

A similar analysis can account for the bonding in all hypervalent molecules. Note, though, that it does not rule out the involvement of d orbitals, for their inclusion may well lower the energy. However, it does show that d orbitals are not a *necessary* component of the explanation of hypervalence. Size—the ability to pack more atoms around the central atom without suffering from strong ligand–ligand repulsions—may be the more important component of the explanation.

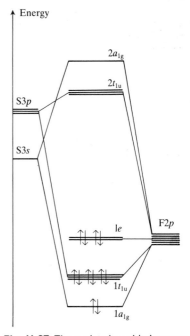

Fig. H.27 The molecular orbital energy levels of sulfur hexafluoride. Note that even without invoking S3d orbitals, there are enough bonding and nonbonding orbitals to accommodate the 12 valence electrons needed to account for the structure of the molecule.

Further information

Many examples of hypervalent compounds will be found in Shriver, Atkins, and Langford (1990), Cotton and Wilkinson (1988), and Greenwood and Earnshaw (1984). For a quantitative discussion of the role of d orbitals in the 'expansion of the octet', see the paper on sulfur hexafluoride by Reed and Weinhold (1986). For a stimulating icon-cracking account of hypervalence in terms of valence bond theory see Cooper, Gerratt, and Raimondi (1989).

Insulator

An electrical **insulator** is a substance that does not conduct electricity; however, strictly speaking, no solid is an electrical insulator. A metallic conductor is a substance with an electrical conductivity that decreases as the temperature is raised, a °semiconductor is a substance with an electrical conductivity that increases as the temperature is raised, and all solid substances fall into one or other of these two classes. However, it is conventional to regard semiconductors with very low electrical conductivities as 'insulators'. Nevertheless, the conductivities of solid 'insulators' increase with temperature, and hence they are properly regarded as semiconductors. Solid 'insulators' are semiconductors with large °band gaps and hence a very low concentration of charge carriers at normal temperatures.

Further information

An introduction to the classification of solids as insulators, etc. will be found in Shriver, Atkins, and Langford (1990); see also Ridley (1988) for a discussion of quantum processes in semiconductors. For more information see the entry on °band theory and the references given there, particularly Anderson (1984), Kittel (1987), and Duffy (1990). For more general references see Parker (1987).

Intermolecular forces

An **intermolecular force** is a force between molecules that have completed their valence requirements and have no further tendency to form chemical bonds. The forces may be either attractive or repulsive, and they typically vary with separation as indicated in Fig. I.1. Intermolecular forces are often called **van der Waals forces** after the Dutch physicist Johannes van der Waals (1837–1923) who investigated their effects on the properties of gases. The term **Keesom force** denotes the interactions between polar molecules, but is now rarely used. The term **London force** refers specifically to the °dispersion force.

Repulsive forces have a very short range, and reflect the increase in energy that occurs when two closed-shell species are forced into contact. For instance, suppose two He atoms approach. (The two atoms are analogues of any pair of closed-shell species.) Their 1s orbitals start to overlap, with the formation of bonding and antibonding σ orbitals (Fig. I.2). There are four electrons to accommodate in the two orbitals. However, the Pauli principle forbids all four to occupy the bonding orbital, and two of the electrons must occupy the antibonding orbital. Since an °antibonding orbital is more antibonding than a bonding orbital is bonding, the energy of the two atoms is now higher than when they are far apart and it increases sharply as the atoms are

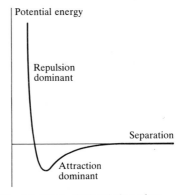

Fig. I.1 The typical variation of an intermolecular potential energy with separation. At short distances (essentially, when the molecules are in contact) the repulsion always dominates.

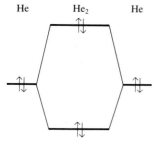

Fig. I.2 When two closed-shell species (represented here by two helium atoms) come close enough together for their orbitals to overlap, one pair of electrons occupies the bonding combination and the other pair occupies the antibonding combination. The net effect is antibonding, with an increase in energy relative to the widely separated species.

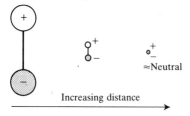

Fig. I.3 The electric field from a dipole decreases rapidly with distance because not only does the strength of the field from each charge decrease, but to a distant observer the positive and negative charges appear to be increasingly coincident and hence cancel.

pressed ever more closely together. This increasing energy is interpreted as the result of a repulsive force between the two atoms. It follows that intermolecular repulsive forces are a consequence of the °Pauli exclusion principle, which prevents more than two electrons occupying the same orbital. Thus, the Pauli principle stops all matter from blending together into one huge cosmic molecule (or even to a point).

The existence of attractive forces is shown by the occurrence of condensed phases of matter. The strongest attractive interactions are **ion–ion interactions** between charged species, which vary inversely with the separation R (Box I.1); their energies are generally of the order of 400 kJ mol^{-1} for typical separations of ions in ionic solids.

Polar molecules may interact with ions by an **ion–dipole interaction**, the potential energy for which varies as $1/R^2$. The shorter range of the interaction (that is, the more rapid decrease in its strength with increasing separation than for the ion–ion interaction) is a consequence of the two charges of the dipole appearing to blend into an electrically neutral point as the separation increases (Fig. I.3). Ion–dipole interactions are of the order of 20 kJ mol^{-1}; one of their important consequences is the hydration of ions in water.

Two polar molecules may interact by a **dipole–dipole interaction**. The potential energy of this interaction varies as $1/R^3$ in a solid (when the molecules are not rotating) but as $1/R^6$ in a fluid (in which the molecules rotate). The short range of the dipole–dipole interaction reflects the fact that at large separations *both* molecules appear to each other to be neutral points. The shorter range ($1/R^6$ as compared with $1/R^3$) of the interaction in a fluid arises from the near cancellation of attractive

Box I.1 A summary of intermolecular forces

Interaction type	Distance dependence of potential energy	Typical energy kJ mol^{-1}	Comment
Ion–ion	$1/r$	250	Only between ions
Ion–dipole	$1/r^2$	15	
Dipole–dipole	$1/r^3$	2	Between stationary polar molecules
	$1/r^6$	0.3	Between rotating polar molecules
London (dispersion)	$1/r^6$	2	Between all types of molecules

The energy of a hydrogen bond A–H⋯B is typically 20 kJ mol^{-1} and occurs on contact for A, B = N, O, or F.

and repulsive orientations that occur as the molecules rotate: the small surviving interaction stems from a slight (Boltzmann) favouring of the attractive (low energy) orientations. Not only does the interaction between the dipoles decrease sharply with distance (as for stationary dipoles), but the favouring of attractive orientations is also less at large separations; hence, the net, surviving interaction weakens more sharply with increasing separation than for stationary dipoles.

A polar molecule near a polarizable molecule (which may or may not be polar) can interact by a **dipole–induced dipole interaction**. In this interaction, the dipole of the polar molecule interacts with the dipole it induces in the polarizable molecule. Since the magnitude of the induced dipole moment is proportional to the °polarizability α of the second molecule, the strength of interaction is proportional to α. The overall $1/R^6$ dependence can be traced to the joint effect of two $1/R^3$ terms. First, the polar molecule generates an electric field that decreases with distance as μ/R^3. That field induces a dipole moment μ' proportional to α and the electric field, and therefore proportional to $\mu\alpha/R^3$. This induced dipole then interacts with the original dipole with a strength proportional to $\mu\mu'/R^3$. Overall, therefore, the interaction varies as $\mu^2\alpha/R^6$.

One of the most important forces, which occurs between polar, nonpolar, and charged species, is the London °dispersion interaction. As explained in that entry, the potential energy of this interaction is proportional to $1/R^6$, and it arises from the correlation of the fluctuations in electron density on adjacent molecules.

Another strong interaction is the °hydrogen bond between two strongly electronegative atoms (N, O, and F). Its strength is typically about 20 kJ mol^{-1}, and so, if it is present, it dominates all the other types of interaction except ion–ion interactions.

The distance dependence of intermolecular interactions and the angular dependence of those between non-spherical molecules are normally expressed in terms of an empirical formula for the potential energy of interaction. Some of the more common intermolecular potentials are illustrated in Fig. I.4 and Box I.2. All the empirical formulae make a crude parametrization of the repulsive interaction, such as expressing it as proportional to $1/R^n$, as in the **Lennard–Jones (n,6)-potential**, with n about 12, or setting it equal to an exponential function, as in the **Buckingham (exp,6)-potential**. The latter is more plausible because the interaction is proportional to the overlap of the outer reaches of the wavefunctions of the two molecules, and (at large distances) all atomic and molecular wavefunctions decrease exponentially with distance. The **Keesom potential** is an expression for the interaction of non-spherical molecules, treating them as cylinders capped by hemispheres. The **Stockmayer potential** adds to the Lennard–Jones potential a term

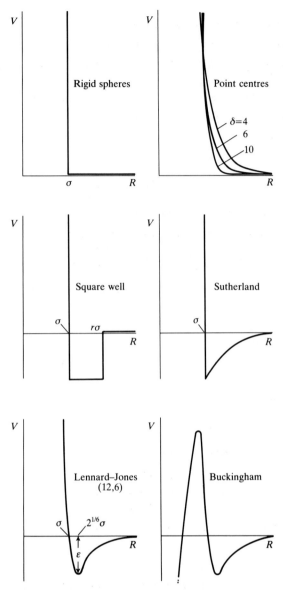

Fig. I.4 Some typical model intermolecular potentials and their parametrizations.

representing the interaction of two dipoles placed at the centres of the molecules.

Other contributions to intermolecular forces include interactions between higher electric multipoles, such as between two °quadrupolar CO_2 molecules or two hexadecapolar CH_4 molecules. These multipolar interactions are of progressively shorter range as the order of the multipole increases. The physical reason is that the more elaborate the

Box I.2 Intermolecular potential energies

Hard spheres: $V = \infty$ for $R < \sigma$ and 0 for $R > \sigma$.

Point centres: $V = DR^{-n}$ with $9 < n < 12$ typically. A 'Maxwellian potential' is one with $n = 4$.

Square well: $V = \infty$ for $R < \sigma$, $-\varepsilon$ for $\sigma < R < r\sigma$, and 0 for $R > r\sigma$.

Sutherland: $V = \infty$ for $R < \sigma$ and $-CR^{-n}$ for $R > \sigma$.

Lennard–Jones: $V = DR^{-m} - CR^{-n}$.

A special case is the '(12, 6)-potential':

$$V = 4\varepsilon \left\{ \left(\frac{\sigma}{R} \right)^{12} - \left(\frac{\sigma}{R} \right)^{6} \right\}$$

Buckingham: $V = Be^{-bR} - CR^{-6} - C'R^{-8}$.

Modified Buckingham (6–exp): $V = \infty$ for $R < R_m$ and

$$V = \frac{\varepsilon}{1 - 6a} \left\{ \frac{6}{a} e^{a(1 - R/R_m)} - \left(\frac{R_m}{R} \right)^{6} \right\} \text{ for } R > R_m$$

where R_m is the value of R for which the latter expression reaches a maximum.

Stockmayer: $V = V_{LJ} + V(\text{dip–dip})$

where V_{LJ} is the Lennard–Jones potential and $V(\text{dip–dip})$ is the electric dipole–dipole interaction:

$$V(\text{dip–dip}) = -\frac{1}{4\pi\varepsilon_0 R^3} \left\{ \boldsymbol{\mu}_1 \cdot \boldsymbol{\mu}_2 - 3\boldsymbol{\mu}_1 \cdot \frac{\boldsymbol{R}\boldsymbol{R}}{R^2} \cdot \boldsymbol{\mu}_2 \right\}$$

For two parallel dipoles (1):

$$V(\text{dip–dip}) = -\frac{\mu_1 \mu_2}{4\pi\varepsilon_0 R^3} (1 - 3\cos^2 \theta)$$

Axilrod–Teller (for three-body interactions):

$$V = -CR_{AB}^{-6} - CR_{BC}^{-6} - CR_{CA}^{-6}$$
$$+ \frac{a(3\cos\theta_A \cos\theta_B \cos\theta_C + 1)}{(R_{AB} R_{BC} R_{CA})^3}$$

where the angles are specified in (2). Typically, $a \approx \frac{3}{4}\alpha' C$.

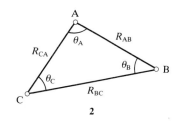

1

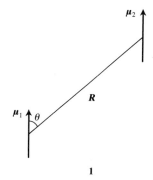

2

arrangement of partial charges, the more rapidly they seem to cancel as the observer moves to greater distances.

The total potential energy of a system of molecules is only approximately the sum of the pairwise interactions discussed so far. In non-ionic condensed media, three-body and higher interaction terms are significant. Thus, the interaction between three Ar atoms differs from the sum of three pairwise interactions by the **three-body potential**, $V(ABC)$, because the presence of A affects the ability of B to interact with C. For instance, the presence of A slightly modifies the polarizabilities of B and C, so their dispersion interaction is slightly different from the dispersion interaction when A is absent. The three-body potential contributes about 10 per cent of the total cohesive energy of solid Ar. One model three-body potential is the **Axilrod–Teller potential**, which is specified in Box I.2.

Further information

The origin of intermolecular forces is described in Atkins (1990) and, in more quantitative detail, in MQM Chapter 13. For more thorough treatments see Kihara (1978), Maitland *et al.* (1981), and the simpler version of this monograph by Rigby *et al.* (1986), and Hirschfelder, Curtiss, and Bird (1954). The volume edited by Hirschfelder (1967) is well worth reading. See also Buckingham (1967).

Internal conversion

The process called **internal conversion** is the radiationless transition from one molecular state to another of the same °multiplicity. An example is the transition from one excited singlet state to another singlet state (Fig. I.5). A radiationless transition between states of *different* multiplicity is called °intersystem crossing.

Internal conversion occurs between vibrational states close to the intersection of the two molecular potential energy curves (as depicted in Fig. I.5), where a molecule in the initial state A has the same energy as in the final state B. As remarked in the entry on °degeneracy, two states of the same energy are mixed strongly if a perturbation is present, and hence the initial state can be nudged into the final state by even a weak perturbation. The perturbation that drives internal conversion is the breakdown of the Born–Oppenheimer approximation, for the electrons do not exactly follow the changing locations of the nuclei.

As in the classical interpretation of the °Franck–Condon principle, the intersection of the potential energy curves marks the point at which the nuclei have the same locations in the two states and at which they are stationary at the turning points of their respective oscillations. Hence, a molecule in state A may swing up to its turning point, where a perturbation flips it into state B with a different arrangement of electrons (and hence with the nuclei experiencing a different force field) but the same arrangement of the nuclei. When the molecule resumes its oscillation it finds itself in the potential well corresponding to state B.

The quantum mechanical version of this description also resembles that for the Franck–Condon principle. Thus, the vibrational wave-functions of the two electronic states resemble each other most closely (in the sense of having the greatest overlap) in the states that occur close to the intersection (Fig. I.5). For these states, the dynamical state

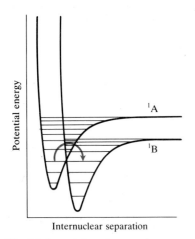

Fig. I.5 In an internal conversion, a molecule makes a non-radiative transition from one electronic state to another of the same multiplicity. The transition occurs between vibrational states close to the point of intersection of the two molecular potential energy curves.

of the nuclei in state B resembles that of the nuclei in state A most closely.

Further information

See MQM Chapter 12 for more information and Hollas (1983, 1987), Struve (1989), and Steinfeld (1985) for applications in spectroscopy. For a detailed theoretical treatment, and further references, the book by Englman (1976) and the volume edited by Lin (1980) are good starting points. For chemical consequences of internal conversion see Wayne (1988), Barltrop and Coyle (1975, 1988), and Calvert and Pitts (1966).

Intersystem crossing

The process called **intersystem crossing** is the radiationless conversion of an atom or molecule into a state of different °multiplicity. An example is the conversion of an excited °singlet state into a triplet state (Fig. I.6).

As explained in the entry on °singlet and triplet states, different multiplicities correspond to different relative orientations of electron spins. Hence, to achieve the conversion of a singlet state into a triplet state it is necessary to change the *relative* orientation of the spins of two electrons (Fig. I.7). This relative reorientation can be achieved by making one electron precess faster than the other by applying different magnetic fields to each. Thus, an in-phase (triplet) orientation is gradually turned into an out-of-phase (singlet) orientation.

A field from a laboratory magnet cannot cause intersystem crossing (unless the two electrons have different °g-factors) because it is homogeneous on a molecular scale, affects both spins equally, and cannot twist one spin relative to the other. However, a magnetic field from within the molecule may be able to rephase the spins. For example, the °spin–orbit coupling experienced by the two electrons may differ if they are in different parts of a molecule, and the different local magnetic fields result in the realignment of the two spins. This is the reason why molecules containing heavy atoms (with large spin–orbit couplings) undergo efficient intersystem crossing (see °phosphorescence).

As in the case of °internal conversion, and for the reasons explained there, intersystem crossing occurs for vibrational states close to the intersection of the potential energy curves of the two electronic states (Fig. I.6).

Further information

See MQM Chapter 12 and McGlynn, Azumi, and Kinoshita (1969) for more information; see Hollas (1983, 1987), Struve (1989), and Steinfeld (1985) for

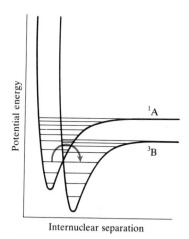

Fig. I.6 In an intersystem crossing, a molecule makes a non-radiative transition from one electronic state to another of a different multiplicity (here, from singlet to triplet). The transition occurs between vibrational states close to the intersection of the two molecular potential energy curves.

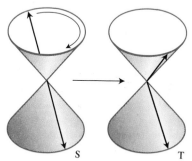

Fig. I.7 Intersystem crossing involves the relative reorientation of electron spins within the molecule. This illustration shows the relative reorientation required if a singlet state is to convert into the $M_S = 0$ component of the triplet. (Different relative motions result in crossing into the $M_S = \pm 1$ components; see °singlet and triplet states.)

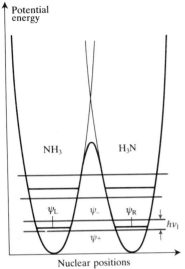

Fig. I.8 The degenerate vibrational states of the double well potential (black horizontal lines) are split into non-degenerate pairs (black lines) because the molecule is able to tunnel from one well to the other. The strongest interaction is between states of the same energy, so to a first approximation we need consider only the interaction between states that have the same energy in each well. (For example, the two zero-point vibrational states interact to give the states labelled ψ_+ and ψ_-.) The splitting between the doubled states, $h\nu_I$, determines the 'inversion frequency' of the molecule.

applications in spectroscopy. For information on radiationless transitions see Englman (1976), Lin (1980), and the entry on °phosphorescence. For the chemical consequences of intersystem crossing see Wayne (1988), Barltrop and Coyle (1975, 1988), and Calvert and Pitts (1966).

Inversion doubling

The trigonal pyramidal NH_3 molecule can vibrate in a symmetrical bending mode, rather like an umbrella being shaken dry. Like an umbrella, the molecule may undergo inversion into another equivalent configuration (Fig. I.8) where it continues to vibrate. Classically, the two configurations are identical and vibrate at exactly the same frequency. **Inversion doubling** is the removal of the degeneracy of the vibrational states of the two individual configurations as a result of the molecule's ability to °tunnel quantum mechanically through the barrier between them.

The fact that the molecule can invert from NH_3 to H_3N implies that the °vibrational wavefunctions of one well continue through the barrier and have non-zero amplitudes in the other well. It follows that the wavefunctions of the left-hand well, ψ_L, have non-vanishing overlap with those of the right-hand well, ψ_R. Just as in the formation of a °molecular orbital (where the wavefunction on one atom overlaps with a wavefunction on the other atom), it follows that the correct description of the vibrational state of the molecule is in terms of the °superpositions $\psi_+ = \psi_L + \psi_R$ and $\psi_- = \psi_L - \psi_R$ of the functions of each well (Fig. I.9).

The two superpositions have different energies, because they have different kinetic energies and different amplitudes in regions of high potential. Why ψ_+ lies lower in energy may be appreciated by imagining the effect of slowly lowering the barrier and in the limit letting it

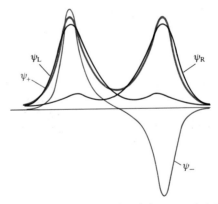

Fig. I.9 The linear combinations ψ_+ and ψ_- of the ground state vibrational wavefunctions of the left and right potential wells, ψ_L and ψ_R. The nodeless wavefunction ψ_+ has the lower energy (see Fig. I.8).

disappear completely to leave a single broad well. In this limit, ψ_+ turns into the ground state, bell-shaped wavefunction, and ψ_- turns into the first excited state. Thus, even though ψ_+ gives a greater probability of finding the molecule inside the barrier (that is, as flat, trigonal planar NH_3), it has a lower energy than ψ_- (which has a node at the centre of the barrier and is never found as flat NH_3). The explanation of why ψ_+ has the lower energy is that it is also necessary to take into account the different kinetic energies of the two combinations, and ψ_-, which is more sharply curved, has a greater kinetic energy than ψ_+.

It follows from the need to form linear combinations of ψ_L and ψ_R that, instead of having pairs of °degenerate vibrational states in the ammonia molecule (one for the vibrations of NH_3 and the other for the vibrations of H_3N), the inversion causes the degeneracy to be removed. Each pair of levels splits into two: each level is **doubled** (Fig. I.8). The energy separation, which is expressed as the **inversion frequency** v_I by division by Planck's constant, depends on the difference in overlap between the vibrational wavefunctions in the two wells, and hence on the height of the potential barrier and the mass of the atoms that have to migrate through it (strictly, on the effective mass of the °normal mode corresponding to inversion). The inversion frequency of NH_3 is 23.786 GHz (0.793 cm^{-1}, an inversion time of 21 ps) for the lowest level; the inversion frequency of ND_3 is 1.600 GHz (an inversion time of 310 ps). On a classical picture of inversion doubling, the inversion frequency would be interpreted as the rate at which NH_3 changes to H_3N and back to NH_3. In quantum mechanics, we could picture the molecule as initially confined to one well with the wavefunction ψ_L, corresponding to the ground vibrational state of NH_3. That wavefunction (which is not a stationary state of the molecule on account of the finite barrier and the accessible other well) evolves into ψ_R, the ground vibrational state of H_3N, and then back to ψ_L, with a frequency v_I.

Further information

For the spectroscopic consequences of inversion doubling see King (1964), Townes and Schawlow (1955), Hollas (1983, 1987). Inversion and other large amplitude motions are discussed in detail by Lister, Macdonald, and Owen (1978). The inversion mode of ammonia was utilized for the forerunner of the laser, the ammonia 'maser'; see Siegman (1986).

Ionic model

The formation of an **ionic bond** involves the transfer of electrons from atoms of one element to those of another, and the subsequent Coulombic attraction between the resulting cations and anions. The feasibility of the transfer (that is, whether it leads to a net reduction in energy) is assessed by considering a gas of atoms of the two elements A and B and the net energy change for:

- Ionization of A, which requires the °ionization energy $I(A)$ of A.

- Electron attachment to B, which releases an energy equal to the °electron affinity $E_{ea}(B)$ of B.

- Coulombic attraction between the ions in the solid aggregate they form.

The last contribution depends on the separation of the ions in the aggregate (the ionic crystal) and their arrangement. It is normally expressed in terms of a scale length d of the crystal lattice and the **Madelung constant** $\mathscr{A}$, which depends on the symmetry of the lattice (Table I.1). An ionic compound may form if the Coulombic attraction between the ions in the solid is greater than the net energy investment $I(A) - E_{ea}(B)$ needed to form the ions initially.

Ionic bonds will form only if the ionization energy of the element A is low, for otherwise the energy investment could not be recovered from the Coulombic attraction. This effectively confines ionic bond formation to the elements of Groups 1 and 2 (and to some extent the members of the d block and Group 13/III) of the Periodic Table. Moreover, cation formation by these elements does not extend beyond loss of their valence electrons (so Na forms Na^+ and Ca forms Ca^{2+}) because the loss of core electrons requires too much energy.

Similarly, the only elements that commonly form anions are those with reasonably high electron affinities, which are those towards the top right of the Periodic Table. However, even these elements will not form anions in which electrons need to occupy a new shell outside the valence shell, for that leads to ions that are too big for the resulting Coulombic attraction in the lattice to overcome the energy investment required for ion formation. Thus O will form O^{2-}, despite the overall

Table I.1 Madelung constants*

Structural type	$\mathscr{A}$
Caesium chloride	1.763
Fluorite	2.519
Rock salt	1.748
Rutile	2.408
Sphalerite	1.638
Wurtzite	1.641

* Some sources cite values that include the charge numbers of the ions (so, for instance, the value for CaF_2 is quoted as 5.039), and it is necessary to verify the definition before using them.

electron affinity being negative, because the oxide ion is small and highly charged, and leads to a very strong Coulombic attraction in the lattice that it ultimately forms. However, F will not form F^{2-} because the resulting anion is so large that the energy needed to go from F^- to F^{2-} cannot be recovered.

The **lattice enthalpy** of a solid is the enthalpy change that accompanies its conversion into a gas of ions. The **ionic model** of bonding in a crystal is valid if the lattice enthalpy calculated on the assumption that the solid consists of ions is in good agreement with the experimental value. Thus, NaCl is reasonably well described as 'ionic', but AgCl is not. The ionic model is reasonably correct for compounds formed between elements on the left and right of the periodic table (such as NaCl). However, there is no such thing as *pure* ionic bonding, since all ions are slightly °polarizable, and the polarization of an anion by a cation is the incipient formation of a °covalent bond. Thus, all 'ionic' solids have some covalent character, but it is least for compounds of elements with widely different electronegativities.

Further information

The ionic model of solid compounds is within the domain of inorganic chemistry, so see Shriver, Atkins, and Langford (1990) where structural and thermodynamic aspects are considered. For more detailed analysis and criticism see Dasent (1982) and Johnson (1982). Ionic bonding is one limit of the °band theory of solids, and that entry should be consulted for a more general view of the structure of solid compounds; see Cox (1987), West (1984), and Duffy (1990) in this connection. See also the suggestions for further reading in the entry on the chemical °bond and the survey by Sacks (1986).

Ionic radius

The **ionic radius** of an element is the contribution of an ion of the element to the internuclear separation in an ionic solid compound (**1**). Unfortunately, there is no unambiguous way of identifying the individual contributions to the internuclear separation (we cannot simply take half the separation as in a °metallic element), and it is necessary to impose a convention. One widely adopted convention is to take the radius of an O^{2-} ion in a six-coordinate environment as equal to 140 pm. Then a separation of 250 pm in an ionic oxide would imply that the ionic radius of the cation present is 110 pm, and so on (Table I.2). However, the choice of 140 pm for the ionic radius of O^{2-} is by no means sacrosanct, and other choices have been made.

One consequence of the arbitrary character of the definition of ionic radii is that one should be very cautious when using values from

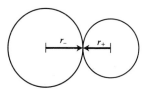

1 Ionic radii

Table I.2 Ionic radii, r/pm

Li^+	Be^{2+}	B^{3+}	O^{2-}	F^-
59 (4)	27 (4)	12 (4)	135 (2)	128 (2)
76 (6)			138 (4)	131 (4)
			140 (6)	133 (6)
Na^+	Mg^{2+}	Al^{3+}		
99 (4)	49 (4)	39 (4)		
102 (6)	72 (6)	53 (6)		
116 (8)	89 (8)			
K^+	Ca^{2+}	Ga^{3+}		
138 (6)	100 (6)	62 (6)		
151 (8)	112 (8)			
159 (10)	128 (10)			
160 (12)	135 (12)			
Rb^+	Sr^{2+}	In^{3+}	Sn^{2+}	Sn^{4+}
149 (6)	116 (6)	79 (6)		69 (6)
160 (8)	125 (8)	92 (8)	122 (8)	
173 (12)	144 (12)			

Numbers in parentheses are the coordination number of the ion.
Data: Shannon and Prewitt (1969).
The radius of the NH_4^+ ion is approximately 146 pm.

different sources, since they may have been defined using different conventions. Moreover, since the values do not have an absolute significance, conclusions based on them, such as those drawn from applying the radius-ratio rule for crystal structures, can be misleading.

Ionic radii vary with the coordination number of the ions. In general, the larger the coordination number, the larger the ionic radius. This dependence probably stems from the geometrical and electrostatic problem of packing large numbers of ions of like charge around a single central ion of opposite charge.

The general trends in ionic radii may be expressed and explained quite simply. Thus, all cations are smaller than their parent atoms since the expanding effects of electron repulsion is reduced slightly when electrons are removed from the neutral atom. All anions are larger than their parent atoms, since the increased electron repulsions puff the ions up to a larger size. For a given electron configuration, ionic radius increases down a group and decreases from left to right across a period. The increase down a group stems from the increasing numbers of electrons needing to occupy shells with higher principal quantum number. The decrease across a period stems from the increasing charge of the nucleus and its ability to draw the surrounding electrons together more compactly. The decrease across Period 6 is enhanced by the °lanthanide contraction.

Further information

See Shriver, Atkins, and Langford (1990) and Wells (1984) for descriptions of the chemical applications of ionic radii. The most complete critical analyses and tabulations of atomic radii are those by Shannon and Prewitt (1969) and Shannon (1976). For surveys of trends in ionic radii through the periodic table see Mason (1988) and Lloyd (1986). The latter looks at the evidence for lanthanide and 'scandinide' contractions.

Ionization energy

The **first ionization energy**, I, of a substance is the minimum energy required to remove an electron from a gas phase atom or molecule:

$$A(g) \rightarrow A^+(g) + e^-(g) \quad \text{Energy required} = I$$

The **second ionization energy**, I_2, of the substance is the minimum energy required to remove a second electron:

$$A^+(g) \rightarrow A^{2+}(g) + e^-(g) \quad \text{Energy required} = I_2$$

Further ionization energies may be defined similarly. In some cases it is convenient to deal with the ionization of an electron from a specified orbital, not necessarily the highest occupied orbital (the °HOMO). Then the ionization energy is denoted I_i, where i labels the orbital originally occupied by the ejected electron. According to the approximation known as °Koopmans' theorem, the ionization energy from an orbital i is equal to (the negative of) the one-electron energy of that orbital: $I_i = -\varepsilon_i$. First ionization energies of atoms vary periodically, as shown in Fig. I.10. In general, I increases from left to right across a period and decreases down a group. The increase across a period is a result of the contraction in the size of the atom as electrons are added to the valence shell (see °atomic radius). The electron is closer

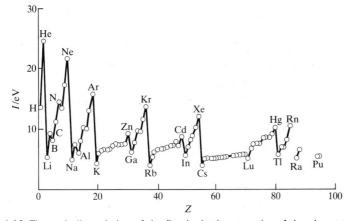

Fig. I.10 The periodic variation of the first ionization energies of the elements.

to a more highly charged nucleus on the right of a period than on the left, and hence is more difficult to remove. On descending a group, the outermost electron is in a shell of increasing principal quantum number; hence it is further away from the attracting nucleus and easier to detach.

Small blips in the smooth increase across a period can be explained in terms of the precise identity of the orbital from which the electron is removed. Thus, the decrease in I between Be and B is because the electron is removed from a $2s$ orbital in Be but from a $2p$ orbital in B (a $2p$ orbital has a higher energy on account of °penetration and shielding). The decrease between N and O occurs because the electron removed from N comes from a singly occupied $2p$ orbital, but in O it comes from a doubly occupied $2p$ orbital, and so is helped on its way by a stronger repulsion from its partner.

Second and higher ionization energies mirror the periodicity of first ionization energies. For ionizations from the valence shell it is reasonably valid to take the nth ionization energy to be approximately n times the first ionization energy of the element.

Table I.3 Ionization energies, $I/(\text{kJ mol}^{-1})$

H							He
1312.0							2372.3
							5250.4
Li	Be	B	C	N	O	F	Ne
513.3	899.4	800.6	1086.2	1402.3	1313.9	1681	2080.6
7298.0	1757.1	2427	2352	2856.1	3388.2	3374	3952.2
Na	Mg	Al	Si	P	S	Cl	Ar
495.8	737.7	577.4	786.5	1011.7	999.6	1251.1	1520.4
4562.4	1450.7	1816.6	1577.1	1903.2	2251	2297	2665.2
		2744.6		2912			
K	Ca	Ga	Ge	As	Se	Br	Kr
418.8	589.7	578.8	762.1	947.0	940.9	1139.9	1350.7
3051.4	1145	1979	1537	1794	2044	2104	2350
		2963	2735				
Rb	Sr	In	Sn	Sb	Te	I	Xe
403.0	549.5	558.3	708.6	833.7	869.2	1008.4	1170.4
2632	1064.2	1820.6	1411.8	1794	1795	1845.9	2046
		2704	2943.0	2443			
Cs	Ba	Tl	Pb	Bi	Po	At	Rn
375.5	502.8	589.3	715.5	703.2	812	930	1037
2420	965.1	1971.0	1450.4	1610			
		2878	3081.5	2466			

Data: Emsley (1989).

Further information

A short summary of ionization energies is given in Table I.3. A reasonably extensive table of ionization energies is given in Appendix 2 of Shriver, Atkins, and Langford (1990). It has been compiled from the very extensive information published by Moore (1971) and Martin *et al.* (1974, 1978). See Emsley (1989) for a handy compilation. For a survey of trends see Agmon (1988).

Isolobality

A ubiquitous concept in chemistry is the substitution of one group or atom for another in a molecule. Thus, we may view $N(CH_3)_3$ as derived from NH_3 by substitution of a CH_3 fragment for each H atom. In current terminology, the structurally analogous fragments are said to be **isolobal** and we express the relationship by the symbol ⟷ as in H ⟷ CH_3. Two fragments are isolobal if their uppermost orbitals have:

- the same symmetry (such as the σ symmetry of the H1s and the Csp^3 orbital);
- similar energies;
- the same electron occupation (1 in each case in H1s and Csp^3).

Two typical families of isolobal fragments are

We can anticipate that we may be able to form molecules such as H_3C—Br and $(OC)_5Mn$—CH_3 by analogy with H—H. The existence of these families suggests that we should encounter molecules such as *cyclo*-C_4H_8, $O(CH_3)_2$, and $N(CH_3)_3$; these compounds are indeed all known. However, isolobal analogies must be used with care, for they may also tempt us into postulating the existence of $(OC)_5Mn$—O—$Mn(CO)_5$ and $(OC)_5Fe$=$Fe(CO)_5$, but both are unknown. Isolobal analogies—like molecular orbitals themselves—provide useful correlations and hints, but they are not definitive.

Further information

Isolobality is discussed and illustrated throughout Shriver, Atkins, and Langford (1990). For a very thorough treatment of the subject see Albright, Burdett, and Whangbo (1985).

Jablonski diagram

A **Jablonski diagram** is a schematic portrayal of the relative positions of the electronic and vibrational levels of a molecule without any attempt being made to show the relative geometrical locations in terms of molecular potential energy curves. Even the vibrational levels are represented only schematically. The diagrams currently used are modified versions of those introduced by A. Jablonski in 1935 in his discussion of the mechanism of °phosphorescence.

An example of a Jablonski diagram is shown in Fig. J.1. The electronic ground state and its stack of vibrational levels are shown schematically on the right of the diagram (the actual vibrational levels are not uniformly spaced). The label S denotes singlet states of the molecule and T denotes triplet states. Straight arrows represent radiative transitions, those upwards correspond to absorption and those downwards to emission. Wavy lines between vertical stacks of levels represent radiationless transitions. Those between terms of the same multiplicity are °internal conversions and those between terms of different multiplicities are °intersystem crossings. Vertical wavy lines within a stack represent relaxation to thermal equilibrium of the vibrational energy as a result of collisions with the surrounding medium.

Further information

For illustrations of the use of Jablonski diagrams in photochemistry see Wayne (1988) and Calvert and Pitts (1966). The original proposals, in a very primitive form, can be found in Jablonski (1935).

Jahn–Teller theorem

The **Jahn–Teller theorem**, which was formulated by H.E. Jahn and Edward Teller in 1937, states that:

> In a non-linear molecule, there is always a distortion that removes any orbital degeneracy of its electronic state.

Thus, if a non-linear molecule is found by calculation to be orbitally degenerate, the actual molecule will be distorted into a non-degenerate, less symmetrical shape. The **static Jahn–Teller effect** is the distortion of orbitally degenerate molecules to achieve a lower energy. The **dynamic Jahn–Teller effect** is the hopping of the distortion from one orientation to another.

The $[CuL_6]^{2+}$ complex (with L a ligand) provides an example of the static Jahn–Teller effect. According to °ligand field theory, the

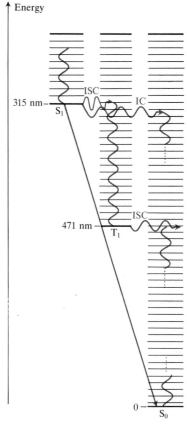

Fig. J.1 An example of a Jablonski diagram (for naphthalene).

1 $t_{2g}^6 e_g^3$

configuration of the octahedral complex is $t_{2g}^6 e_g^3$ (**1**). However, this configuration is orbitally degenerate because $e_g^3 = d_{z^2}^2 d_{x^2-y^2}^1$ and $e_g^3 = d_{z^2}^1 d_{x^2-y^2}^2$ have the same energy. The Jahn–Teller theorem therefore predicts that the complex will distort in such a way that the orbital degeneracy is eliminated (Fig. J.2).

The physical reason for the distortion of $[CuL_6]^{2+}$ can be understood by considering the forces operating on the ligands. In the configuration $d_{z^2}^2 d_{x^2-y^2}^1$ there is more electron density along the z axis than in the equatorial plane; the opposite is true in the configuration $d_{z^2}^1 d_{x^2-y^2}^2$. In the former, there is a tendency for the ligands on the z axis to move away from the central ion, and for the equatorial ligands to move in. In the second configuration, the opposite shifts are to be expected. If the molecule distorts by stretching along the z-axis, $d_{z^2}^2 d_{x^2-y^2}^1$ will have a lower energy than $d_{z^2}^1 d_{x^2-y^2}^2$ (Fig. J.2). If the complex becomes compressed along the z axis, the configuration $d_{z^2}^1 d_{x^2-y^2}^2$ will have the lower energy. It follows that we should expect the molecule to be found in either the elongated or the compressed form, but which of these it actually adopts is very difficult to predict.

The high-spin d^4 configuration $t_{2g}^3 e_g^1$, the low-spin d^7 configuration $t_{2g}^6 e_g^1$, and the d^9 configuration $t_{2g}^6 e_g^3$ are all cases in which there is degeneracy because a single electron (or a single hole) is present in two degenerate e_g orbitals (d_{z^2} and $d_{x^2-y^2}$). Examples include Cu^{2+} (d^9) and

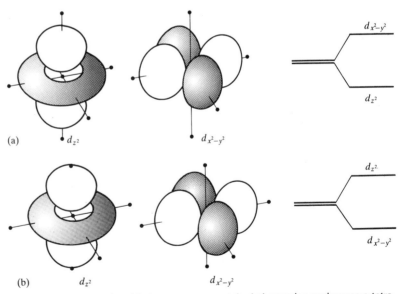

Fig. J.2 The change in orbital energy as an octahedral complex undergoes a tetragonal distortion by (a) elongation along z and (b) compression along z. The removal of degeneracy results in a lower energy for the distorted conformation so long as there are fewer than four electrons to be accommodated in the e_g orbitals of the octahedral complex.

Cr^{2+} (d^4) (which form complexes with closely related distorted structures), and Mn^{3+} (d^4); other examples are the rare low-spin Co^{2+} (d^7), low-spin Ni^{3+} (d^7), and Ag^{2+} (d^9) complexes. Jahn–Teller distortions are not expected for d^3, high-spin d^5, and low-spin d^6 or d^8.

Jahn–Teller distortions are expected for configurations that give rise to t_{2g} orbital degeneracy, such as d^1, d^2, high-spin d^6, and high-spin d^7. In practice the t_{2g} orbitals are directed between (rather than along) the metal–ligand axes, and the weaker effect of changes in electron density results in distortions that in practice are too small to see, or which are averaged to zero by the fluxional motion of the complex.

The experimental detection of a true Jahn–Teller distortion is very difficult as there are other reasons why a complex may be distorted. In particular it is difficult to distinguish it from a distortion imposed by packing requirements within the crystal.

Further information

For a simple, qualitative discussion of the Jahn–Teller theorem and the accompanying effects see Williams (1979), Ballhausen (1979), and Bunker (1979). A more thorough treatment is given by Griffith (1964), and a full and interesting discussion of the spectroscopic consequences will be found in §1.2 of Herzberg (1966). A full discussion has been given by Englman (1972). For spectroscopic consequences see Lever (1984). The original papers are by Jahn and Teller (1937) and Jahn (1938).

Kinetic energy

The **kinetic energy**, T, of a particle is the energy it possesses on account of its motion. In classical mechanics, the kinetic energy is related to the linear momentum p by $T = p^2/2m$ (and by $T = \frac{1}{2}mv^2$, where v is the velocity). In quantum mechanics the kinetic energy is proportional to the second derivative of the wavefunction (Box K.1). As the wavefunction becomes more sharply curved, so the kinetic energy of the particle increases. Conversely, a particle with a wavefunction that is almost flat everywhere has almost zero kinetic energy.

The relation between wavelength and kinetic energy (and thence momentum) of free particles is the basis of the °de Broglie relation. If the particle is bound to a potential centre (as for an electron in an atom) the wave may not spread far enough through space for a wavelength to be a meaningful concept, but the relation between curvature and kinetic energy remains valid and helpful. Thus, in the °hydrogen atom the 1s orbital is a simple exponentially decaying function which never passes through zero (and has no definable wavelength), yet we know that the electron has kinetic energy by virtue of the curvature of the wavefunction.

Further information

See MQM Chapters 2 and 3. All standard texts on quantum mechanics provide examples of wavefunctions that can be interpreted in terms of the relation between curvature and kinetic energy, so see Bransden and Joachain (1989), Das and Melissinos (1986), and Davydov (1976). For pictures of wavefunctions for a number of systems see Brandt and Dahmen (1985).

Box K.1 The kinetic energy

The classical expression for the kinetic energy is

$$T = \tfrac{1}{2}mv^2 = p^2/2m$$

It follows that the operator for kinetic energy is

$$T = -\frac{\hbar^2}{2m}\nabla^2$$

where ∇^2 is the °laplacian. For motion in one dimension,

$$T = -\frac{\hbar^2}{2m}\frac{\mathrm{d}^2}{\mathrm{d}x^2}$$

and the kinetic energy is proportional to the second derivative of the wavefunction.

Koopmans' theorem

In 1933 the Dutch theoretical chemist T. Koopmans proposed the following theorem:

> The °ionization energy of an atom or molecule is equal to the energy of the orbital from which the electron is ejected.

Formally: $I_i = -\varepsilon_i$, where ε_i is the energy of the electron in the orbital i (Fig. K.1). Thus, since the first ionization energy of the Na atom is 5.1 eV, we know that the energy of the 3s orbital (from which the least tightly bound electron is ejected) is − 5.1 eV. The theorem is exact only for hydrogenic atoms. For many-electron atoms and molecules the theorem is an approximation because it assumes that the remaining electrons and bond lengths do not reorganize to take advantage of the absence of the ejected electron. For instance, when °self-consistent field orbital energies are calculated, the ionization energies calculated by Koopmans' theorem may be inaccurate because the theorem assumes that the electrons in the ion occupy the same orbitals as they did in the atom. Furthermore, the Hartree–Fock scheme neglects electron °correlation effects, and is non-relativistic. The neglect of relativistic effects can lead to serious errors for electrons that are strongly bound and hence subject to high accelerations, such as the core electrons of heavy atoms.

Further information

Koopmans' theorem (note the location of the apostrophe) is now widely used in elementary discussions of °photoelectron spectroscopy, so see Ballard (1978). The original paper is Koopmans (1933), and a critical discussion will be found in Richards and Cooper (1983). This theorem is relevant to the discussions of °ionization energy, °electron affinity, °electronegativity, and °hardness.

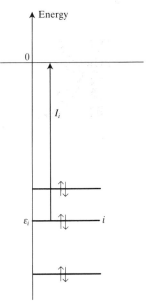

Fig. K.1 According to Koopmans' theorem, the ionization energy of an electron in orbital *i* is equal to the (negative) of the one-electron orbital energy.

Lanthanide contraction

The **lanthanide contraction** is the striking reduction in atomic and ionic radius that occurs across the first row of the f block and which leaves the subsequent elements of Period 6 with smaller radii than would have been expected by extrapolation from previous periods. Similar contractions occur in the d block, and have been named (but apparently not entirely seriously) 'scandinide contractions'. The origin of the lanthanide contraction is the poor shielding provided by the f electrons. Although the poor shielding is conventionally ascribed to the highly angular character of the orbitals, a closer analysis shows that the true explanation lies in their radial character. First, we should note that the $4f$ orbitals make only a minor contribution to the electron density in the outer reaches of the atom, so the electron density in the $5s$ and $5p$ orbitals determines the atomic radius. Next, we note that the single maximum of a $4f$ orbital lies just outside the fourth and third radial maxima of the $5s$ and $5p$ orbitals respectively, so the latter orbitals penetrate the $4f$ orbitals quite effectively and experience only weak °shielding. As a result, their *overall* extent, which determines the radius of the atom, is reduced. The $3d$ electrons do make a substantial contribution to the electron density in the outer parts of the atoms of the first row of the d block. The radial maximum of the $3d$ orbitals lies outside the $3p$ orbitals at the start of the row, and both sets of orbitals contract as Z increases. However, the $3d$ set contracts to a greater extent, and moves inside the $3p$ orbitals for the later members of the series. The greater contraction of the $3d$ orbitals is ascribed to their high angular dependence, which enables them to shrink without unduly increasing their interaction with each other. Because they shrink, and because they contribute directly to the atomic radius, the atoms contract.

Further information

The consequences of the lanthanide contraction are explored in Shriver, Atkins, and Langford (1990), Puddephatt and Monaghan (1986), and Mason (1988). The analysis of the effect, and its extension to the d block, is based on the helpful article by Lloyd (1986).

Laplacian

The **laplacian operator**, ∇^2, is the generalization to several dimensions of the differential operator d^2/dx^2. Its explicit form is given in Box L.1. The symbol ∇^2 is read 'del squared'. An informal interpretation of the significance of the laplacian is that it measures the sharpness of the curvature of a function of several variables. The greater the curvature

Box L.1 The laplacian and the legendrian

Cartesian coordinates:

$$\nabla^2 = \frac{\partial^2}{\partial x^2} + \frac{\partial^2}{\partial y^2} + \frac{\partial^2}{\partial z^2}$$

Spherical polar coordinates:

$$\nabla^2 = \frac{\partial^2}{\partial r^2} + \frac{2}{r}\frac{\partial}{\partial r} + \frac{1}{r^2}\Lambda^2$$

$$\Lambda^2 = \frac{\partial^2}{\partial \theta^2} + \frac{\cos\theta}{\sin\theta}\frac{\partial}{\partial\theta} + \frac{1}{\sin^2\theta}\frac{\partial^2}{\partial\phi^2}$$

Cylindrical polar coordinates:

$$\nabla^2 = \frac{\partial^2}{\partial r^2} + \frac{1}{r}\frac{\partial}{\partial r} + \frac{1}{r^2}\frac{\partial^2}{\partial\theta^2} + \frac{\partial^2}{\partial z^2}$$

of a function f (with respect to any or all of the coordinates), the greater the value of $\nabla^2 f$ (Fig. L.1). At a 'saddle point', where the curvature is positive (**1**) in one direction and negative (**2**) in a perpendicular direction, the total curvature may be zero, in which case $\nabla^2 f = 0$ (Fig. L.2).

It is inconvenient to work with the laplacian expressed in cartesian coordinates when the system has spherical or cylindrical symmetry; the appropriate expressions are given in Box L.1. The part of the laplacian involving the angular derivatives is called the **legendrian**, Λ^2. It measures the curvature of a function relative to the surface of a sphere. Thus, when Λ^2 is applied to an °s orbital, the outcome is zero because, relative to a spherical surface, the spherical s orbital is 'flat'.

Pierre Simon de Laplace (1749–1827), after whom the laplacian operator is named, was a notable French mathematician, and his

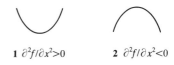

1 $\partial^2 f/\partial x^2 > 0$ **2** $\partial^2 f/\partial x^2 < 0$

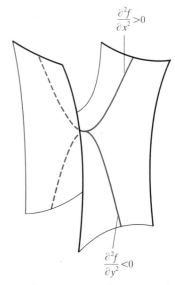

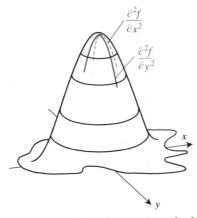

Fig. L.1 The laplacian operator measures the curvature of a function relative to all the axes that define it. In this illustration, the function depends on two variables (x and y), and the laplacian gives the sum of the two second derivatives with respect to each axis.

Fig. L.2 At a saddle point the function has a positive curvature in one direction but a negative curvature in the perpendicular direction. The sum of the two second derivatives is zero if the curvatures are equal but opposite.

formidable *Mécanique céleste* is a profound and worthy memorial to his work. Among its many pages of mathematics is the **Laplace equation**, $\nabla^2 f = 0$. Although the equation was formulated to describe the properties of gravitational fields, it has turned out to be applicable to a wide variety of phenomena. That it governs properties like the flow of incompressible fluids, gravitational and electromagnetic fields, and thermal conduction, is probably due to the fact that it is an equation that expresses the tendency of distributions towards flatness (that is, for wrinkles to spread).

Further information

For an account of the transformation of the laplacian from cartesian to polar coordinates, see Kyrala (1967), Riley (1974), and Arfken (1985). A brief biography of Laplace is available in Newman (1954).

Level

In general, an **energy level** is a permitted value of the energy of a system. More precisely, it is a permitted °eigenvalue of the °hamiltonian of a system. Thus the energy levels of the H atom are the values of $-hcR/n^2$ and those of the harmonic oscillator are the values of $(v + \frac{1}{2})\hbar\omega$.

The word 'level' signifies something more specific in atomic spectroscopy. An **atomic level** consists of all the states of a configuration with a given L, S, and J. For example, the °configuration $1s^2 2s^2 2p^1$ gives rise to a ^2P °term with the two levels $^2P_{1/2}$ and $^2P_{3/2}$

where the subscript denotes the two possible values of the total electronic angular momentum J (see the °Clebsch–Gordan series). The number of levels of a term is the **multiplicity** of the term. For instance, the multiplicity of a ^2P term is 2 and that of a ^3D term is 3 ($^3D_1$, $^3D_2$, and $^3D_3$). The $^2P_{1/2}$ level of the ^2P term consists of two individual states of the atom (with $M_J = +\frac{1}{2}$ and $-\frac{1}{2}$) and the $^2P_{3/2}$ level consists of four individual states (with $M_J = +\frac{3}{2}, +\frac{1}{2}, -\frac{1}{2}$, and $-\frac{3}{2}$).

Further information

See the entry on °atomic spectra and the bibliography there. Levels, multiplicity, and fine structure are discussed by Gerloch (1986), Cowan and Duane (1981), and Woodgate (1980). For a thorough theoretical treatment see Condon and Odabaşi (1980).

Lewis structures

The great American chemist Gilbert Lewis (1875–1946) proposed one of the most important of chemistry's collection of concepts when he identified the central importance of the electron pair in bond formation. In 1916 he published his proposal that identified a °covalent bond as a shared electron pair, a double bond as two shared electron pairs, and so on. ('And so on' does not go very far here, for chemists are rarely concerned with anything more elaborate than triple bonds, or three shared electron pairs, and never more than four shared pairs.) Lewis combined the identification of a shared electron pair and a chemical bond with the view that atoms tended to complete their °octet (for reasons explored in that entry). Thus, he devised a way of representing and deducing the electronic structures of molecules by drawing structures, which are now called **Lewis structures**, that showed how electron pairs are shared and how atoms complete their octets by pooling their electron resources with other atoms.

To construct a Lewis structure for a given molecule:

● Write down the symbols of the atoms to show which are neighbours.

For acetic acid we would write

$$\begin{array}{ccc} H & & O \\ H & C & C \\ H & & O \quad H \end{array}$$

(The arrangement of atoms is usually known from other considerations.)

● Count up the total number of valence electrons that must be accommodated in the structure.

In CH_3COOH, there are 24 valence electrons ($2 \times 4 = 8$ from the C, $4 \times 1 = 4$ from the H, and $2 \times 6 = 12$ from the O).

● One pair of electrons is then drawn between each pair of immediately neighbouring atoms (for they must be bound by at least a single bond).

For acetic acid we would write (**3**). At this stage we have accounted for 14 electrons (in seven pairs).

● Distribute the remaining electrons in such a way that the octet of each atom (the duplet of H) is completed, either by forming a double bond or by leaving the atoms with one or more nonbonding lone pairs.

Now we have arrived at (**4**), and have accounted for all 24 electrons. The appearance of a Lewis structure is simplified by representing each shared electron pair by a line (a chemical bond), so the final Lewis

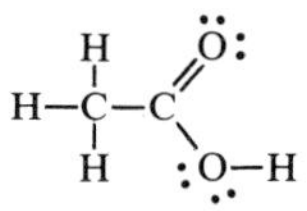

5

structure in this example is shown as (**5**). The Lewis structures for a number of common molecules are shown in Table L.1.

In a number of cases there are several ways of fulfilling the requirements of Lewis's theory. For instance, in the case of the ozone molecule, O_3, we can write two Lewis structures:

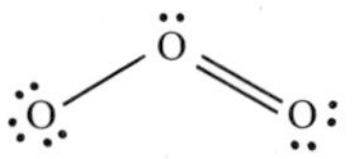

and

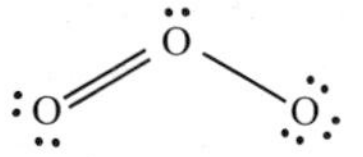

In such cases the actual structure is a °resonance hybrid—a blend—of the alternative structures.

Lewis structures are portrayals of the topology of a molecule: specifically, they portray its **connectivity**, the arrangement of links between neighbouring atoms. Except in simple cases they do not portray the geometry of the molecule, its 'shape'. Thus, the Lewis structure of the SO_4^{2-} ion correctly shows that each O atom is linked to a central S atom, but it does not show that the ion is tetrahedral. The development of Lewis's approach to account for the shapes of molecules is the domain of °VSEPR theory.

Lewis's ideas were wonderful, but not unassailably so. His ideas work only under certain circumstances (which we understand, particularly in the light of the development of °molecular orbital theory). For instance, certain atoms can become °hypervalent and act as though they are able to tolerate an expansion of their octet to ten or even

Table L.1 The Lewis octet structures of some molecules and ions

Molecule	Lewis structures*		
H_2	H—H		
N_2, CO	:N≡N:	:C≡O:	
O_3, SO_2, NO_3^-			
NH_3, SO_3			
PO_4^{3-}, SO_4^{2-}, ClO_4^-			

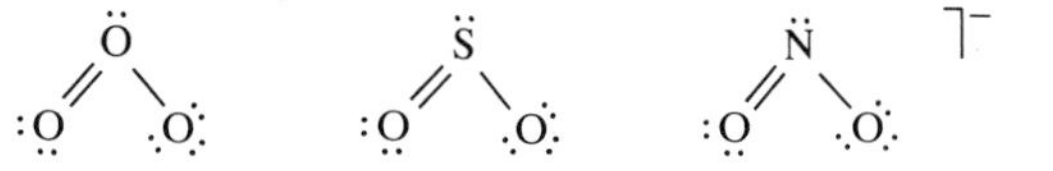

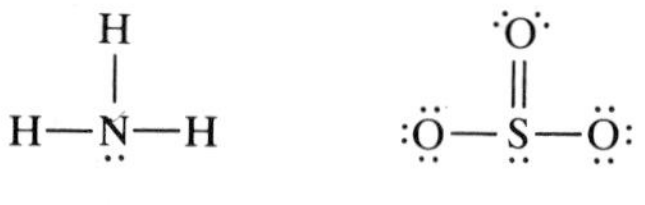

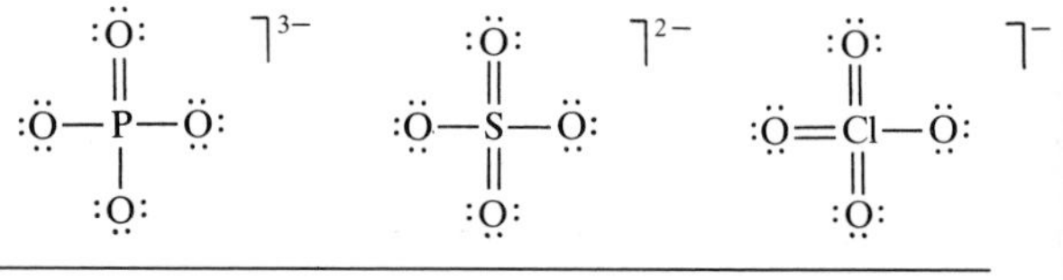

* Only representative resonance structures are given. Shapes are indicated only for diatomic and triatomic molecules.

twelve electrons. The most common atoms that behave in this way are Si, P, S, and Cl, and some of the 'expanded octet' structures that result are show in Table L.2. The rule to remember is that Period 2 elements do not normally become hypervalent but that elements in subsequent periods can do so.

The conventional explanation of hypervalence is the availability of *d* orbitals and their use to accommodate the additional electrons of the expanded octet. However, such an explanation is more a rule of thumb than a deep understanding, and for what may be the true explanation of °hypervalence see that entry. A second class of compounds that are not accommodated by Lewis's concepts are **electron-deficient compounds**, which are compounds that have too few electrons for a structure to be written at all. The most common introductory example is B_2H_6, in which there must be at least seven bonds: this requires 14 electrons, but there are only 12 valence electrons available. The explanation of the existence of such compounds is to be found in °molecular orbital theory, and there is nothing mysterious about them.

The modern explanation of the Lewis approach to the electronic structure of molecules can be found in the °molecular orbital theory of chemical bonding, in which electrons occupy orbitals that spread over the entire molecule. According to the °Pauli exclusion principle, no more than two electrons can occupy an orbital, and the Lewis structure is a form of accounting in which, in effect, each electron pair denotes an occupied bonding orbital.

Table L.2 Lewis expanded octet structures

Further information

The construction of Lewis structures is a field that is well trodden in elementary chemistry texts, so see Atkins (1989) for an introduction. For more information, see the entry on the °chemical bond. For a survey of Lewis's contribution to chemistry, and a biographical note, see Stranges (1984), Jensen (1984), Pauling (1984), and Kasha (1984).

Lifetime broadening

When a state has a finite lifetime τ, its energy is uncertain to an extent δE, where $\delta E \approx \hbar/\tau$. If the state in question is a participant in a spectroscopic transition, then the transition will spread over a frequency corresponding to the uncertainty in energy. In terms of frequencies and wavenumbers, the corresponding spectral linewidths are

$$\delta v = \frac{1}{2\pi\tau} \text{ which implies that } \delta v/\text{Hz} = \frac{0.1592}{\tau/\text{s}}$$

$$\delta\tilde{v} = \frac{1}{2\pi c\tau} \text{ which implies that } \delta\tilde{v}/\text{cm}^{-1} = \frac{4.771}{\tau/\text{ns}}$$

Lifetime broadening is widely ascribed to the °uncertainty principle, but for technical reasons (see that entry) it is better to regard it as a direct consequence of the Schrödinger equation. Lifetime broadening is the outcome of identifying the energy of a state with the frequency at which its wavefunction oscillates, and realizing that a decaying wavefunction corresponds to the superposition of a range of oscillating functions.

For a justification of this interpretation, we note that if the time-dependent wavefunction of a system is $\Psi = \psi e^{-iEt/\hbar}$, it oscillates in time with a single frequency $E/\hbar$ and hence corresponds to the precise energy E (Fig. L.3a). For such a wavefunction the probability that the system will be found in the state is $|\Psi|^2 = |\psi|^2$, which is independent of the time. Therefore, if the state persists for ever, its energy is infinitely precise. However, if the state decays with a time-constant τ, the probability $|\Psi|^2$ decays exponentially as $|\psi|^2 e^{-t/\tau}$, and hence the wavefunction itself has the form $\Psi = \psi e^{-(iE/\hbar + 1/2\tau)t}$ as depicted in Fig. L.3b. Because the state is now no longer simply oscillating but is also decaying, we can no longer identify its energy as E.

To make progress we use the fact that an oscillating, decaying function Ψ can be expressed as a sum of many oscillating functions centred on the frequency $E/\hbar$ (Fig. L.4). Furthermore, the shorter the lifetime τ, the wider the range of frequencies needed to recreate the decay. Since the presence of each oscillating function implies that the

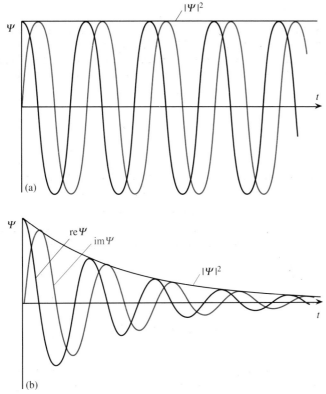

(a)

(b)

Fig. L.3 (a) The wavefunction corresponding to an eigenstate of the hamiltonian oscillates for ever, and its square modulus is constant: an infinite lifetime is associated with a precise energy. (b) If the wavefunction oscillates and decays, so that the probability of the system being found in the state decays with time, the energy is ill-defined. The faster the decay, the more uncertain the energy.

system may be found with that energy, it may be found with any of a range of energies, and the width of the range increases as the lifetime shortens.

Further information

For an analysis of lifetime broadening along the lines outlined here see §8.9 of MQM. The full discussion of the widths of levels and the relation to their lifetime can be found in Heitler (1954) and Craig and Thirunamachandran (1984); see also Davydov (1976) and Bransden and Joachain (1989).

Ligand field theory

The collection of concepts called **ligand field theory** is used to describe the electronic structure of *d*-metal complexes. Ligand field theory is essentially a °molecular orbital theory in which orbitals that spread

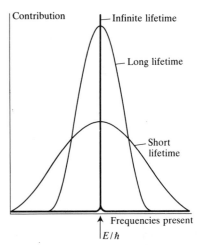

Fig. L.4 A representation of the energies (expressed as frequencies) present in wavefunctions with different lifetimes. The shorter the lifetime, the greater the frequency spread in the components needed to recreate the wavefunction.

over the entire complex are built from the d orbitals of the central atom and °symmetry adapted linear combinations of the ligand orbitals. A characteristic feature of the theory is the **ligand field splitting parameter**, Δ, which is used to correlate and rationalize the spectroscopic, magnetic, and thermodynamic properties of complexes.

We shall illustrate the approach by considering an octahedral complex of a d^n ion, in which the ligands have σ lone pairs of electrons directed towards the central ion. From the six ligand σ orbitals we can form six symmetry adapted linear combinations (Fig. L.5) of symmetry classification a_{1g} (one), e_g (two), and t_{1u} (three). The two e_g orbitals have a net °overlap with two of the d orbitals of the central ion, and form two bonding and two °antibonding molecular orbitals with them. The other three d orbitals, which are classified as t_{2g}, do not have ligand partners of the same symmetry, and so remain as nonbonding

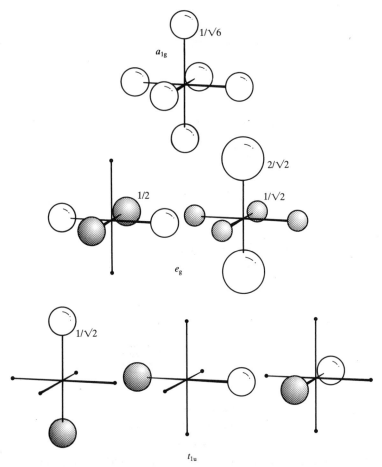

Fig. L.5 The symmetry adapted linear combinations of ligand σ orbitals in an octahedral complex. Tints denote relative signs; numbers denote the coefficients in the linear combinations.

orbitals on the metal atom. The resulting molecular orbital diagram is shown in Fig. L.6. As the ligand orbitals are lower in energy than the metal d orbitals, the bonding combinations are largely ligand in character and the antibonding combinations are largely metal atom in character.

Since the metal ion supplies n electrons and the six ligand lone pairs provide 12, there are $12 + n$ electrons to accommodate in the molecular orbitals. According to the normal °building-up principle, the first 12 electrons occupy the a_{1g}, e_g, and t_{1u} bonding orbitals (which are largely ligand in character). The next n electrons must compete for places in the nonbonding t_{2g} and antibonding e_g orbitals, both of which are predominantly metal in character.

A crucial aspect of ligand field theory is that the gap between the t_{2g} and e_g orbitals, which is denoted Δ, is generally quite small. There-

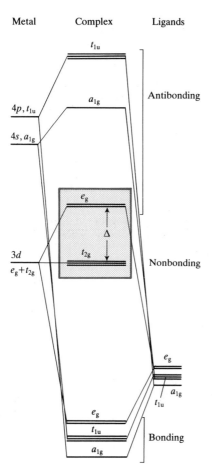

Fig. L.6 The molecular orbital energy levels of an octahedral complex; only local σ ligand orbitals are taken into account. The electrons notionally supplied by the ligands occupy and fill the six bonding orbitals. The remaining electrons are distributed over the frontier orbitals, which are shown surrounded by a box.

fore, we must modify the normal rules of the building-up principle. In particular, there may now be a competition between all n electrons occupying the t_{2g} orbitals or being distributed over the e_g orbitals too. The advantage of the former arrangement is that the high energy e_g orbitals are not occupied. The advantage of the latter arrangement (in which e_g orbitals are occupied) is that it may be possible to avoid having two electrons in one t_{2g} orbital. The outcome of the competition depends on the relative sizes of Δ and the **pairing energy** P, which is the repulsion energy between two paired electrons in the same d orbital.

In the **strong-field case**, with $\Delta > P$, it is energetically advantageous for as many electrons as possible to occupy the t_{2g} orbitals (**6**). In the **weak-field case**, with $\Delta < P$, so little energy is expended in occupying the upper orbitals that the electrons minimize their repulsions by occupying all the orbitals singly (and with parallel spins, see °Hund rules) before completing the t_{2g} orbitals (**7**). The resulting energetically most favourable configurations of d^n ions are given in Table L.3.

Where two configurations can occur (according to whether the ligand field is strong or weak), the configuration with the greater number of unpaired electrons is called the **high-spin case** and that with the lower number of unpaired spins is called the **low-spin case**. Thus, a d^4 complex with the configuration t_{2g}^4 is low spin (two unpaired electrons, **6**), but if it has the configuration $t_{2g}^3 e_g^1$ it is high spin (and has four unpaired electrons, **7**). High-spin complexes have a higher magnetic moment than low-spin complexes, and the two may be distinguished by their different paramagnetic susceptibilities, which are proportional to

6 Strong field low-spin d^4

7 Weak field high-spin d^4

Table L.3 Ligand field stabilization energies of d^n ions

n	Example	Octahedral				Tetrahedral	
		Strong-field		Weak-field			
		N	LFSE	N	LFSE	N	LFSE
d^0	Ca^{2+}, Sc^{3+}	0	0	0	0	0	0
d^1	Ti^{3+}	1	0.4	1	0.4	1	0.6
d^2	V^{3+}	2	0.8	2	0.8	2	1.2
d^3	Cr^{3+}, V^{2+}	3	1.2	3	1.2	3	0.8
d^4	Cr^{2+}, Mn^{3+}	2	1.6	4	0.6	4	0.4
d^5	Mn^{2+}, Fe^{3+}	1	2.0	5	0	5	0
d^6	Fe^{2+}, Co^{3+}	0	2.4	4	0.4	4	0.6
d^7	Co^{2+}	1	1.8	3	0.8	3	1.2
d^8	Ni^{2+}	2	1.2	2	1.2	2	0.8
d^9	Cu^{2+}	1	0.6	1	0.6	1	0.4
d^{10}	Cu^+, Zn^{2+}	0	0	0	0	0	0

N is the number of unpaired electrons; LFSE is in units of Δ; the calculated relation is $\Delta_T \approx 0.44 \Delta_0$.
Source: Shriver, Atkins, and Langford (1990).

$S(S+1)$, with S (the total spin) equal to $\frac{1}{2}N$, where N is the number of unpaired electrons.

So far we have considered only σ bonding between the central atom and the ligands. However, some ligands have orbitals of π symmetry with respect to the M—L axes and can overlap the appropriate combinations of d orbitals. If there are two π orbitals on each ligand, there are 12 orbitals in all (Fig. L.7), and they form °symmetry adapted combinations of symmetry t_{1g}, t_{1u}, t_{2g}, and t_{2u}. The $t_{2g}\pi$ orbitals have non-zero overlap with the t_{2g} d orbitals (which are nonbonding in the absence of π overlap). If the ligand π orbitals are full (as in Cl^-), they lie below the metal orbitals, and the effect of the bonding is to decrease Δ (Fig. L.8a). On the other hand, if the ligand π orbitals are initially empty (as in CO), they lie above the d electron energies, and the effect of π overlap is to increase Δ (Fig. L.8b).

The strength of the ligand field depends on the identities of the ligands and the central metal ion and on the presence or otherwise of π overlap. The variation of Δ with ligand is expressed by the **spectrochemical series**:

$$I^- < Br^- < S^{2-} < SCN^- < Cl^- < NO_3^- < F^- < C_2O_4^{2-} < H_2O < NCS^-$$
$$< NH_3 < en < bipy < phen < NO_2^- < PPh_3 < CN^- < CO$$

which broadly follows the sequence

π bases < weak π bases < no π character < weak π acids < π acids

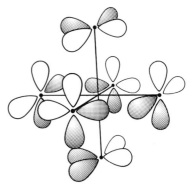

Fig. L.7 The 12 ligand orbitals of local π symmetry that are used to discuss the role of metal–ligand π bonding.

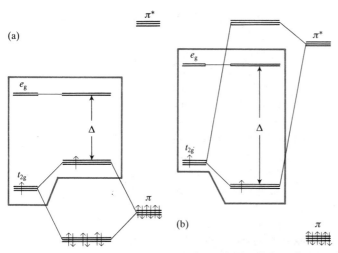

Fig. L.8 The effect of π orbital overlap on the ligand field splitting of an octahedral complex. (a) If the ligands supply electrons (are Lewis π bases), the ligand field splitting Δ is decreased. (b) The opposite holds for ligands with empty (and typically antibonding) π acceptor orbitals (Lewis π acids).

To explain the order, it is necessary to consider the influence of π bonding on the separation of the d orbitals in the complex as well as the influence of σ bonding, particularly where π bonding is irrelevant (in the middle of the series). The variation of Δ with the metal ion is summarized by the series:

$$Mn^{2+} < V^{2+} < Co^{2+} < Fe^{2+} < Ni^{2+} < Fe^{3+} < Co^{3+} < Ir^{3+} < Pt^{4+}$$

The ligand field splitting increases with oxidation number and on descending a group: the latter trend stems from the improvement in σ bonding that is achieved for the more diffuse $4d$ and $5d$ orbitals.

Further information

Ligand field theory is described in Chapters 7 and 14 of Shriver, Atkins, and Langford (1990). Texts that cover the subject, in approximate order of difficulty, are Figgis (1966), Ballhausen (1979), and Griffith (1964). For a thorough survey of the chemistry of d-metal complexes see Cotton and Wilkinson (1988).

Linear combination of atomic orbitals

The **linear combination of atomic orbitals** (LCAO) approximation expresses a °molecular orbital as a sum of atomic orbitals centred on each nucleus. Thus, the σ and σ^* orbitals of the H_2 molecule are expressed as a superposition of the two H1s orbitals. In general, N molecular orbitals can be created from a superposition of N atomic orbitals.

The accuracy of the representation of the true molecular orbital as an LCAO increases with the size of the **basis set**, the number of atomic orbitals employed in the superposition. In the elementary discussion of H_2 the basis set consists of the two H1s atomic orbitals, and this gives a reasonable first approximation to the bond. However, the description can be improved considerably by expanding the basis to include other s orbitals and higher p, d, f, and other orbitals. The use of a small basis set in the LCAO description of molecular bonding is one of the most severe sources of error in the method.

Further information

The LCAO approximation is one of the foundations of the theory of the chemical bond, and more information will be found under the appropriate entries, especially °molecular orbital theory and °symmetry-adapted orbitals. See in particular DeKock and Gray (1980), Coulson (1979), Szabo and Ostlund (1982), and Hehre *et al.* (1986). Then see Cook (1978), Clark (1985), and the useful discussion and set of references given in Hirst (1990).

Linear momentum

The **linear momentum**, p, is formally the product of the mass and velocity of a particle: a heavy particle moving rapidly has a higher linear momentum than a light particle moving slowly because it is both heavier and travelling faster. The greater the rate of change of momentum in a collision, the stronger the force a particle exerts. In quantum mechanics, the linear momentum is inversely proportional to the wavelength of the particle, as expressed by the °de Broglie relation $p = h/\lambda$. A wave of short wavelength (corresponding to the particle having a high linear momentum) is like a compressed spring in the sense that it has a high tendency to spread, and hence it exerts a strong force on the boundaries it encounters (Fig. L.9).

Linear momentum parallel to an axis is °complementary to position along that axis. That is, there is an °uncertainty relation between momentum along x and position along x (and likewise relative to the other axes). Thus, although in classical mechanics a particle may simultaneously have an exact linear momentum and position on each point of its trajectory, according to quantum mechanics the closer we specify the linear momentum, the more uncertain the position (and vice versa).

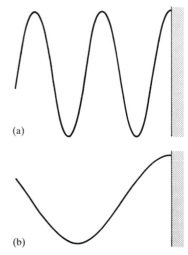

(a)

(b)

Fig. L.9 A pictorial interpretation of the de Broglie relation. (a) A short wavelength, high momentum wave is like a compressed spring in the sense that it has a high tendency to spread, and hence it exerts a strong force on the boundaries it encounters. (b) A long wavelength, low momentum wave exerts a weaker force.

Further information

The classical role of momentum (including its more general formulation in terms of the Lagrangian and Hamiltonian formulations of classical mechanics) is described in Goldstein (1982). The quantum mechanical description of particles in states of specified momentum is discussed in Chapter 3 of MQM and almost all books on quantum mechanics; see, for instance, Bransden and Joachain (1989) and Das and Melissinos (1986). For pictures see Brandt and Dahmen (1985). In the presence of a magnetic field the definition of linear momentum must be treated with care: see Chapter 14 and Appendix 21 of MQM.

Localized orbitals

A **localized orbital** is a °linear combination of molecular orbitals that is localized almost wholly on a single group of atoms in a molecule. Thus, a localized orbital in the H_2O molecule may be an orbital that is almost completely confined to one of the two O—H bond regions rather than spreading over the entire molecule.

The physical and chemical properties of many molecules suggest that to a significant extent electrons may be regarded as belonging to different parts of the molecule. For example, that bond lengths are largely transferable from one molecule to another suggests the existence

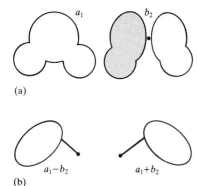

Fig. L.10 (a) Two delocalized molecular orbitals of an AB$_2$ molecule and (b) their sum and difference, which are orbitals localized in each A—B bonding region.

of a localized A—B bond. Similarly, the spectroscopic and chemical properties of the carbonyl group are similar whatever the identity of the rest of the molecule. In the 'natural' approach to the description of the electronic structure of molecules, the distribution of electrons is expressed in terms of molecular orbitals that spread throughout the molecule, and hence appears to run counter to the existence of localized units. However, an equally valid description of a molecule may be found by forming linear combinations of the delocalized orbitals, and if the combinations are well chosen, they correspond to electron density localized in specific regions of the molecule. This procedure is illustrated in Fig. L.10.

The molecular properties for which it seems (admittedly subjectively) natural to use localized or delocalized orbitals are summarized in Table L.4. Broadly speaking, delocalized orbitals are best used for the description of electronic excitations, since the whole molecule participates. Localized orbitals are convenient when discussing the properties of individual bonds or lone pairs. However, either description may be used in principle, since they are mathematically equivalent to each other. An example from solid state chemistry is the description of an ionic solid in terms of °band theory (the delocalized orbital description) or localized ions (the localized orbital description). In this case, we can think of the electrons either as occupying bands formed from the anion orbitals, or as occupying the atomic orbitals of the individual anions.

The mathematical basis of the equivalence of delocalized and localized orbitals in the framework of molecular orbital theory is that the many-electron wavefunction of a molecule satisfies the °Pauli principle if it is written as a °Slater determinant. However, it is a property of determinants that rows or columns may be added to other rows or columns without changing the value of the determinant. Therefore, rows and columns of a Slater determinant may also be added and subtracted without changing the overall wavefunction. These manip-

Table L.4 A general indication of the properties when localized and delocalized descriptions are appropriate

Localized appropriate	Delocalized appropriate
Bond strengths	Electronic spectra
Force constants	Photoionization
Bond lengths	Electron attachment
Brønsted acidity	Magnetism
VSEPR description of	Walsh description of
molecular geometry	molecular geometry
	Reduction potentials

Source: Shriver, Atkins, and Langford (1990).

ulations correspond to taking linear combinations of the original delocalized orbitals, and lead to a determinant expressed in terms of localized orbitals.

In the spin-coupled °valence bond description of molecules, the molecular wavefunction is not invariant to linear transformations of the kind that leave invariant the molecular orbital description, so their localized form is not arbitrary. That is, the theory leads inescapably to a localized bond description (even for °benzene).

Further information

Good accounts of the formation of localized bonds are those by Boys (1960) and Steiner (1976). For an analysis see Edmiston and Ruedenberg (1963, 1965), and for elementary descriptions see Coulson (1979) and Pilar (1990). For the °valence bond account see that entry and Cooper, Gerratt, and Raimondi (1986).

Lone pair

A **lone pair** of electrons is a pair of electrons in the valence shell of an atom that is not engaged in bonding. For example, the NH_3 molecule (**8**) has one lone pair on the N atom, and the H_2O molecule (**9**) has two lone pairs on the O atom.

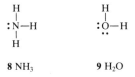

8 NH₃   9 H₂O

Lone pairs are important both structurally and chemically. They influence the shape of a molecule by exerting strong repulsive forces on the electron pairs in neighbouring bonds and on other lone pairs. For example, the trigonal pyramidal shape of the NH_3 molecule may be ascribed in part to the effect of the single lone pair on the N atom exerting a repulsive force on the three N—H bonding pairs. The analysis and prediction of the influence of lone pairs on the shapes of molecules is the province of °VSEPR theory. Lone pairs participate in the chemical properties of molecules on account of their ability to act as Lewis bases (electron pair donors), Brønsted bases (proton acceptors), and nucleophiles (agents that attack regions of low electron density). In each case, the action of the lone pair is a consequence of its ability to form a bonding molecular orbital with an empty orbital of the appropriate symmetry (Fig. L.11); since the lone pair provides two electrons and the acceptor orbital supplies none, there is a net lowering of energy when the complex forms.

Since lone pairs are not tied down into a bonding region by a second nucleus they can contribute to moderately low energy °electronic transitions. For instance, the $\pi^* \leftarrow n$ transition in carbonyl compounds is a major mode of excitation (see °colour).

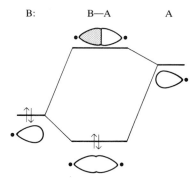

Fig. L.11 The formation of a complex from a Lewis acid A and a Lewis base B: can be expressed in terms of the formation of molecular orbitals from the orbital occupied by the lone pair on B and the empty orbital on A. Only the bonding orbital is occupied in the complex, which therefore has a lower energy than the individual components.

Further information

Further information about lone pairs and some references will be found in the entries on °Lewis structures, °VSEPR theory, and °colour. Aspects of Lewis acid and base character are discussed in Shriver, Atkins, and Langford (1990) and, in more detail, in Jensen (1980).

Magnetic dipole and electric quadrupole radiation

The most intense transitions in molecules are due to °electric dipole transitions. When these transitions are forbidden by °selection rules other mechanisms may have sufficient strength to cause a transition, but at a much lower intensity.

A **magnetic dipole transition** generates a photon by stimulating the magnetic component of an electromagnetic field. The weakness of the interaction of the molecule and field can be traced to the fact that the displaced charge generates a magnetic field only if it moves in a curved path (Fig. M.1b), but on the scale of a molecule any rotational component of charge displacement is very small. For an °electric dipole transition there is no need for curvature in the charge displacement (Fig. M.1a), and the transition couples much more strongly to the field.

An **electric quadrupole transition** arises from a displacement of charge that has a °quadrupolar character in that it resembles two dipolar transitions back-to-back (Fig. M.1c). The °selection rule for quadrupole transitions, $\Delta l = 0, \pm 1, \pm 2$, appears to conflict with the conservation of angular momentum, for the spin of a photon is 1. However, when the photon is generated it is flung off the molecule with a spatial variation that corresponds to its having an *orbital* angular momentum; the different relative orientations of the spin and orbital angular momenta of the photon can result in it providing or carrying away 0, 1, or 2 units of angular momentum.

Further information

Magnetic dipole transitions are nicely discussed in §3.7 of Craig and Thirunamachandran (1984), §IIID.2b of Kuhn (1962), §3.2.2 of Griffith (1964), and §7.6 of Hameka (1965). See also Heitler (1954) and Berestetskii, Lifshitz, and Pitaevskii (1971). If a molecule can undergo a transition to the same state by both electric and magnetic dipole transitions, it is optically active (see °chirality and °optical activity). Magnetic resonance (see °nuclear magnetic resonance and °electron spin resonance) relies on magnetic dipole transitions in which the magnetic component of the electromagnetic field interacts with a spin magnetic moment.

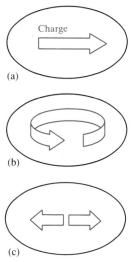

Fig. M.1 The motion of electronic charge in the course of (a) electric dipole, (b) magnetic dipole, and (c) electric quadrupole transitions.

Magnetic moment

see °Bohr magneton and °g-value.

Magnetic properties

The density of magnetic lines of force inside a **diamagnetic** substance placed in a magnetic field is less than it would be in a vacuum

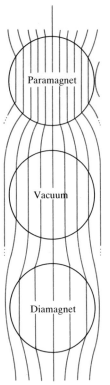

Fig. M.2 The magnetic induction can be pictured as a measure of the density of lines of force. In a paramagnetic sample, the density is greater than in a vacuum; in a diamagnetic sample, the density is less.

(Fig. M.2). That is, the magnetic **induction** B is lower inside a diamagnetic substance than in a vacuum. In a **paramagnetic** substance the density of lines of force is greater than it would be in a vacuum. The reduction in the induction inside a diamagnetic substance arises because the applied field induces a **magnetization** M (a magnetic dipole moment density) that opposes the applied field. The increase in induction inside a paramagnetic substance arises because the induced magnetization adds to the applied field. A physical manifestation of diamagnetism and paramagnetism is that a diamagnetic sample tends to move out of a region of magnetic field and a paramagnetic sample tends to move into it. The effect is employed in the 'Gouy balance', which is used to determine magnetic susceptibilities by measuring the apparent weight of a sample with and without a magnetic field present.

The ratio of the induced magnetization M to the strength of the applied field H is the **magnetic susceptibility**, χ, of the sample. (see Box M.1 and Table M.1). The susceptibility is negative for a diamagnetic material and positive for a paramagnetic material. The susceptibility of any substance can be written as the sum of a paramagnetic contribution χ_p and a (negative) diamagnetic contribution χ_d. All molecules have a diamagnetic contribution to their susceptibility because an applied field can always induce a current within their ground state molecular orbitals. This induced circulating current generates a magnetic field in opposition to the field applied and hence reduces the magnetic induction in the sample.

Table M.1 Magnetic susceptibilities at 298 K

	$\chi/10^{-5}$	$\chi_m/(10^{-4}\ cm^3\ mol^{-1})$
Water	−9.0	−16
Benzene	−0.72	−6.4
Cyclohexane	−0.79	−8.5
Carbon tetrachloride	−0.89	−8.4
$NaCl\,(s)$	−1.39	−3.75
$Cu\,(s)$	−9.6	−6.8
$S\,(s)$	−1.29	−2.0
$Hg\,(l)$	−2.85	−4.2
$CuSO_4\cdot 5H_2O\,(s)$	+17.6	+192
$MnSO_4\cdot 4H_2O\,(s)$	+264	+2790
$NiSO_4\cdot 7H_2O\,(s)$	+41.6	+600
$FeSO_4(NH_4)_2SO_4\cdot 6H_2O\,(s)$	+7.55	+1510
$Al\,(s)$	+2.2	+2.2
$Pt\,(s)$	+2.62	+23.8
$Na\,(s)$	+0.073	+1.7
$K\,(s)$	+0.56	+2.5

Data: Kaye and Laby (1973) and $\chi_m = \chi M/\rho$.

Box M.1 Magnetism

The **magnetization**, M, of a sample is related to the applied field, H, by

$$M = \chi H$$

where χ is the dimensionless **volume magnetic susceptibility**. The latter is related to the **molar magnetic susceptibility**, χ_m by

$$\chi_m = \chi V_m$$

where V_m is the molar volume. The **magnetic flux density**, B, is related to the applied field by

$$B = \mu_0 (H + M) = \mu_0 (1 + \chi) H$$

where μ_0 is the vacuum permeability. A substance for which $\chi < 0$ is **diamagnetic** and one for which $\chi > 0$ is **paramagnetic**.

The **Langevin–Pauli formula** for the diamagnetic molar susceptibility is

$$\chi_m = -\frac{e^2 \mu_0 N_A \langle r^2 \rangle}{6 m_e}$$

The temperature dependence of the paramagnetic magnetization of a sample composed of spins S with magnetic moment μ is given by the **Brillouin function**, $\mathscr{B}$:

$$M = N\mu \mathscr{B}(x), \qquad x = \mu H / kT$$

$$\mathscr{B} = \frac{2S+1}{2S} \coth\left\{ \left(\frac{2S+1}{2S} \right) x \right\} - \frac{1}{2S} \coth\left\{ \frac{x}{2S} \right\}$$

At high temperatures ($\mu H / kT \ll 1$) the paramagnetic molar susceptibility is given by the **Curie law**:

$$\chi_m = \frac{C}{T} \qquad C = \frac{N_A g_e^2 \mu_0 \mu_B^2 S(S+1)}{3k}$$

A useful relation is that

$$N_A g_e^2 \mu_0 \mu_B^2 / 3k = 6.3001 \text{ cm}^3 \text{ K}^{-1} \text{ mol}^{-1}$$

If the substance undergoes a ferromagnetic transition at the Curie temperature T_C, its susceptibility is given by the **Curie–Weiss law**:

$$\chi_m = \frac{C}{T - T_C}$$

The **temperature-independent paramagnetic susceptibility** of a substance is given by the expression

$$\chi_m = \frac{N_A e^2 \mu_0}{6 m_e^2} {\sum_n}' \frac{|\langle 0 | l | n \rangle|^2}{E_n - E_0}$$

For most substances the diamagnetic contribution is dominant and the material is diamagnetic. However, if the sample possesses unpaired electrons, the paramagnetic contribution dominates the diamagnetic and the sample is paramagnetic overall. It is usually found that χ_p decreases as the temperature is raised, and that in the vicinity of room temperature this **spin paramagnetism** obeys the **Curie law**, $\chi_p \propto 1/T$. In a few cases the paramagnetic term dominates the diamagnetic even

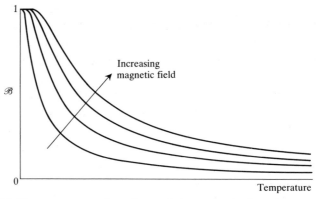

Fig. M.3 The temperature dependence of the magnetization of a paramagnetic sample expressed in terms of the Brillouin function.

though all the electrons are paired. In such cases it is also found that this weak paramagnetism is independent of temperature; hence this spin-independent paramagnetism is referred to as **temperature-independent paramagnetism** (TIP).

Spin paramagnetism arises from the net °magnetic moment of a collection of unpaired electrons. Thus, the populations of α and β spin states differ in a magnetic field (because they have different energies), and there are more electrons with β spin than with α spin. Since the spin magnetic moments do not cancel, the sample has a net magnetic moment and the field this moment generates adds to the field applied. As the temperature is raised, the populations of the α and β states become progressively more equal and the net moment decreases. Close to room temperature, where the populations are similar, the calculated magnetization decreases as $1/T$, in accord with the Curie law. At lower temperatures the magnetization is given by the **Brillouin function**, $\mathscr{B}$ (Box M.1 and Fig. M.3).

Temperature-independent paramagnetism arises from °orbital angular momentum induced by the applied field. In contrast to diamagnetism, in which currents are induced within the ground state of the molecule, the currents arise because electrons are forced to migrate through the molecular framework by making use of excited states (Fig. M.4). The circulation gives rise to a magnetic field that adds to the applied field, and hence contributes to the paramagnetism of the sample.

Many paramagnetic materials undergo a transition to a phase in which the spins align cooperatively and strongly enhance the magnetic properties of the material. When the neighbouring spins align in the same direction throughout a **domain**, a region of the sample, a strong magnetization is obtained that persists even after the magnetizing field

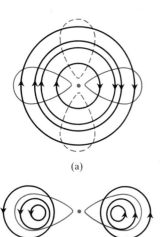

(a)

(b)

Fig. M.4 (a) A paramagnetic current (which is responsible for temperature-independent paramagnetism) arises when an applied field induces a circulation of electrons that makes use of excited state orbitals (shown with a broken line). (b) A diamagnetic current (which is responsible for diamagnetism) is a current confined to the ground state orbitals.

is removed (Fig. M.5). This aligned phase is called the **ferromagnetic phase**. It occurs below the **Curie temperature**, T_C, which is characteristic of the substance (Table M.2). Above the Curie temperature the paramagnetism of a potentially ferromagnetic substance follows the **Curie–Weiss law**, $\chi \propto 1/(T_C + T)$.

Another group of materials, of which NiO is an example, undergo a transition to an **antiferromagnetic phase** below the **Néel temperature**, T_N (Table M.2). In this phase neighbouring spins are antiparallel (Fig. M.5c) and the spin paramagnetism is locked into a very low value. In a **ferrimagnetic phase** there are two types of ion with different spin magnetic moments that are locked together antiferromagnetically (Fig. M.5d); since the moments are different, they do not cancel completely.

The spin–spin interactions responsible for cooperative magnetization are electrostatic in origin and related to the °exchange energy; see *Further information*.

Further information

See Shriver, Atkins, and Langford (1990), Atkins (1990), and MQM Chapter 14 (in that order) for more detailed summaries of the magnetic properties of materials. For more information try White (1983), Carlin (1986), Kittel (1986), and the useful set of references provided by Hatfield (1987). The standard work on magnetic susceptibilities is van Vleck (1932), but as its date suggests it is now more revered than used. The texts by Earnshaw (1968) and Hinchcliffe and Munn (1985) are useful elementary introductions; Davies (1967) is more advanced. For experimental procedures see Zijlstra (1967).

Fig. M.5 (a) A paramagnetic state of a solid: there is a random distribution of α and β spins (or of the projection of the spin magnetic moment if $S > \frac{1}{2}$). (b) A ferromagnetic substance, in which spins are parallel over an extensive domain. (c) An antiferromagnetic substance, in which neighbouring spins are antiparallel and the net magnetic moment is zero. (d) A ferrimagnetic substance, in which neighbouring spins are antiparallel but of different magnitudes, so the net magnetic moment is non-zero.

Matrix

A **matrix** *M* is a rectangular array of numbers (**1**). The numbers making up a matrix are called the **matrix elements**. A matrix is a generalization of the concept of 'number' in the sense that an ordinary number is a 1×1 matrix and is therefore a special case of a general $n \times m$ matrix. The matrix elements, which are denoted M_{rc}, are labelled by the row r and column c they each occupy. In a square $n \times n$ matrix the number of rows is equal to the number of columns, and there are n^2 elements. Square matrices are very important in quantum mechanics (although oblong matrices also occur), and our comments will be confined to them. Some matrices with special properties and names are given in Box M.2.

Matrices may be combined by a generalization of the rules of addition and multiplication used for simple numbers (Box M.2). A particu-

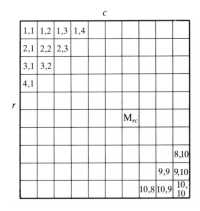

1

Table M.2 Curie (T_C) and Néel (T_N)
temperatures of solids

	T_C/K	T_N/K
Fe	1043	
Co	1404	
Ni	631	
Dy	85	179
Cr		475
Mn		100
Fe_3O_4	858	
Fe_2O_3		960
Mn_3O_4	43	
CuF_2		69
$CoCO_3$		18

Data: Gray (1972).

Box M.2 Matrices

A **square matrix M** is an array of n^2 **elements**, M_{rc}, where r labels the row and c the column:

$$M = \begin{pmatrix} M_{11} & M_{12} & M_{13} & \dots & M_{1n} \\ M_{21} & M_{22} & M_{23} & \dots & M_{2n} \\ . & . & . & \dots & . \\ . & . & . & \dots & . \\ M_{m1} & M_{m2} & M_{m3} & \dots & M_{nn} \end{pmatrix}$$

The **rules of combination** of matrices are:

Addition: $M + N = S$, where $S_{rc} = M_{rc} + N_{rc}$

Multiplication: $MN = P$, where $P_{rc} = \sum_q M_{rq} N_{qc}$

Specific types of matrix include the following:

Diagonal matrix, all $M_{rc} = 0$ expect for $r = c$

Unit matrix, 1, a diagonal matrix with all $M_{rr} = 1$

Inverse matrix, M^{-1}, such that $MM^{-1} = M^{-1}M = 1$

Transposed matrix, M^T, for which $(M^T)_{rc} = M_{cr}$

Complex conjugate matrix, M^*, for which $(M^*)_{rc} = (M_{rc})^*$

Adjoint matrix, M^+, defined as $M^+ = (M^T)^*$

Hermitian, self-adjoint matrix: a matrix for which $M^+ = M$

larly important point, the most significant difference from simple numbers, is that matrix multiplication is in general **non-commutative**. That is, it is not generally true that the product MN is equal to the product NM. The difference $MN - NM$ is the °commutator of M and N.

Matrices are fundamental to °group theory and to Heisenberg's formulation of quantum mechanics (see °matrix mechanics). However, the language of matrices has spread into the Schrödinger formulation of quantum mechanics. Thus, it is common to refer to an integral of the form $\int \psi_r^* \Omega \psi_c \mathrm{d}\tau$ as a 'matrix element' of the operator Ω and to write it Ω_{rc}.

Further information

For reasonably thorough introductory accounts of the properties of matrices, see Stephenson (1973), Riley (1974), and Arfken (1985). Starzak (1989) is a fairly comprehensive and wide ranging source of applications to all manner of problems. For some of the applications of matrices in chemistry and physics see the entries on °matrix mechanics and °group theory.

Matrix mechanics

The formalism of quantum mechanics initiated in 1925 by Werner Heisenberg (1901–1976) and expressed formally by him in collaboration with Max Born and Pascual Jordan in the following year, takes as its starting point the view that we should not seek to build visualizable models of atoms. Instead, we should deal directly with observable quantities, such as the frequencies of the radiation they emit. Heisenberg, Born, and Jordan's approach consisted of representing observables by °operators and looking for the equation that expressed their time-dependence. The contrast with °Schrödinger's formulation is sharpest here, for the latter presumed that the operators are independent of time but act on functions that evolve. He then formulated an equation that expressed the evolution of these functions. In due course, the two approaches were seen to be mathematically equivalent, differing only in the point of view (Fig. M.6).

The operator formulation of quantum mechanics turned out to hinge on one principal postulate:

> The position of a particle along a coordinate q and its linear momentum p along that coordinate must obey the rule $qp - pq = i\hbar$.

$\hbar$ is Planck's constant h divided by 2π. This fundamental postulate of quantum mechanics marks a profound break with classical mechanics, for it shows that observables cannot be represented by simple numbers (which commute) but must be represented by °matrices, which do not necessarily commute.

The error introduced by using the wrong rule, $qp - pq = 0$, and therefore treating q and p as ordinary numbers (as 'c-numbers', where c stands for classical), is only of the order of $\hbar$. Therefore classical calculations are acceptable in practice when inaccuracies of the order of Planck's constant can be tolerated.

Further information

Interesting, succinct accounts of matrix mechanics are given in the short books by Green (1965) and Jordan (1986); see also Das and Melissinos (1986). The original papers are those by Heisenberg (1925), Born and Jordan (1925), and Born, Heisenberg, and Jordan (1926). English translations are available in van der Waerden (1967). Born's contribution is nicely illustrated in the collection of correspondence between him and Einstein (Born 1970). Jammer (1966) puts Heisenberg's contribution into perspective.

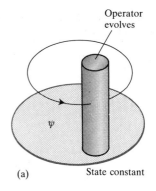

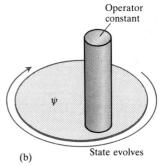

Fig. M.6 A diagrammatic representation of the difference between the Heisenberg and Schrödinger pictures of quantum mechanics. (a) In the Heisenberg picture, the state is fixed but the operator (represented by the column) moves. (b) In the Schrödinger picture, the operator is constant and the state moves. The outcomes are equivalent.

Metallic radius

The **metallic radius** of an element that forms a metallic phase is half the internuclear separation of nearest neighbour atoms, adjusted to correspond to 12-coordination.

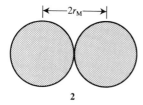

2

The first part of the definition is straightforward: a metallic element is pictured as built from identical spheres in contact, and the radius of the spheres is half the separation of the centres of those in contact (2). The adjustment to 12-coordination (the coordination number of a close-packed structure) is made because the internuclear separation is found empirically to increase with increasing coordination number. Average values fall in the following ratios:

Coordination number:	4	6	8	12
Relative radius:	0.88	0.96	0.97	1

Since it is desirable to disentangle the intrinsic size of an atom from the effects of packing, the metallic radius puts all atoms on the same footing. Thus, the metallic radius of sodium is reported as 191 pm; but as the room temperature form of the metal has 8-fold coordination, its actual radius in the solid is 0.97×191 pm $= 185$ pm.

Table M.3 gives a list of metallic radii, and their periodic variation is shown in Fig. M.7. The trends are similar to those of °ionic radii, and metallic radii increase down a group and decrease from left to right across a period. The increase down a group stems from the increasing numbers of electrons needing to occupy more distant shells. The decrease across a period stems from the increasing charge of the nucleus and its ability to draw the surrounding electrons together more compactly. The decrease across Period 6 is enhanced by the °lanthanide contraction.

Further information

See Shriver, Atkins, and Langford (1990) for a more detailed account, and Wells (1984). The periodicity of atomic radii is described by Puddephatt and

Table M.3 Metallic radii, r_M/pm

Li	Be												
157	112												

Na	Mg											Al	
191	160											143	

K	Ca	Sc	Ti	V	Cr	Mn	Fe	Co	Ni	Cu	Zn	Ga	
235	197	164	147	135	129	137	126	125	125	128	137	153	

Rb	Sr	Y	Zr	Nb	Mo	Tc	Ru	Rh	Pd	Ag	Cd	In	Sn
250	215	182	160	147	140	135	134	134	137	144	152	167	158

Cs	Ba	Lu	Hf	Ta	W	Re	Os	Ir	Pt	Au	Hg	Tl	Pb	Bi
272	224	172	159	147	141	137	135	136	139	144	155	171	175	182

The values refer to coordination number 12.
Data: Wells (1984).

Monaghan (1986) and by Mason (1988) and the variation of radius with coordination number and the Goldschmidt correction is discussed in Chapter 29 of Wells (1984).

Molecular orbital theory

In the **molecular orbital (MO) theory** of chemical bonding, an electron is supposed to occupy an orbital that spreads throughout the molecule. The square of the amplitude of the orbital at any point (more precisely, $\psi^*\psi$) is proportional to the electron density there. It is common practice to construct molecular orbitals from a °linear combination of atomic orbitals (LCAO), and we shall adopt this approach in the following.

In the LCAO approximation it is supposed that a molecular orbital can be expressed as a sum of the atomic orbitals on the constituent atoms of the molecule. The central feature of the method is the construction of molecular orbitals from all the atomic orbitals (typically of the valence shells of the atoms) that have the appropriate symmetry to have a net °overlap with each other.

The main features of the approach are as follows:

● Molecular orbitals are formed from all the valence orbitals (but without reference to the number of electrons that are to be accommodated).

● N atomic orbitals overlap to form N molecular orbitals.

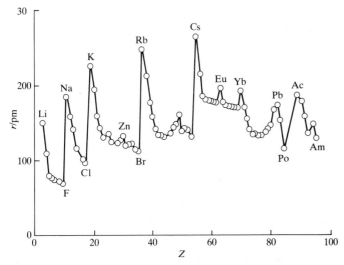

Fig. M.7 The variation of metallic radius through the Periodic Table. The atomic radii of non-metallic elements have been included to show the periodicity more clearly.

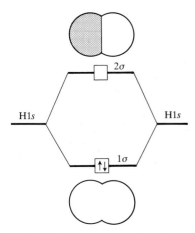

Fig. M.8 The formation of bonding and antibonding σ orbitals in H_2 and their occupation in the ground state of the molecule.

- For a given set of contributing atomic orbitals, the greater the number of interatomic nodes, the higher the energy.

- The appropriate number of electrons is accommodated in accord with the rules of the °building-up principle (with no more than two electrons in any one orbital).

We shall illustrate the approach in three stages, starting with molecular hydrogen. The molecular orbitals of H_2 are built from the H1s orbitals of the two atoms. The sum of the two orbitals gives one molecular orbital and the difference gives another (Fig. M.8); we shall denote them 1σ and 2σ respectively. In general, a σ **orbital** is a molecular orbital with cylindrical symmetry around the internuclear axis. Since 1σ has no interatomic node, it lies lower in energy than 2σ which has one node where the two atomic orbitals cancel. Each atom provides one electron, and both can be accommodated (with paired spins) in the lower 1σ orbital. The configuration of the molecule is therefore $1\sigma^2$.

The conventional physical reason for the difference in energy between 1σ and 2σ is that the sum of two atomic orbitals results in constructive interference and an accumulation of electron density in the internuclear region whereas the difference results in destructive interference and the exclusion of electron density from the internuclear region. (See °bond for a deeper analysis of this point.) If 1σ is occupied, the energy of the molecule is lower than the two separated atoms; hence 1σ is classified as a **bonding orbital**. If 2σ were occupied, the energy of the molecule would be higher than the two separated atoms; hence 2σ is an **antibonding orbital**. The H_2 molecule has a lower energy than the separated atoms because only the bonding orbital is occupied.

The eight molecular orbitals that can be built from the two 2s and six 2p orbitals of Period 2 homonuclear diatomic molecules are shown in Fig. M.9. Note that they fall into two distinct sets. Four of the orbitals have σ symmetry and are built from the two 2s and two $2p_z$ orbitals (where z is the internuclear direction). The other four orbitals, which are built from the two $2p_x$ and two $2p_y$ orbitals, have a nodal plane along the internuclear axis, and are called π **orbitals**. It is difficult to predict the relative energies of the eight orbitals reliably, but two alternatives are shown in the illustration. Figure M.10 shows the calculated positions of the orbitals, and we see that Fig. M.9a applies as far as N_2 but Fig. M.9b applies to O_2 and F_2. (The distinction hinges on the separation of the 2s and 2p orbitals in the atoms, which increases across the period.)

The configurations of the molecules are now readily found by applying the building-up principle in the usual way. Thus, for O_2 there are 12 valence electrons to accommodate, and the resulting configuration (which is illustrated in Fig. M.9b) is

$$1\sigma^2 2\sigma^2 3\sigma^2 1\pi^4 2\pi^2$$

The two outermost electrons occupy *different* 2π orbitals, and have parallel spins (see the °Hund rules). Hence, O_2 is predicted to be a °paramagnetic molecule, which is the case in practice.

The final example shows how readily molecular orbital theory takes into its stride the existence of molecules that presented grave difficulties to °Lewis (see also the discussion of °hypervalence). The Lewis electron pair theory of the chemical bond found difficulty with explaining the existence of **electron-deficient compounds**, those having too few electrons for a Lewis structure to be written. An example is diborane, B_2H_6, which has 12 valence electrons but needs at least seven bonds to hold the eight atoms together. Molecular orbital theory simply notes that from the 14 atomic valence orbitals that are available we can build 14 molecular orbitals, of which approximately seven will be bonding and seven will be antibonding (Fig. M.11). The 12 electrons can be accommodated in the lowest six orbitals, which are net bonding. hence the molecule has a lower energy than the separated atoms.

The point of the diborane example is that it shows that the bonding ability of an electron pair is *distributed* over several internuclear regions

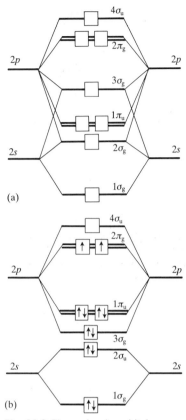

Fig. M.9 The molecular orbital energy level diagram for the Period 2 homonuclear diatomics. (a) For lithium to nitrogen, (b) for oxygen and fluorine. The reversal of the $1\pi_u$ and $3\sigma_g$ orbitals between the two sets of molecules reflects the interaction between orbitals of the same symmetry, which has the effect of pushing $3\sigma_g$ higher when the $2s,2p$ separation is small (on the left of the period). The configuration completed in (b) is that for O_2.

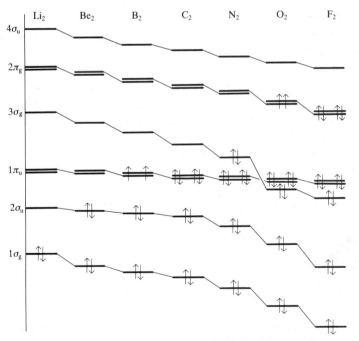

Fig. M.10 The molecular orbital energy levels of the Period 2 homonuclear diatomic molecules from Li_2 to F_2 and their ground state electron configurations.

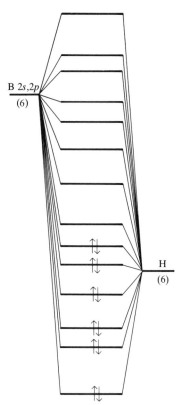

B 2s,2p
(6)

H
(6)

Fig. M.11 A schematic version of the molecular orbital energy levels of diborane. The 12 electrons occupy the lowest six orbitals, which are of predominantly bonding character. Hence the energy of the molecule is lower than that of the individual atoms even though, in Lewis terms, it does not have enough electrons for electron pair bonds between each pair of atoms.

and that Lewis's requirement that an electron pair be shared between a given pair of atoms is too stringent. A similar explanation applies to °hypervalent compounds, and that entry should be consulted.

Further information

See MQM Chapter 10 for details. Numerous books deal with the molecular orbital theory of the chemical bond; see in particular DeKock (1987), DeKock and Gray (1980), Coulson (1979), Albright, Burdett, and Whangbo (1985), Hehre *et al.* (1986), Szabo and Ostlund (1982), and McWeeny (1989). For graphical images of orbitals see Tedder and Nechvatal (1985), Jørgensen and Salem (1973), Verkade (1986), and Hout, Pietro, and Hehre (1984). More information, and further references, can be found in the entries on °*ab initio* techniques, °band theory, °bond, °Hückel method, °linear combinations of atomic orbitals, °self-consistent fields, °semiempirical methods, °symmetry-adapted orbitals, and °Walsh diagrams, as well as elsewhere throughout the book.

Molecular orbital and valence bond theories: a comparison

The details of molecular orbital (MO) and valence bond (VB) theories are given under their separate headings. In this synopsis we summarize their similarities and differences.

Both theories seek to describe the electronic structures, shapes, and energies of molecules. Both theories, at least in their simplest interpretation, account for bonding in terms of an accumulation of electron density in the internuclear region. Both achieve that accumulation by recognizing that electrons cannot be localized on a single atom when that atom is part of a molecule.

Although both methods emphasize the role of the electron pair, in practice they do so in different ways. The MO theory starts by formulating all the molecular orbitals that may be built from the available atomic orbitals; it then inserts the electrons into the molecular orbitals in accord with the °Pauli principle. The VB method concentrates on electron pairs from the outset, and calculates the energies of various structures that have all the electrons in the molecule paired in all possible ways. Then, having set up all these 'perfect pairing' structures, it expresses the true wavefunction as a °superposition of them. This process introduces the concept of °resonance, which is a central aspect of VB theory.

Electrons are allowed to spread over the whole molecule in the MO method, but these delocalized orbitals may be transformed into a collection of °localized orbitals. In VB theory attention is concentrated on individual bonds, and this feature is largely preserved in the final superimposed wavefunction. The MO concept of delocalization, as in

°benzene, has °resonance as its counterpart in VB theory, as between equivalent Kekulé structures.

Although both methods give a similar distribution of electrons, there are notable differences. The MO theory writes the wavefunction of an electron as $A(1) + B(1)$, and the overall wavefunction of two electrons as the product $[A(1) + B(1)][A(2) + B(2)]$. The VB theory writes their wavefunction as $A(1)B(2) + B(1)A(2)$. Expansion of the two-electron MO leads to $A(1)A(2) + B(1)B(2) + A(1)B(2) + B(1)A(2)$. This differs from the VB function by the presence of the first two terms. These terms can be interpreted as the contribution of a state in which both electrons are on the same atom, which may be either A or B. As these extra terms appear with the same weight as the other terms, we conclude that the MO theory does not fully take into account the effect of **electron correlation**, the tendency of electrons to keep apart. The VB theory moves to the opposite extreme and forbids both electrons to be on A or B simultaneously; therefore it overestimates the role of electron correlation. It can be improved by adding ionic terms to the original covalent wavefunction (which leads to the concept of ionic–covalent °resonance).

Molecular orbital theory has received far more attention than VB theory quantitatively because self-consistent field methods are relatively easily programmed for computers. The difficulty in dealing with VB calculations has been the very large number of canonical and ionic structures that had to be taken into account; however, this is no longer a problem with the development of 'spin-coupled' valence bond theory. Molecular orbital theory, in contrast to VB theory, is particularly inadequate for the prediction of dissociation energies, for it is poor at large distances.

Further information

See the individual entries on °molecular orbital and °valence bond theories for their respective details. Coulson (1979) and Murrell, Kettle, and Tedder (1970, 1985) provide further comparisons of the two approaches. The review by Cooper, Gerratt, and Raimondi (1990) gives valence bond theory an attractive face.

Morse potential

The **Morse potential**, which is specified in Box M.3 and depicted in Fig. M.12, is a convenient mathematical model of a molecular potential energy curve. It fails at very short distances (where it approaches a finite value), but is a reasonable representation of the shape of such curves. It was proposed by the American theoretical physicist P.M. Morse in 1929.

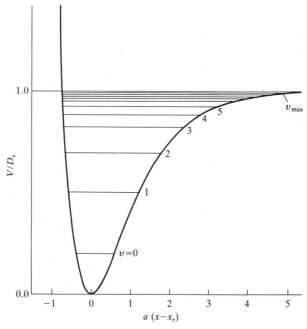

Fig. M.12 The Morse potential, its parametrization (*a* and D_e), and its finite number of bound-state energy levels. Note the convergence of the levels as the unbounded continuum is approached.

The importance of the Morse potential stems from the fact that the Schrödinger equation may be solved exactly for a particle confined by it. The solutions are given in Box M.3. There are only a finite number of them, and their separations decrease as the quantum number increases (as in an actual molecule).

Further information

The original proposal, and the explicit solution, is in Morse (1929). A critical discussion with further references is given by Herzberg (1950), who also describes its applications to spectroscopy. For a very general method of solving Schrödinger equations with Morse-like potentials (and, in fact, a broad class of differential equations in general) see Wybourne (1974).

Box M.3 The Morse potential

The **Morse potential** is the expression

$$V = D_e \{1 - e^{-a(x-x_e)}\}^2$$

where D_e is the depth of the minimum and $a = (\mu/2D_e)^{1/2}\omega$. The eigenvalues are

$$E = (v + \tfrac{1}{2})\hbar\omega - (v + \tfrac{1}{2})^2 \hbar\omega x_e \qquad x_e = \frac{a^2\hbar}{2\mu\omega}$$

with $v = 0, 1, 2, \ldots$.

Multiplicity

The **multiplicity** of a °term is equal to $2S+1$, where S is the total spin angular momentum quantum number. When $L \geq S$, the multiplicity is equal to the number of °levels the term possesses. That is, it is the number of different values of the total electronic angular momentum quantum number J that may be ascribed to the term. The numerical value of $2S+1$ is denoted by an upper left superscript on the term symbol. For example, a ^{2}D ('doublet D') term has two levels, namely ^{2}D$_{5/2}$ and ^{2}D$_{3/2}$. The multiplicity of a term is equal to $2S+1$ when $L \geq S$ because $2S+1$ values of J may be formed by coupling the spin and orbital °angular momenta together (see the °Clebsch–Gordan series). When $L \leq S$, only $2L+1$ values of J may be formed, and so such a term has only $2L+1$ levels. As an example, a ^{3}P term (a 'triplet P term') has a multiplicity of three, and its levels are ^{3}P$_2$, ^{3}P$_1$, and ^{3}P$_0$. A ^{2}S term (a 'doublet S term') has $L=0$ and $S=\frac{1}{2}$, and although it is a doublet term it has only one level (that with $J=\frac{1}{2}$).

Further information

See MQM Chapter 9 for further details about the classification of atomic states and the nomenclature that is used. Full details will be found in books on atomic spectra, notably Herzberg (1944), Woodgate (1980), Shore and Menzel (1968), and Condon and Odabaşi (1980). Stevenson (1965) gives a moderately simple yet complete account of the multiplet structure of atoms and molecules. See Wayne (1988) for a description of how the multiplicity of a species determines its chemical properties.

Node

A **node** is a point, line, or surface at which a wavefunction is zero and which separates regions of opposite sign of a wavefunction. The condition that the wavefunction should change sign at a node implies that the points at infinity where ψ approaches zero are not counted as nodes, nor are the boundaries of a well, at which the wavefunction is zero. As a general rule, the greater the number of nodes in a wavefunction, the higher its energy.

The real and imaginary components of the (°complex) wavefunction of a free particle each have an infinite number of evenly spaced nodes (Fig. N.1; ψ itself actually has no nodes), and the closer the nodes the higher the linear momentum. The momentum increases with decreasing nodal spacing because the wavelength of the wavefunction progressively decreases as the nodes are brought closer together (see the °de Broglie relation). The wavefunctions of a °particle in a one-dimensional square well are standing waves with nodes at a number of regularly spaced points between them (Fig. N.2). The walls, at which $\psi = 0$, are not nodes because the wavefunction does not change sign there. The more nodes there are, the higher the °kinetic energy, because the curvature of the wavefunction increases (and the wavelength shortens) as more nodes are added.

An atomic °s orbital may have a **radial node**, a spherical surface on which $\psi = 0$ everywhere. The $1s$ orbital has no radial node, a $2s$ orbital has one, a $3s$ orbital has two, and in general an ns orbital has $n-1$ radial nodes. The energy of the orbital increases as the number of radial nodes increases on account of the changes in kinetic and potential energy that accompany the changes in the shape of the wavefunction. However, the increasing number of radial nodes actually leads to a *decrease* in kinetic energy (see the °virial theorem) because, although the nodes increasingly buckle the wavefunction, they force it to spread

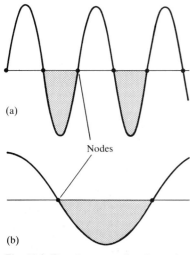

(a)

Nodes

(b)

Fig. N.1 The closer together the nodes in a wavefunction, the shorter the wavelength and hence the higher the linear momentum. (a) High momentum, (b) low momentum.

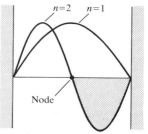

$n=2$ $n=1$

Node

Fig. N.2 The greater the number of nodes, the greater the energy. This illustration shows two wavefunctions of a particle in a box. The function for $n=1$ is the ground state and has no nodes; the function for $n=2$ is the first excited state and has one node.

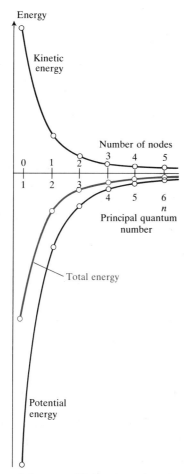

Fig. N.3 The variation of the potential, kinetic, and total energy with the number of nodes (and the principal quantum number) for a hydrogenic atom. Note that the kinetic energy decreases as the number of nodes increases but the increase in potential energy dominates it, and there is a net increase in energy.

over a progressively greater region of space and the net outcome is a decrease in its curvature. Since the nodes force the orbital to be more diffuse, its *potential* energy becomes less negative (Fig. N.3), and overall the total energy rises.

All (real) °p orbitals have one **angular node,** a plane through the nucleus on which $\psi = 0$ everywhere. All (real) °d orbitals have two angular nodes, all (real) °f orbitals have three, and so on. In general, an atomic orbital with °angular momentum quantum number l has l angular nodes. The increasing number of angular nodes effectively shortens the wavelength with respect to a circular path around the nucleus, and hence the momentum with respect to that angular coordinate increases (Fig. N.4).

The energy of molecular orbitals increase as the number of inter-

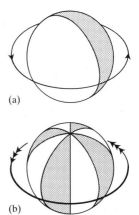

Fig. N.4 The greater the number of angular nodes, the greater the orbital angular momentum. (a) A low momentum state, $l=1$, (b) a higher momentum state, $l=4$. Note that only the real and the imaginary components of the complex wavefunction have nodes; the overall wavefunction is nodeless.

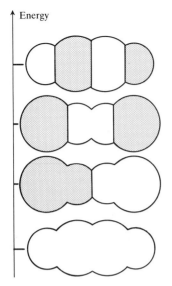

Fig. N.5 In general, the energy of
molecular orbitals increases with the
number of internuclear nodal planes,
which introduce a progressively larger
degree of antibonding character into
the wavefunction.

nuclear nodes increases (Fig. N.5), and the greater the number of internuclear nodes, the greater the °antibonding character of an orbital. Each new node excludes more electron density from the internuclear region where it lies, and leads to an energetically less favourable electron distribution (but see °bond for a deeper analysis).

Further information

See Chapters 3 and 4 of MQM for many examples of wavefunctions with nodes. The pictures in Brandt and Dahmen (1985) may be helpful. For the nodal structure of molecular orbitals, see the picture books referred to in the entry on °molecular orbitals. Since the nodes are places where a function is zero, the tables of zeros of functions in Abramowitz and Stegun (1965) can be useful for locating them. A general theorem states that the ground state of any one-particle system is nodeless: see §18 of Landau and Lifshitz (1958a).

Non-crossing rule

Consider two states of an atom or molecule with energies that depend on some parameter P (for example, a bond length or an external field). The **non-crossing rule** states that:

The energies of two states with the same symmetry never cross.

It follows that a variation in P leads to the energy variation illustrated in Fig. N.6.

The non-crossing rule is used in the construction of °correlation diagrams, such as °Tanabe–Sugano diagrams and °Walsh diagrams, where the parameter P corresponds respectively to the strength of the electron–electron repulsion and to the bond angle.

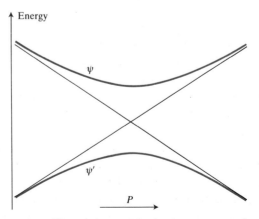

Fig. N.6 As a parameter P is varied, two states begin to converge, but do not cross if they have the same symmetry.

The time-dependence of a system in which there are non-crossing states is relevant to the discussion of °adiabaticity. Thus, if a system is initially in state ψ (Fig. N.6) and P is varied very slowly, it will remain in state ψ. However, if the change in P is rapid, the system may cross into state ψ', and hence emerge as an excited state of the new system.

Further information

A diagram like the one shown here can be obtained by solving the Schrödinger equation for a two-level system: this is done in Chapter 8 of MQM. Such an analysis formed the basis of Teller's deduction of the rule: see Teller (1937). A more abstract and earlier deduction of the rule is that given by von Neumann and Wigner (1929). For simpler, more conventional approaches, see Coulson (1979) and Herzberg (1950).

Normal modes

A molecule built from N atoms requires $3N - 6$ ($3N - 5$ if it is linear) independent coordinates to specify its internal motions. The **normal modes** of vibration are the linear combinations of these displacements that are dynamically independent for a quadratic potential. That is, so long as the vibrational motion is simple °harmonic, an excited normal mode persists indefinitely and does not stimulate the motion of another normal mode. Any complex vibrational mode of a polyatomic molecule may be expressed as a linear combination of its normal modes so long as the vibrations remain harmonic.

First we show how the formulae $3N - 6$ and $3N - 5$ are obtained, and then we discuss how the modes are chosen.

The location of N atoms requires the specification of $3N$ coordinates (x_1, y_1, and z_1 for atom 1, likewise for atom 2, and so on). A change in any of these coordinates corresponds to a change in the molecule's location, orientation, or shape. We could gives the values of all the coordinates relative to x, y, and z axes set in the laboratory (Fig. N.7a). However, it proves more sensible to choose them differently. Thus, three of the coordinates may be chosen to be the position of the centre of mass of the molecule, and they specify the location of the molecule as a whole (Fig. N.7b). The remaining $3N - 3$ coordinates therefore specify the positions of the atoms relative to the centre of mass. Of these remaining coordinates, three (two if the molecule is linear) are needed to specify the orientation of the molecule. A change in any of these three (or two) coordinates corresponds to a rotation of the molecule about its centre of mass. The remaining $3N - 6$ coordinates specify the relative positions of the atoms *within* the stationary, fixed-orientation molecule. A change in any of these coordinates corresponds to a distortion of the molecule without moving its centre of mass or

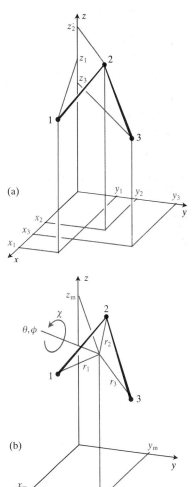

(a)

(b)

Fig. N.7 Two ways of specifying the locations of the atoms of a triatomic angular molecule. (a) The nine coordinates of the three atoms are reported. (b) The three coordinates of the centre of mass, the three angles that specify the orientation of the molecule, and the three locations of the atoms relative to the centre of mass are given.

resulting in an overall rotation. Hence $3N - 6$ (or $3N - 5$) coordinates are vibrational modes of the molecule.

Now consider the linear CO_2 molecule and its $3 \times 3 - 5 = 4$ vibrational modes. We could identify one vibrational mode with the stretching of one of the CO bonds. However, such a mode is not dynamically independent of the stretching of the other CO bond, for as soon as the first bond is excited it starts to shake the other bond, and hence excites that bond too. Thus, vibrational energy flows to and fro between the two bonds: they are not dynamically independent.

Now suppose that instead of exciting one of the bonds we excite the joint **symmetric stretch** (Fig. N.8a) of them both. The C atom is buffeted equally from both sides, and the vibration persists until it is terminated by a collision or some other external cause. In particular, the symmetric stretch is dynamically independent of the alternative combination of the two CO bond vibrations, the **antisymmetric stretch** (Fig. N.8a). If the latter is excited, it persists until it is deactivated by an external cause. Moreover, the two stretches are independent of the two **bending** modes of the molecule. Thus we see that by a judicious choice of the modes of vibration of the molecule we may obtain a set of independent motions: these are the four normal modes of the molecule.

The same analysis may be applied to other molecules. In every case, so long as the vibrations are strictly °harmonic, it is possible to form

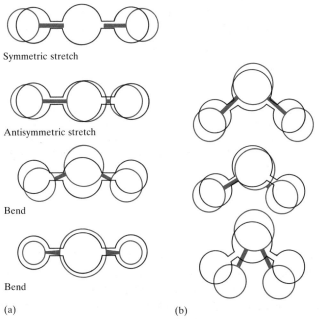

Symmetric stretch

Antisymmetric stretch

Bend

Bend

(a)   (b)

Fig. N.8 (a) The four normal modes of a linear triatomic molecule. (b) The three normal modes of an angular triatomic molecule.

$3N - 6$ dynamically independent combinations of the $3N - 6$ bond stretches and bends of a molecule. The three normal modes of H_2O (or of any angular AB_2 molecule) are shown in Fig. N.8b. Three of the 30 normal modes of a benzene molecule are shown in Fig. N.9. The collective motion of the atoms that corresponds to a normal mode is called a displacement of the molecule along the specified **normal coordinate**.

Each normal mode is equivalent to a single °harmonic oscillator of a certain mass m and °force constant k. The effective mass is a combination of the masses of the individual atoms weighted by the amount each atom moves when the molecule is displaced along the normal coordinate of the mode. For example, since the C atom does not move during the symmetrical stretch of CO_2, it makes no contribution to the effective mass of that vibration. Similarly, the effective force constant of a normal mode is a combination of the force constants of all the contributing stretches and bends. The allowed vibrational energy levels are then given by the expression obtained for a °harmonic oscillator, and are equal to $(v + \frac{1}{2})\hbar\omega$, with $\omega = (k/m)^{1/2}$.

The excitation of a normal mode by the absorption of electromagnetic radiation is governed by °selection rules. In particular, a mode can be excited by an °electric dipole transition only if the displacement along the normal coordinate results in a change in electric dipole moment. If the dipole moment does change, the mode is said to be **infrared active**, for such modes are responsible for the infrared absorption spectra of molecules. The two bending modes of CO_2 are infrared active, as is the antisymmetric stretch; the symmetric stretch is infrared inactive (but can be excited by collisions).

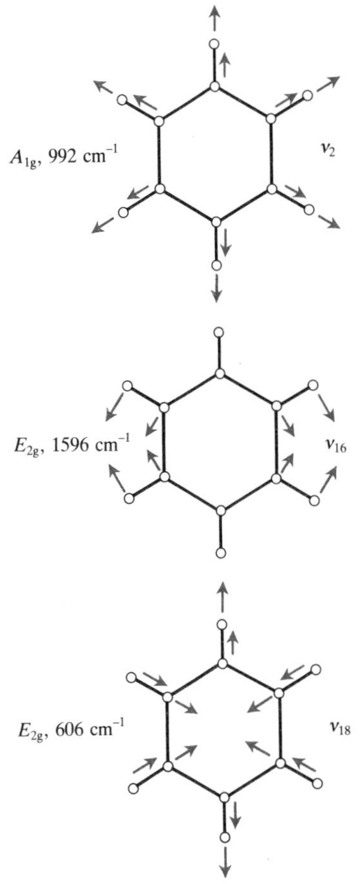

A_{1g}, 992 cm^{-1} v_2

E_{2g}, 1596 cm^{-1} v_{16}

E_{2g}, 606 cm^{-1} v_{18}

Fig. N.9 Three of the 30 normal modes of the benzene molecule, their symmetry designation, and their wavenumbers.

Further information

See Chapter 11 of MQM for an introduction to normal modes and their symmetry properties. For a more thorough introduction, see Woodward (1972), Hollas (1983), Flygare (1978), Herzberg (1966), and Bunker (1979). An excellent description of all aspects of normal mode analysis is that by Wilson, Decius, and Cross (1955).

Normalized function

A bound-state wavefunction ψ is **normalized** to 1 if $\int \psi^* \psi \, d\tau = 1$, where the integration is over all space (and includes summation over spin states if ψ includes spin). The Born interpretation of a normalized °wavefunction is that $\psi^* \psi \, d\tau$ is equal to the probability that a particle is in the volume element $d\tau$. Two functions that are both normalized and mutually orthogonal are said to be **orthonormal**.

A general function ψ may be normalized to 1 by multiplying it by the **normalization constant** N, where

$$N = \left\{ \frac{1}{\int \psi^* \psi \, d\tau} \right\}^{1/2}$$

for this guarantees that the integral over the square of $N\psi$ is equal to 1.

Further information

Simple examples of the normalization of functions will be found throughout MQM. A particular problem arises with the normalization of infinitely extended wavefunctions, such as $\sin kx$. The technique of 'δ-function normalization' is then sometimes used. The procedure is described in Davydov (1976), Goldberger and Watson (1964), Bransden and Joachain (1989), and Das and Melissinos (1986).

Nuclear magnetic resonance

The technique of **nuclear magnetic resonance** (NMR) is the observation of the resonant absorption of radiofrequency radiation by magnetic nuclei in the presence of an externally applied magnetic field. Current techniques use superconducting magnets to achieve fields of up to about 14 T, corresponding to a proton resonance frequency of up to about 600 MHz.

A nucleus with spin I has $2I+1$ permitted orientations relative to an applied magnetic field. Each orientation has a different energy, with the lowest level (marginally) the most highly populated. The energies of the transitions between these spin states are monitored by exciting nuclei in the sample with a pulse of radiofrequency radiation and then observing the return of the nuclear magnetization back to thermal equilibrium. After data processing (Fourier transformation), the data are displayed as an absorption spectrum (Fig. N.10), with peaks at frequencies corresponding to transitions between the different nuclear energy levels.

The technique provides three principal items of information:

- the **shielding constant**, σ, from the **chemical shift**, δ;
- the **spin–spin coupling constant**, J, from the **fine structure**;
- motional information from the nuclear spin **relaxation times**, T_1 and T_2.

The frequency of an NMR transition depends on the local magnetic field, B', which is the sum of the applied field, B, and the field induced in the molecule. The induced contribution is written $-\sigma B$, and hence

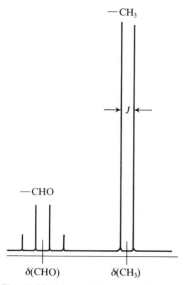

Fig. N.10 A typical simple nuclear magnetic resonance signal showing the chemical shifts (δ) and the spin–spin coupling constant (J). The intensities of the resonances are proportional to the numbers of resonant protons.

$B' = (1 - \sigma)B$. In practice, the position of an NMR signal is expressed as a chemical shift, which is defined in terms of the difference between the resonance frequency of the nucleus in the sample and the same nucleus in a reference compound:

$$\delta = \frac{\nu - \nu_{ref}}{\nu_{ref}} \times 10^6$$

When $\delta < 0$, the nucleus is said to be **shielded** relative to the reference; when $\delta > 0$ the nucleus is **deshielded** with respect to the reference. Differences in shielding constants (and hence chemical shifts) arise because the applied field induces different fields in different chemical groups. However, as several factors contribute to the shielding, a simple physical interpretation of chemical shifts in terms of electron density is generally not possible. More information will be found in the entry on the °chemical shift.

Spin–spin coupling of magnetic nuclei gives rise to multiplets of lines (Fig. N.10). The magnetic moments of the other nuclei modify the local magnetic field at the nucleus of interest. As a result, the energies of its spin states are modified and it gives a signal at a frequency that depends on the orientation of nearby nuclear spins. The spin–spin coupling within a **magnetically equivalent** group of nuclei (a group of identical nuclei with the same chemical shift, such as a —CH$_3$ group) may be large, but selection rules forbid its detection in the spectrum. For more information, including the mechanism of the interaction of nuclear spins, see °spin–spin coupling.

The strength of coupling, which is reported as the spin–spin coupling constant J, decreases rapidly with distance through chemical bonds, and in many cases is greatest when the two atoms are directly bonded to each other. For simple **first–order spectra**, those for which chemical shifts are much larger than spin–spin multiplet splittings, J is equal to the separation of adjacent lines in a multiplet. The properties of spin–spin coupling are such that a multiplet of $2I + 1$ lines appears in the spectrum of a spin-$\frac{1}{2}$ nucleus when that nucleus (or a group of magnetically equivalent spin-$\frac{1}{2}$ nuclei) is coupled to a nucleus of spin I. The relative line intensities of a multiplet arising from the coupling of the observed nucleus to N equivalent spin-$\frac{1}{2}$ nuclei are given by Pascal's triangle, which begins as follows:

Intensity ratio	Coupling to
1	0 other nuclei
1 1	1 spin-$\frac{1}{2}$ nucleus
1 2 1	2 spin-$\frac{1}{2}$ nuclei
1 3 3 1	3 spin-$\frac{1}{2}$ nuclei
1 4 6 4 1	4 spin-$\frac{1}{2}$ nuclei

The shape of the resonance lines, and specifically their width, is determined by °relaxation processes. In brief, local fluctuations in the magnetic field induce transitions between nuclear spin states and hence reduce their lifetime. The °lifetime broadening effect then results in a broadened line. Similarly, local inhomogeneities in the magnetic field (which can arise from molecular interactions as well as inhomogeneities in the applied field) result in imprecisely defined energy levels and hence broader lines. The determination of line widths and relaxation times is an important technique for the study of molecular motion in fluid solutions.

Further information

See Atkins (1990) for an introduction to NMR. More detailed introductions are those by Harris (1986) and Sanders and Hunter (1987). For applications to organic chemistry, see Derome (1987), and for applications to inorganic chemistry see Ebsworth, Rankin, and Cradock (1987). The handbook compiled by Freeman (1987) along the lines of this handbook, is a useful *aide memoire* for more advanced techniques. The dictionary compiled by Homans (1989) is also very useful for a wider range of terms. The book by Abragam (1961) remains authoritative but is dated. See the entries on the °chemical shift, °spin–spin coupling, and °relaxation for more information and further references to the electronic origin of these properties.

Nuclear magneton

The **nuclear magneton** μ_N is used for reporting the magnetic moments of nuclei, and is the analog of the °Bohr magneton for electrons. It is defined in the same way as the Bohr magneton, but with the proton mass in place of the electron mass:

$$\mu_N = \frac{e\hbar}{2m_p} = 5.051 \times 10^{-27} \text{ J T}^{-1}$$

Since the proton is not a truly fundamental particle (it is composed of three quarks), the nuclear magneton is not as fundamental a quantity as the Bohr magneton. It is a convenient quantity, however, because the sizes of most nuclear magnetic moments are close to μ_N and can be reported as a small numerical multiple of μ_N (see °g-value and Table G.1). For example, the magnetic moment of the proton itself is $2.79\mu_N$.

The nuclear magneton is about a thousand times smaller than the Bohr magneton. Therefore, a nucleus gives rise to a magnetic field that is about a thousand times weaker than that of an electron with the same angular momentum. The classical explanation is that a heavy particle need rotate at a lower speed than a light particle in order to achieve the same angular momentum. Therefore, the heavy particle gives rise to a weaker electric current even though it has the same angular momentum as the light particle. Because the current associated with the heavy particle is weaker, the magnetic field it generates is also weaker.

The neutron, despite its zero electric charge, has a magnetic moment of $-1.91\mu_N$. The explanation is that a neutron is a composite particle (like the proton, it is composed of three quarks), and that each of the charged components has a magnetic moment. Although the charges of the three quarks cancel, resulting in a neutral composite particle, their magnetic moments do not cancel, and the neutron has a non-zero magnetic moment overall.

Further information

See the entry on the °g-value for more information and further references. The relation of magnetic moment to nuclear structure (to the small extent that it is understood) is described by Jelley (1990).

Nuclear statistics

The requirement that nuclei must satisfy the °Pauli principle has consequences for the permitted °rotational states of a molecule. Since the rotation of a molecule may interchange equivalent nuclei, the overall wavefunction must change sign if the nuclei are °fermions but retain its sign if the nuclei are bosons. As a result, a symmetrical molecule can occur in only some of all possible rotational states. The effect of the Pauli principle on the population of rotational states of molecules is called **nuclear statistics**.

We shall consider the H_2 molecule, but similar arguments apply to other molecules. The first point to note is that since protons are fermions, the only acceptable states of the molecule are those with wavefunctions that change sign when the labels of the two protons are interchanged (Fig. N.11).

Next, we note that the interchange of the labels can also be achieved by a rotation of the molecule through 180° followed by an interchange of the spins of the protons (Fig. N.11). The rotation of the molecule results in a change of sign if the rotational quantum number J is odd, but there is no change in sign if J is even. This behaviour can be understood by noting that rotational wavefunctions are the same as the angular factors of atomic orbitals, with l replaced by J. As s orbitals do not change sign under a 180° rotation, p orbitals do change sign, d orbitals do not, and so on, we conclude that the sign is $(-1)^l$ for atomic orbitals and hence $(-1)^J$ for rotational wavefunctions.

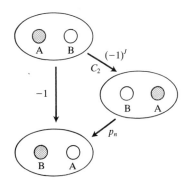

Fig. N.11 The relabelling of the two protons of H_2 (vertical arrow) must result in a change of sign of the overall wavefunction of the molecule. A rotation by 180° (C_2) results in a change in sign of $(-1)^J$. The change in sign of the nuclear spin permutation (p_n) depends on the nuclear spin state.

Now we need to consider the effect of interchanging the nuclear spins. As explained and illustrated in the entry on °singlet and triplet states, there are three wavefunctions corresponding to parallel spins and one corresponding to paired spins:

Parallel spins	Paired spins
$\alpha_1\alpha_2$	
$\alpha_1\beta_2 + \beta_1\alpha_2$	$\alpha_1\beta_2 - \beta_1\alpha_2$
$\beta_1\beta_2$	

All three states corresponding to parallel spins are even under spin interchange but the paired spin state is odd. For example, when we interchange α and β spins, the paired state becomes $\beta_1\alpha_2 - \alpha_1\beta_2$, which is the same as $-\{\alpha_1\beta_2 - \beta_1\alpha_2\}$.

At this stage, we have arrived at the result that the *overall* change of sign is $(-1)^J$ if the proton spins are parallel and $(-1)^{J+1}$ if they are paired. However, we have also seen that the Pauli principle requires that the overall wavefunction must change sign. Hence:

If the proton spins are parallel, J must be odd.
If the proton spins are paired, J must be even.

Hydrogen with parallel spins (and therefore confined to rotational states with odd J values) is called **ortho-hydrogen**. Hydrogen with paired nuclear spins (and therefore in rotational states with even J values) is called **para-hydrogen**.

At thermal equilibrium at high temperatures, a sample of molecular hydrogen consists of *para*-H_2 and *ortho*-H_2 in the ratio 1:3 (because there are three ways of having the proton spins parallel but only one way of having them paired). However, at low temperatures, where all the gas tends to occupy the lowest rotational level, for which $J = 0$ (an even value), a sample at thermal equilibrium contains only *para*-H_2. The two forms of hydrogen have different thermal properties (particularly heat capacities) on account of the differences in their available rotational energy levels. They also have different spectroscopic properties, since only certain states are allowed.

The attainment of thermal equilibrium can be very slow because the relative orientation of the nuclear spins must change. A sample of *para*-H_2 will change only slowly at room temperature to the thermal equilibrium mixture, because many of the nuclear spins must be reorientated to make them parallel to their partners. This reorientation can be achieved more rapidly in the presence of a catalytic surface on which the molecules may dissociate and then recombine with random partners. Alternatively, it may be brought about by a paramagnetic molecule (such as O_2 or a complex) that provides a magnetic field which can interact more strongly with one proton than the other and so drive them into new relative orientations (see Fig. S.11).

Nuclear statistics affect the spectra and thermal properties of all molecules containing equivalent nuclei, such as O_2, HCCH, H_2O, CH_3CH_3, and CH_4. However, the analysis gets quite complicated, particularly when the electronic state is not Σ_g^+, because its symmetry must also be taken into account. The same is true of excited vibrational states of the molecule, for they too must be included in the analysis.

Further information

See MQM Chapter 11 for an introduction to the more general case. A simple discussions will be found in Struve (1989), Graybeal (1988), and Hollas (1983, 1987). all of which discuss the spectroscopic consequences of nuclear statistics. The thermal consequences of nuclear statistics are explained by Davidson (1962) and Hecht (1990). See Townes and Schawlow (1955) and Hollas (1983) for extensions to more complex molecules.

Operators

According to quantum mechanics, all the dynamical information about a system is contained in its °wavefunction. The outcome of an observation of a particular property can be predicted by performing the appropriate mathematical operation on the wavefunction. Each physical observable is therefore represented by an **operator**, a mathematical action such as multiplication or differentiation. The operator corresponding to the observable Ω is often denoted $\hat{\Omega}$, but we shall not make that distinction here. Different choices of operators for observables correspond to the different **representations** of quantum mechanics. The conditions such operators must satisfy are as follows:

- The operators must be linear and °hermitian.

A **linear operator** Ω is one for which, if a is a constant, $\Omega(af) = a\Omega f$. Differentiation is a linear operation, taking logarithms is not. The property of hermiticity (see that entry) ensures that physical observables have real values.

- The operators used for position x and linear momentum p must obey the °commutation rule $xp - px = i\hbar$.

This criterion is one of the fundamental postulates of °matrix mechanics. The latter condition expresses the essential difference between classical and quantum mechanics.

In the **position representation** the operator for position on the x axis is multiplication by the coordinate x. The commutation rule then implies that the operator for linear momentum parallel to the x axis is differentiation with respect to x (the explicit form of the operator is given in Box O.1). Once we know the operators for position and

---

Box O.1 Operators

The operators for position, q, and momentum, p_q, along q must satisfy the **commutation relation**

$$[q, p_{q'}] = i\hbar\delta_{q,q'}$$

Different choices that satisfy this relation correspond to different **representations** of quantum mechanics. Two common representations are:

The **position representation**, in which

$$q = q \times, \qquad p_q = \frac{\hbar}{i}\frac{\partial}{\partial q}$$

The **momentum representation**, in which

$$p_q = p_q \times, \qquad q = -\frac{\hbar}{i}\frac{\partial}{\partial p_q}$$

momentum, we can set up the operators for all physical observables (other than those for the 'internal' properties of a particle, such as its electric charge and its spin) because they can be expressed in terms of the two basic observables. For example, the kinetic energy in classical mechanics is equal to $p^2/2m$, and hence the corresponding operator in the position representation is proportional to differentiation followed by differentiation. That is, the kinetic energy operator is proportional to second differentiation, d^2/dx^2 (Box O.1).

The possible outcomes of measuring an observable are the °eigenvalues of the corresponding operator. The mean outcome of many measurements is the °expectation value of the operator. Both those entries should be consulted for more details.

Further information

Operators are at the very heart of quantum theory, and so books that deal with the subject's fundamentals deal with operators at length. For a simple account of basic theory see MQM Chapter 5. For a classic account of operators and observables see Dirac (1958); for a résumé see Bransden and Joachain (1989). More mathematical accounts of operators will be found in von Neumann (1955), Jordan (1969), and Jauch (1968). An introduction to representation theory is given in Davydov (1976).

Orbital

See °Atomic orbital and °Molecular orbital.

Orbital angular momentum

See °Angular momentum.

Orbital approximation

The wavefunction of a many-electron system is a function $\psi(r_1, r_2, \ldots)$ of all the coordinates of all the electrons. According to the **orbital approximation**, this very complicated function is approximated as a product of one-electron wavefunctions—°orbitals—and written $\psi(r_1)\psi(r_2)\ldots$. For example, the ground state of a He atom is a very complicated function of the coordinates of both electrons, but is approximated as the product of two hydrogenic $1s$ orbitals. A many-electron wavefunction written as a product of one-electron orbitals does not satisfy the °Pauli principle, but it may be made to do so by an appropriate antisymmetrization procedure: see °Slater determinants.

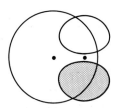

Fig. O.1 An *s* orbital and a *p* orbital in this orientation are orthogonal because the integral over the region where their product is positive is cancelled by the integral over the region where their product is negative.

Further information

See Richards and Cooper (1983) for a simple introduction, and Cook (1978), Szabo and Ostlund (1982), and McWeeny (1989) for more details. Further information will also be found in the references cited in °*ab initio*, °molecular orbitals, and °self-consistent fields.

Orthogonality

Two wavefunctions ψ_1 and ψ_2 are **orthogonal** if

$$\int \psi_1{}^* \psi_2 \, d\tau = 0$$

Two examples are the orthogonality of the 1*s* and 2*s* orbitals of the same atom and the orthogonality of a 1*s* orbital on one atom and a 2*p*π orbital on a neighbouring atom (Fig. O.1). Two functions that are both °normalized (to 1) and mutually orthogonal are said to be **orthonormal**. Sets of functions that are not mutually orthogonal may be orthogonalized by the **Schmidt procedure** (Box O.2).

The °eigenfunctions corresponding to different eigenvalues of a °hermitian operator are mutually orthogonal. Thus, the wavefunction of a harmonic oscillator with $v = 0$ is orthogonal to all other wavefunctions of the oscillator. Eigenfunctions with degenerate eigenvalues need not be mutually orthogonal, but the Schmidt procedure can be used to produce combinations of them that are orthogonal. In °group theoretical terms, two functions are orthogonal if they belong to different irreducible representations of the point group of the system.

Further information

For the proof that the eigenfunctions of hermitian operators are orthogonal see Chapter 5 of MQM. The group-theoretical aspects of orthogonality are described in Chapter 7 of MQM, and lucidly in Tinkham (1964). The Schmidt procedure is described in McGlynn *et al.* (1972) and Steiner (1976).

Box O.2 Schmidt orthogonalization

If the normalized wavefunctions *a*, *b*, *c*, ... are not orthogonal, form the combinations

$$a' = a$$

$$b' = b - S_{ab}a$$

$$c' = c - S_{ac}a - S_{b'c}b'/S_{b'b'}$$

and so on, where $S_{gh} = \int g^* h \, d\tau$, then the functions *a'*, *b'*, *c'* are mutually orthogonal (but not normalized). For the general case, see Steiner (1976).

Oscillator strength

The **oscillator strength** f of a transition is a measure of its intensity. It may be interpreted as the ratio of the observed intensity to the intensity of radiation absorbed or emitted by a harmonically bound electron in an atom (one bound to a nucleus by a Hooke's law force, and hence able to oscillate with simple harmonic motion in three dimensions); for such an electron $f = 1$. For strongly allowed transitions f is close to 1; for symmetry-forbidden transitions f is often close to 10^{-1}, and for spin-forbidden transitions f is close to 10^{-5}. Typical values for several types of transition are given in Table O.1.

The theoretical expression for f in terms of the °electric dipole transition moment is given in Box O.3, which shows that f increases as the square of the transition dipole moment. The oscillator strength is measured experimentally by determining the integrated °absorption intensity, as specified in Box O.3: the greater the area under the absorption curve, the greater the oscillator strength of the transition.

The **Kuhn–Thomas sum rule** states that:

The sum of the oscillator strengths for all the transitions of an N-electron system is equal to N.

Thus, *overall* an H atom behaves like a harmonically bound electron, because the sum of the oscillator strengths for all transitions away from the ground state is 1.

Table O.1 Oscillator strengths

	f
Electric dipole allowed	1
Magnetic dipole allowed	10^{-5}
Electric quadrupole allowed	10^{-7}
Spin forbidden (S–T)	10^{-5}
Parity forbidden	10^{-1}

Box O.3 The oscillator strength

The **definition** of the oscillator strength f for a transition that has transition dipole moment μ and occurs at the frequency ν is

$$f = \frac{8\pi^2 m_e \nu |\mu|^2}{3he^2}$$

A useful relation is

$$\frac{3he^2}{8\pi^2 m_e} = 7.095 \times 10^{-43}\,(\text{C m})^2\,\text{Hz}$$

The oscillator strength is related to the **integrated absorption coefficient** $\mathscr{A}$ by

$$f = \frac{4m_e c\varepsilon_0 \ln 10}{N_A e^2} \times \mathscr{A}$$

$$= 1.44 \times 10^{-19} \times \mathscr{A}/(\text{cm}^2\,\text{mmol}^{-1}\,\text{s}^{-1})$$

The **Kuhn–Thomas sum rule** states that in an N-electron molecule,

$$\sum f = N$$

where the sum is over all the transitions.

Further information

For the properties of the oscillator strength, its calculation, its relation to measured intensities, and a proof of the Kuhn–Thomas sum rule, see Appendix 17 of MQM. See also Davydov (1976). More sum rules of a similar kind are given in Bethe (1964).

Overlap integral

The **overlap integral**, S, of two °normalized wavefunctions ψ_1 and ψ_2 is defined as $\int \psi_1^* \psi_2 \, d\tau$. It may be pictured as a measure of the mutual resemblance and coincidence of the two functions. Thus, if the two functions are identical and coincide, $S = 1$; if the two functions are °orthogonal, $S = 0$.

The variation of the overlap integral for two H1s orbitals with internuclear distance is shown in Fig. O.2. At large separations $S \approx 0$ because where one wavefunction has a large amplitude the other is close to zero, their product is therefore everywhere very small, and so too is the sum (integral) of these products. When the separation is almost zero, the two wavefunctions peak in the same region of space, the integral of their product is nearly the same as the integral of the square of either one of them, and now $S \approx 1$.

The variation of the overlap integral for an H1s orbital brought up to an H2p orbital along the latter's axis is also shown in Fig. O.2. The integral is close to zero at large separations for the same reason as before. However, after S passes through a maximum it falls to zero when the separation is zero. When the two nuclei coincide (Fig. O.3), the product of amplitudes is positive in one half of space and negative in the other half of space; hence, the integral, the total area under the

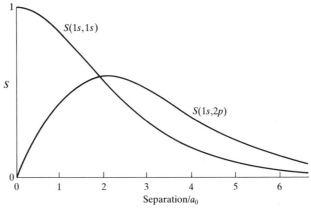

Fig. O.2 The variation of the overlap integral of hydrogen atomic orbitals with distance.

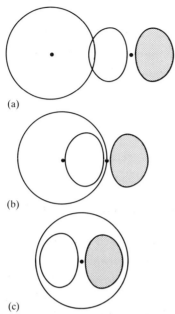

(a)

(b)

(c)

Fig. O.3 (a) At large distances the product of wavefunctions is small and the overlap integral is close to zero. (b) The overlap integral rises to a maximum and then (c) falls to zero again when the positive and negative regions cancel.

curve, is zero. For the same reason, the overlap integral between an H1s orbital and an H2p_x orbital is zero at all separations (Fig. O.1).

Further information

For the evaluation of overlap integrals see McGlynn *et al.* (1972). Extensive tables of overlap integrals have been published, and a good bibliography will be found on p.421 of that reference. Explicit expressions for overlap integrals between various atomic orbitals are given in Box O.4.

Box O.4 Overlap integrals

The following are some overlap integrals between °Slater type orbitals on identical atoms:

$S(1s, 1s) = (1 + \zeta R + \frac{1}{3}\zeta^2 R^2)\, e^{-\zeta R}$

$S(1s, 2p\sigma) = \zeta R(1 + \zeta R + \frac{1}{3}\zeta^2 R^2)\, e^{-\zeta R}$

$S(2s, 2s) = (1 + \zeta R + \frac{4}{9}\zeta^2 R^2 + \frac{1}{9}\zeta^3 R^3 + \frac{1}{45}\zeta^4 R^4)\, e^{-\zeta R}$

$S(2s, 2p\sigma) = \frac{1}{6}\sqrt{3}\zeta R(1 + \zeta R + \frac{7}{15}\zeta^2 R^2 + \frac{2}{15}\zeta^3 R^3)\, e^{-\zeta R}$

$S(2p\pi, 2p\pi) = (1 + \zeta R + \frac{2}{5}\zeta^2 R^2 + \frac{1}{15}\zeta^3 R^3)\, e^{-\zeta R}$

In each case R is the internuclear separation in units of a_0.

For two STOs on the *same* atom:

$$S(1s, 2s) = \frac{24\zeta_1^{3/2}\zeta_2^{5/2}}{3^{1/2}(\zeta_1 + \zeta_2)^4}$$

For more formulae, see Mulliken *et al.* (1949).

Oxidation number

The **oxidation number** (ON) is a formal parameter used to assess the extent to which an atom has gained or lost electrons on formation of a compound. The name is derived from the identification of oxidation with electron loss, and the greater the oxidation number the greater the extent to which electrons have been lost from the atom. Reduction corresponds to electron gain, and the oxidation number of an element is decreased when it is reduced chemically. An element with a particular oxidation number is said to be in that **oxidation state**. Thus, an N atom with oxidation state $+5$ is said to be in its $+5$ oxidation state, and is denoted N(V).

The oxidation numbers ascribed to elements can be illustrated by considering MgO and SO_2. The °ionic compound MgO can be regarded as consisting of Mg^{2+} and O^{2-} ions. As the Mg atom has lost two electrons we ascribe to it the oxidation number $+2$; the O atom has gained two electrons, and so we ascribe to it the oxidation number -2. We treat the covalent compound SO_2 as though the electron transfer is complete (even though it is not), and *formally* regard the molecule as $\{O^{2-}S^{4+}O^{2-}\}$; thus we ascribe the oxidation number $+4$ to the S atom and -2 to each O atom.

Rules have been devised (Box O.5) that extend the concept of oxidation number to compounds that do not contain oxygen, just as the

---

Box O.5 Oxidation numbers

Work through the following rules in the order given. Stop as soon as the oxidation number has been assigned.

- The sum of the oxidation numbers of all the atoms in the molecule or ion is equal to its total charge.
- For atoms in their elemental form, $ON = 0$.
- For elements of

  Group I/1: $ON = +1$

  Group II/2: $ON = +2$

  Group III/13 (other than boron): $ON = +3$ for M^{3+}

  $\qquad\qquad\qquad\qquad\qquad\qquad\quad = +1$ for M^+

  Group IV/14: $ON = +4$ for M^{4+}

  $\qquad\qquad\qquad +2$ for M^{2+}

- *For hydrogen*, $ON = +1$ when combined with a nonmetal

  $\qquad\qquad\qquad\qquad -1$ when combined with a metal

- For fluorine, $ON = -1$ in all its compounds.
- For oxygen, $ON = -1$ in peroxides (O_2^{2-})

  $\qquad\qquad\qquad -\frac{1}{2}$ in superoxides (O_2^-)

  $\qquad\qquad\qquad -\frac{1}{3}$ in ozonides (O_3^-)

  $\qquad\qquad\qquad -2$ unless combined with F

modern concept of 'oxidation' is independent of the presence of oxygen. The feature to bear in mind in general is that the oxidation number is a parameter that summarizes the *exaggeration* of the relocation of electrons in a compound compared to the elements. The exaggeration takes the form of pretending that the atoms do indeed become the ions that their relative electronegativities would lead us to expect, with F becoming F^- in fluorides, and with H becoming H^+ (even in such typically covalent compounds as CH_4).

The usefulness of the oxidation number lies in our ability to use it to recognize oxidation and reduction and to correlate the chemical properties of a substance (for example, the acidic or basic character of an oxide or its likely potency as an oxidizing or reducing agent).

Further information

The oxidation number of elements is treated exhaustively in elementary chemistry, so see Atkins (1989) for an introduction and Shriver, Atkins, and Langford (1990) for some of the chemical correlations of properties and oxidation numbers. One method for the determination of oxidation numbers is explained by Kauffman (1986) and some of the limitations of the concept are described by Woolf (1988). Another assessment of the relocation of electrons (one based on an exaggerated covalent picture in contrast to the exaggerated ionic picture of oxidation states) is the °formal charge on an atom.

p orbital

A **p orbital** is an °atomic orbital with $l = 1$. An electron in a p orbital is called a **p electron**, and has an °orbital angular momentum of magnitude $2^{1/2}\hbar$. Each p subshell of an atom consists of *three* orbitals, corresponding to the three values of m_l (+1, 0, and −1) that are permitted when $l = 1$.

A p orbital has one angular node, which separates the wavefunction into two lobes of opposite sign (Fig. P.1). Since the node passes through the nucleus, there is zero probability of a p electron being found there. In this respect a p orbital differs sharply from an °s orbital, which can be found at the nucleus. The difference can be traced to the effect of the orbital angular momentum, which generates such a strong centrifugal force at small distances from the nucleus (the force varies as $1/r^3$) that it overcomes the Coulombic attraction (which varies as $1/r^2$).

The radial dependence of p orbitals is depicted in Fig. P.2 and given explicitly in Box P.1. In all cases the amplitude is zero at the nucleus,

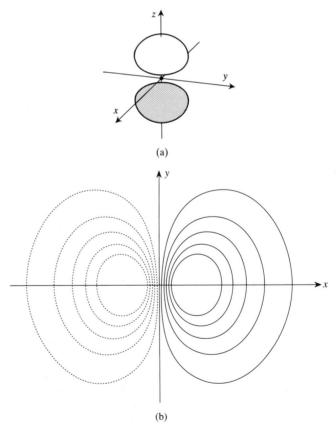

Fig. P.1 (a) The boundary surface of a p orbital (in this case, p_z). The p_x and p_y orbitals differ only in their orientations. (b) Contours of equal amplitude of a hydrogen 2p orbital.

Box P.1 *p* orbitals

$\psi = R(r)\, Y(\theta, \phi)$

Complex polar form:

$m_l = 0,$ $Y = \left(\dfrac{3}{4\pi}\right)^{1/2} \cos\theta$

$m_l = \pm 1,$ $Y = \pm\left(\dfrac{3}{8\pi}\right)^{1/2} \sin\theta\, e^{\pm i\phi}$

Real cartesian form:

p_x $rY = \left(\dfrac{3}{4\pi}\right)^{1/2} x$

p_y $rY = \left(\dfrac{3}{4\pi}\right)^{1/2} y$

p_z $rY = \left(\dfrac{3}{2\pi}\right)^{1/2} z$

Radial wavefunctions:

$n = 2$ $R = \dfrac{1}{2\sqrt{6}}\left\{\dfrac{Z}{a_0}\right\}^{3/2} \rho\, e^{-\rho/2}$ $\rho = Zr/a_0$

$n = 3$ $R = \dfrac{1}{9\sqrt{6}}\left\{\dfrac{Z}{a_0}\right\}^{3/2} (4-\rho)\, e^{-\rho/2}$ $\rho = 2Zr/3a_0$

$n = 4$ $R = \dfrac{1}{32\sqrt{15}}\left\{\dfrac{Z}{a_0}\right\}^{3/2} (20 - 10\rho + \rho^2)\, e^{-\rho/2}$ $\rho = Zr/2a_0$

Average values:

$\langle r \rangle = \dfrac{n^2 a_0}{Z}\left\{\dfrac{3}{2} - \dfrac{1}{n^2}\right\}$ $\langle 1/r \rangle = \dfrac{Z}{a_0 n^2}$

See also Box H.7 for general expressions.

and then oscillates through $n-2$ radial °nodes before decaying exponentially towards zero at large distances.

The orbitals with the two non-zero m_l values are °complex wavefunctions. It is conventional, however, to represent the orbitals by the boundary surfaces of their real linear combinations and to label them according to the axis along which they have maximum amplitude. Thus, the $m_l = 0$ orbital is real and proportional to $\cos\theta$ (where θ is the colatitude in °polar coordinates). Its maximum amplitude occurs for $\theta = 0$ and $180°$, which is on the z axis; hence the $m_l = 0$ orbital is also called the p_z orbital. The p orbitals with $m_l = \pm 1$ are proportional to $\sin\theta$, and hence both have their maximum amplitudes in the equatorial plane (for which $\theta = 90°$). One linear combination is proportional to $\sin\phi$, and hence has its maxima along the y axis; it is therefore called a p_y orbital. The other is proportional to $\cos\phi$, and hence is called a p_x orbital.

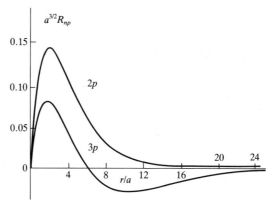

Fig. P.2 The radial functions of the 2*p* and 3*p* orbitals of a hydrogen atom. Note that the *p* orbitals of successively higher shells have one additional radial node.

Apart from their orientations, the three *p* orbitals of a shell are identical, and can be rotated into each other. They are therefore °degenerate in free atoms. In general, and for reasons connected with °penetration and shielding effects, the *p* orbitals of a many-electron atom have a higher energy than the *s* orbitals of the same shell.

Further information

See the entries on °atomic orbitals for references to further information. The representation of complex orbitals (which shows the relation between the orbitals and the value of m_l most clearly) is described by Breneman (1988). See also Gerloch (1986).

π orbital

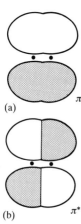

Fig. P.3 (a) A bonding π orbital between two atoms and (b) its antibonding counterpart with an additional internuclear nodal surface.

A **π orbital** is a molecular orbital that, when viewed along the internuclear axis, resembles a *p* orbital. The orbital has two lobes of opposite sign separated by a nodal surface that passes through the internuclear axis (Fig. P.3). Two electrons in a π orbital constitute a **π bond**.

The most common examples of π orbitals are formed from the overlap of *p* orbitals on adjacent atoms (Fig. P.3). The in-phase combination of the *p* orbitals gives rise to a bonding π orbital and the out-of-phase combination gives an antibonding π orbital, which is often denoted π*. The electron configuration $\sigma^2\pi^2$ is the molecular orbital version of a **double bond** (Fig. P.4a). The configuration $\sigma^2\pi^4$ (where the two pairs of π electrons occupy two perpendicular π orbitals formed from p_x and p_y overlap respectively) is a **triple bond** (Fig. P.4b).

It is difficult to predict the relative strengths of single and multiple bonds because the formation of a π bond is accompanied by a shortening of the internuclear distance, which affects the strength of

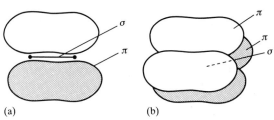

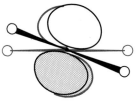

(a)   (b)

Fig. P.4 The orbital composition of (a) a double bond (one σ bond and one π bond) and (b) a triple bond (one σ bond and two π bonds).

Fig. P.5 A π bond brings a degree of torsional rigidity to a molecule, because the overlap of the component p orbitals is decreased if the molecule is twisted.

the σ bond. For example, the enthalpies of CC and NN bonds (in kJ mol^{-1}) vary as follows:

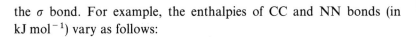

	X—X	X=X	X≡X
CC	348	612	837
NN	163	409	944

The CC triple bond is weaker than three CC single bonds but the NN triple bond is substantially stronger than three NN single bonds.

A π bond is **torsionally rigid** in the sense that it opposes twisting around the internuclear axis. The explanation lies in the decrease in π overlap that occurs as one p orbital is rotated relative to the other (Fig. P.5) and the corresponding rise in energy as the strength of the bond decreases.

Further information

See the entry on °molecular orbitals for more information and further references to the role of π orbitals in molecular structure. Coulson (1970) deals with them explicitly, and DeKock and Gray (1980) give a straightforward introduction. For further information, and contour plots of π orbitals, see Webster (1990).

Pairing

Two electrons are **paired** if their spins are antiparallel (one α and the other β) and their total spin angular momentum is zero (see °singlet and triplet states for a vector diagram).

Electron pairs are of prime importance in the theory of the chemical °bond, and it is sometimes said that bond formation reflects the tendency of electrons to pair. However, that remark is shorthand for the fact that, according to the °Pauli exclusion principle, two electrons must pair if they are both to enter a bonding molecular orbital. The role of electron pairing is different when separate orbitals are occupied by two electrons, for then there is no need for them to pair, and according to °Hund's rule of maximum multiplicity, the lowest energy is achieved if the electrons do not pair.

Further information

See °Lewis structures for the emergence of the realization of the importance of the electron pair in bonding, and the entries on the chemical °bond and °molecular orbital theory for background information. The °valence bond theory of molecular structure captures the essence of Lewis's electron pair approach by focusing on electron pairs from the outset and develops the approach in quantum mechanical terms. That two paired electrons have strictly antiparallel spins but two electrons with 'parallel' spins make an acute angle to each other is explained in the entry on °singlet and triplet states.

Parity

See °*gerade* and *ungerade* (g and u).

Particle in a box

A 'box' is a region of space in which the potential energy of a particle is zero but which rises abruptly to infinity at the walls (Fig. P.6). The abrupt rise is characteristic of a **square well potential**, in which the 'squareness' refers to the perpendicular rise of the potential, not the geometrical shape of the box. The box may be one-dimensional, in which the particle is free to move between 0 and L along the x axis. A two-dimensional box is a plane surface of any geometrical shape. Three-dimensional square wells include the cubic box and the spherical cavity (a 'spherical square well').

In each case the energy of the particle arises entirely from its °kinetic energy inside the box. Because the particle is confined, its energy is °quantized. The permitted energies (which are set out in Box P.2 for various box shapes) are obtained from the °Schrödinger equation and the boundary conditions characteristic of the well (that the wavefunction must be zero at each wall). Within a rectangular box the permitted functions are the same as free-particle functions, but for them to fit between 0 and L their wavelengths must satisfy $n \times \frac{1}{2}\lambda = L$ with $n = 1,2,\dots$ It follows from the °de Broglie relation that the momentum of the confined particle is limited to the values $p = nh/2L$, and hence that the kinetic energy, $p^2/2m$, of the particle is $n^2(h^2/8mL^2)$, as given in Box P.2. The wavefunctions are sine waves of wavelength $\lambda = 2L/n$ (Fig. P.7).

The properties of a particle in a one-dimensional square well can be summarized as follows:

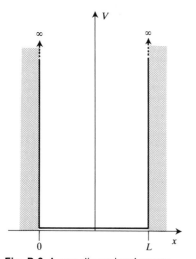

Fig. P.6 A one-dimensional square well potential characteristic of a particle in a box.

● The energies are determined by a single quantum number n and are proportional to n^2 (see Box P.2), with $n = 1,2,\dots$

Box P.2 Particles in boxes

For a **one-dimensional infinite square well**,

$$E_n = \frac{n^2 h^2}{8mL^2} \qquad n = 1, 2, \ldots$$

$$\psi_n = \left\{\frac{2}{L}\right\}^{1/2} \sin \frac{n\pi x}{L}$$

For an **N-dimensional infinite square well**,

$$E_{n_1, n_2 \ldots n_N} = \frac{h^2}{8m}\left\{\frac{n_1^2}{L_1^2} + \frac{n_2^2}{L_2^2} + \ldots + \frac{n_N^2}{L_N^2}\right\} \qquad \text{all } n_i = 1, 2, \ldots$$

$$\psi_n = \left\{\frac{2^N}{L_1 L_2 \ldots L_N}\right\}^{1/2} \sin \frac{n_1 \pi x}{L_1} \sin \frac{n_1 \pi y}{L_2} \ldots \sin \frac{n_N \pi z}{L_N}$$

For a **spherical cavity** of radius a:

$$E_{n, l} = \frac{h^2 \rho_{n, l}^2}{8ma^2} \qquad n = 1, 2, \ldots \text{ and } l = 0, 1, \ldots$$

where the first few ρ have the following values:

l	$n=$	1	2	3	4	5
0		1.0000	2.0000	3.0000	4.0000	5.0000
1		1.4303	2.1590	3.4709	4.4775	5.4816
2		1.8346	2.8950	3.9226	4.9385	5.9189

The wavefunctions are

$$\psi_{n, l, m_l} = N_n j_l (kr) Y_{l, m_l} (\theta, \phi) \qquad k = \pi \rho / a$$

where the j are spherical Bessel functions.

- There is an irremovable °zero point energy of magnitude $h^2/8mL^2$.

- All the states are non-degenerate.

- The lighter the particle, the greater the energy level separation.

- The separation of the energy levels increases as the walls become more confining.

- The wavefunctions are sine functions of wavelength $L/2n$ inside the box and are zero everywhere outside the box.

The zero-point energy corresponds to a state in which the particle is most probably near the centre of the box, and furthest from the walls. At higher quantum numbers the probability density is spread more evenly throughout the region.

Since the wavelength must decrease as the width of the box is decreased (so that the wavefunction fits between the walls, Fig. P.8), the kinetic energy also increases as L is decreased. In the limit of infinite width (effectively, no walls), the levels form a continuum and the energy is no longer quantized.

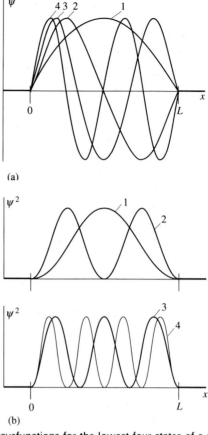

(a)

(b)

Fig. P.7 (a) The wavefunctions for the lowest four states of a particle in a square well. (b) The probability distributions for the same four states. Note how the probability becomes more uniformly distributed over the length of the well as the quantum number increases.

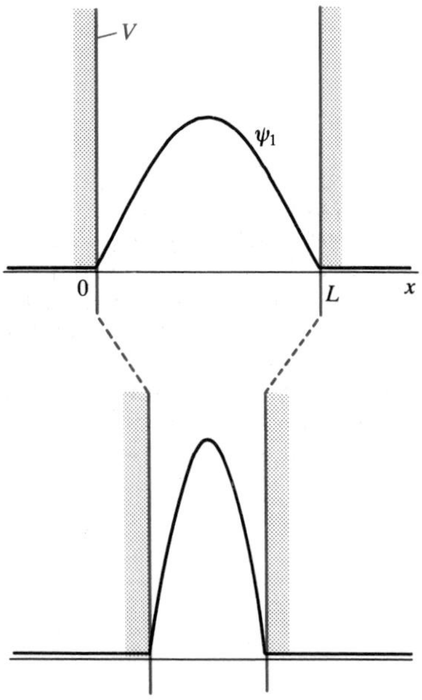

Further information

See Chapter 3 of MQM for detailed solutions of the Schrödinger equation for rectangular boxes. The solutions for a spherical cavity will be found in Chapter 9 of Das and Melissinos (1986) and in Davydov (1976). The book by Kauzmann (1957) is still worth consulting in this connection. The other two references (Das and Melissinos, and Davydov) are good sources of discussion of wells with finitely high walls. See also Johnson and Pedersen (1973). The wavefunctions for finite wells can also be found using the software compiled by Ayscough (1990); see also Albeverio (1988). For an account of the relativistic square well see Gumbs and Kiang (1986) and Coulter and Adler (1971).

Paschen–Back effect

The **Paschen–Back effect**, which was reported by F. Paschen and E. Back in 1921, is the decoupling of spin and orbital angular momenta

Fig. P.8 If a well is compressed, the curvature of the wavefunction is increased, which corresponds to an increase in the kinetic energy (and hence the total energy) of the particle confined in the well.

by an applied magnetic field. The effect is observed when very high fields are used in the °Zeeman effect. Because an electromagnetic field interacts with the spatial distribution of the electrons, in the uncoupled system it causes transitions without affecting the spin. As a result, the anomalous Zeeman effect characteristic of atoms with both spin and orbital angular momenta is replaced in the uncoupled system by the normal Zeeman effect of atoms without spin.

The Paschen–Back effect may be expressed in terms of the °vector model of the atom. The °spin and °orbital angular momenta of an atom are coupled together by °spin–orbit coupling and form a resultant $J = L + S$. Both components °precess around their resultant, and J also precesses about the direction of the applied field (Fig. P.9a). However, as the applied field is increased its interaction with the spin and orbital °magnetic moments becomes so great that it begins to compete with the spin–orbit coupling. Now S and L begin to precess independently around the direction of the field, and the spin–orbit coupling is broken. If the field is made sufficiently strong (of the order of 1 T) the competition is resolved in its favour, and the state is one of almost pure precession of each type of momentum about the field's direction (Fig. P.9b).

Further information

See the entry on the °Zeeman effect for the context in which the Paschen–Back effect was first recognized; see also MQM Chapter 9. The spectroscopic aspects of the effect are described by Herzberg (1944) and Kuhn (1962); see Condon and Odabaşi (1980) for full details.

Pauli exclusion principle

The **Pauli exclusion principle** states:

> No more than two electrons may occupy one orbital, and if two electrons are present in an orbital they must be °paired.

The Pauli exclusion principle is of immense importance in chemistry, for it underlies the °building-up principle used to account for the electron configurations of atoms and molecules. It is also the fundamental reason for the importance of the electron pair in the formation of chemical °bonds. As explained in the entry on °intermolecular forces, the exclusion principle also accounts for the fact that matter does not blend together into a homogeneous, undifferentiated continuum, but consists of discrete atoms and molecules.

The Austrian physicist Wolfgang Pauli (1900–1958) proposed the exclusion principle in 1924 (in terms of Bohr orbits and the old

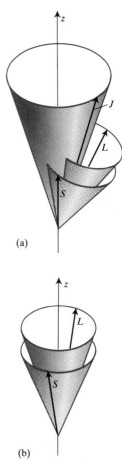

Fig. P.9 (a) The precession of the orbital and spin angular momenta about their resultant (by spin–orbit coupling) and the precession of the resultant about a weak magnetic field along the z-axis. (b) When the field is very strong (as in the Paschen–Back effect), the spin and orbital momenta precess individually about it: the momenta have been uncoupled from each other.

quantum theory) to account for the absence of a number of expected lines in the spectrum of helium. Later it was realized that the exclusion principle is a consequence of a very deep principle of nature, the °Pauli principle, which requires the overall wavefunction to change sign under pairwise interchange of identical fermions.

The connection between the Pauli principle and the Pauli exclusion principle can be appreciated as follows. Suppose that two electrons did both occupy the same orbital ψ with the same spin α. Their joint wavefunction would then be written $\psi^{\alpha}(r_1)\psi^{\alpha}(r_2)$. However, relabelling the electrons changes this wavefunction to $\psi^{\alpha}(r_2)\psi^{\alpha}(r_1)$, which is the same as before apart from the order of the factors. Since the wavefunction does not change sign, the Pauli principle forbids it to occur. Hence, an atom or molecule cannot have two electrons with the same spin in the same orbital.

°Bosons are not subject to an exclusion principle, and any number of bosons can occupy the same state. For example, a large number of °photons may occupy the same state and give rise to an intense, monochromatic light beam.

Further information

The origin of the original version of the exclusion principle will be found in Jammer (1966). The role of the exclusion principle in atomic structure is described in the entry on °building-up principle, and its role in molecular structure is described in the entries on °molecular orbital theory and °intermolecular forces; further references will be found in those entries. Wavefunctions of many-electron systems satisfy the exclusion principle if they are written as °Slater determinants. For the populations of states by particles that must satisfy the exclusion principle, see the entry on the °Fermi–Dirac distribution.

Pauli principle

The **Pauli principle** states that:

> The total wavefunction of a system composed of identical fermions (bosons) must be antisymmetrical (symmetrical) under particle interchange.

An antisymmetrical function is one that reverses its sign under the stated operation. For example, when x is replaced by $-x$ a function f is antisymmetrical if $f(-x) = -f(x)$. A symmetrical function remains unchanged: $f(-x) = f(x)$. A °fermion is a particle with half-integral spin (such as an electron) and a °boson is a particle with integral spin (such as a photon). The °Pauli exclusion principle is a direct consequence of the Pauli principle applied to electrons (which are fermions).

The principle can be seen to be plausible by noting that if two particles are indistinguishable, then when they are interchanged the calculated properties of the system must remain unchanged. In particular, since the particle density is proportional to ψ^2 (see °wavefunction), when two indistinguishable particles are interchanged (so that the labels 1 and 2 are exchanged everywhere that they appear in the wavefunction) the particle density, and hence ψ^2, must not change. This implies that ψ itself can either change sign when the particles are interchanged, or not change sign. That for bosons $\psi \to \psi$ whereas for fermions $\psi \to -\psi$ is consistent with this requirement (but not, of course, explained by it).

Some inkling (but no more than that) of the reason for the distinction can be obtained by noting the analogy between interchanging identical fermions (which leaves a state *apparently* unchanged but actually changes its sign) and the rotation of a spin-$\frac{1}{2}$ particle through $360°$ (which changes the sign of its spin state). There appears to be a general requirement that actions which for spin-1 particles restore original states completely (such as $360°$ rotations and particle interchanges), must be applied *twice* to recover the initial state of a spin-$\frac{1}{2}$ system. Pauli found a theoretical basis for this aspect of the principle in 1940, when he showed that it stemmed from considerations of the requirements of relativistic invariance.

Further information

The relativistic origins of the Pauli principle are explored in Pauli (1940); this paper is present in the volume edited by Schwinger (1958). The reason why the Pauli principle has given rise to less philosophical disquiet than the °uncertainty principle is the subject of a reflective article by Margenau (1944) which has been added to by Jammer (1966, 1974). For the implementation of the Pauli requirement of antisymmetrization see McWeeny (1989), Szabo and Ostlund (1982), and Hehre *et al.* (1986). The entry on °Slater determinants is also relevant.

Penetration and shielding

By **penetration** is meant the probability that an electron will be found inside the region of space occupied by an atom's core. **Shielding** is the reduction in the net strength of the nuclear attraction for an electron by repulsions from other electrons in an atom, particularly the electrons of the core. Penetration and shielding jointly affect the energies of electrons in atoms and result in them having energies that differ from those in hydrogenic atoms. In particular, shielding and penetration result in an *ns* orbital having a lower energy than the *np* orbitals of the same shell.

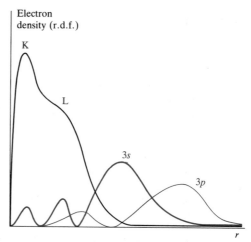

Fig. P.10 The radial distribution functions of the 3*s* and 3*p* orbitals of a sodium atom superimposed on the core electron density. Note how the 3*s* orbital gives a greater probability for the electron to be at short distances from the nucleus than does the 3*p* orbital.

As the amplitude of an $^{\circ}s$ orbital is non-zero at $r=0$, an *s* electron may be found at the nucleus. In contrast, a $^{\circ}p$ orbital has a node at the nucleus, and is very unlikely to be found close to it (Fig. P.10). The physical reason for the difference is that a *p* electron has orbital angular momentum, and hence is prevented from approaching the nucleus by the centrifugal force arising from its motion. An *s* electron has no orbital angular momentum, is not flung away from the vicinity of the nucleus, and hence may penetrate the inner core of electrons.

A spherical charge distribution can be modelled as a point charge at its centre. Moreover, only charge density inside a sphere defined by the distance of the point from the nucleus contributes to the net force at the point (Fig. P.11). Thus, an electron that is typically found at a small distance from the nucleus experiences a smaller repulsion from the core electrons than one found largely outside the core. That is, an electron that penetrates appreciably through a core experiences only a lightly shielded nucleus, but one that penetrates only little experiences strong shielding from the core electrons.

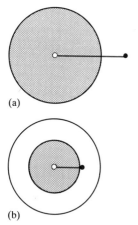

Fig. P.11 (a) The field experienced at a point outside a spherical distribution of charge is equivalent to that generated by a point charge of equal magnitude at the centre of the distribution. (b) If the point is inside the spherical distribution of charge, the field is the same as that of a point charge of magnitude equal to the total charge inside the spherical surface defined by the point.

Shielding reduces the effective nuclear charge experienced by the electron from Ze to $Z_{\text{eff}}e$. The **effective atomic number** Z_{eff} is normally written $Z-\sigma$, where Z is the true atomic number and σ is a **screening constant** (Table P.1). It follows from the effects of penetration and shielding that the effective atomic number for an *s* electron is greater than that for a *p* electron of the same shell. Consequently, an *s* orbital has a lower energy in a many-electron atom than a *p* orbital of the same shell. Similarly, a *p* orbital has a lower energy than a *d* orbital

Table P.1 Effective atomic numbers, Z_{eff}

	H							He
Z	1							2
$1s$:	1							1.69

	Li	Be	B	C	N	O	F	Ne
Z	3	4	5	6	7	8	9	10
$1s$:	2.69	3.68	4.68	5.67	6.66	7.66	8.65	9.64
$2s$:	1.28	1.91	2.58	3.22	3.85	4.49	5.13	5.76
$2p$:			2.42	3.14	3.83	4.45	5.10	5.76

	Na	Mg	Al	Si	P	S	Cl	Ar
Z	11	12	13	14	15	16	17	18
$1s$:	10.63	11.61	12.59	13.57	14.56	15.54	16.52	17.51
$2s$:	6.57	7.39	8.21	9.02	9.82	10.63	11.43	12.23
$2p$:	6.80	7.83	8.96	9.94	10.96	11.98	12.99	14.01
$3s$:	2.51	3.31	4.12	4.90	5.64	6.37	7.07	7.76
$3d$:			4.07	4.29	4.89	5.48	6.12	6.76

Data: Clementi and Raimondi (1963).

of the same shell, since a d orbital penetrates even less than a p orbital, and so is even more shielded from the nucleus.

Further information

Penetration and shielding lie at the heart of the °building-up principle for atoms, so see that entry and the texts—mostly of inorganic chemistry—that pursue the connection, such as Shriver, Atkins, and Langford (1990) and Puddephatt and Monaghan (1986). Screening constants can be estimated using the rules devised by °Slater for the construction of his orbitals; alternatively they can be calculated more precisely from self-consistent field orbitals, as described in the same entry. The articles by Lloyd (1986) and Rich and Suter (1988) are relevant in this connection.

Perturbation theory

Most systems of interest are described by °Schrödinger equations that are too difficult to solve exactly. However, inspection often reveals that a simpler, solvable system closely resembles the true, intractable system. If the °wavefunctions and energies of this simpler system are known it is possible to adjust them so that they are distorted in the direction of the true wavefunctions and energies. If the true system resembles the simpler system very closely, the amount of distortion required is only a very small **perturbation** of their form. The modification of the simple wavefunction can be achieved by mixing into it other wavefunctions of the simple system in the appropriate proportions, and perturbation theory provides the recipe for the mixture. At the same time, perturbation theory shows how to calculate the additional terms that must be added to the energy of the simple system to yield the energy of the true system.

Time-independent perturbation theory deals with perturbations that are constant in time, such as the effect of a steady electric field on an atom. **Time-dependent perturbation theory** is concerned with perturbations that vary with time. For example, a perturbation may be turned on, or it may oscillate.

We shall illustrate some of the ideas of perturbation theory by considering a system composed of two states with wavefunctions ψ and ψ'. The energies of the two states are E and E' respectively. (We shall deal only with perturbation theory for non-degenerate states; special precautions must be taken when the energies of different states coincide.) An example is the distortion of the 1s orbital of an H atom by an applied electric field (the °Stark effect) in which we model the distortion (the °polarization of the atom) by mixing the 2p_z orbital into the ground state (Fig. P.12). A full treatment would include all the orbitals of the atom, and would build up an increasingly accurate version of the true distortion by including more and more orbitals in the superposition. The explicit expressions for the general case are collected in Box P.3. In **first-order perturbation theory** the corrected wavefunction is approximated by a sum $\psi + \lambda\psi'$ of the two unperturbed states of the system, with the mixing coefficient λ given by

$$\lambda = \frac{\text{strength of perturbation}}{\text{separation of energy levels}}$$

We see that the stronger the perturbation (the more intense the electric field), the greater the distortion. The greater the energy separation of the unperturbed system, the smaller the distortion achieved by a given perturbation. Thus, a system with widely spaced energy levels (like the ground state of the H atom) is 'stiff' and unresponsive to perturbations. Conversely, a system with closely spaced energy levels is highly responsive to a perturbation, and might be severely distorted by even a weak perturbation. This dependence on the energy level separation is the underlying reason why classically it is easy for a force to distort a weak spring but not a stiff spring. A weak spring (one with a small °force constant) has closely spaced energy levels, and a perturbation (the force we apply to stretch the spring) can mix the wavefunctions together very strongly: we therefore see the spring stretch appreciably. A stiff spring has widely spaced energy levels, and is much less easily perturbed.

An important feature of perturbation theory is the manner in which the form of the perturbation is impressed on the distorted system. In a sense, the perturbation leaves its footprint on the system. Thus, an electric field in the z direction impresses 'z character' on a spherical

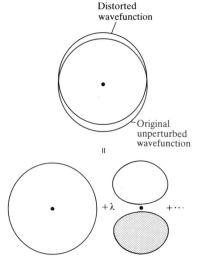

Fig. P.12 The distortion of a spherical atom by an electric field in the z-direction can be modelled by adding some p_z orbital character into the s orbital of the unperturbed atom.

Box P.3 Perturbation theory

In **time-independent perturbation theory**, we suppose that the eigenstates and eigenvalues of the unperturbed hamiltonian $H^{(0)}$ are known:

$$H^{(0)}|n\rangle = E_n|n\rangle$$

and we require the eigenstates and eigenvalues of the hamiltonian

$$H = H^{(0)} + H^{(1)} + H^{(2)}$$

where the $H^{(1)}$ and $H^{(2)}$ are respectively first and second order in a parameter (such as the field strength). We write

$$E'_k = E_k + E_k^{(1)} + E_k^{(2)} + \dots$$

$$|k\rangle' = |k\rangle + |k\rangle^{(1)} + |k\rangle^{(2)} + \dots$$

Then

$$E_k^{(1)} = \langle k|H^{(1)}|k\rangle$$

$$|k\rangle^{(1)} = \sum_n{}' c_n|n\rangle \text{ with } c_n = \frac{\langle n|H^{(1)}|k\rangle}{E_k - E_n}$$

$$E_k^{(2)} = \langle k|H^{(2)}|k\rangle + \sum_n{}' \frac{\langle k|H^{(1)}|n\rangle\langle n|H^{(1)}|k\rangle}{E_k - E_n}$$

In **time-dependent perturbation theory**, where

$$H = H^{(0)} + H^{(1)}(t)$$

we write

$$|k; t\rangle' = \sum_n c_n(t)|n\rangle \, e^{-iE_n t/\hbar}$$

$$c_n(t) = c_n(0) - \frac{i}{\hbar}\sum_m \int_0^t c_m(t')\langle n|H^{(1)}(t')|m\rangle \, e^{i\omega_{nm}t'} \, dt'$$

where $\hbar\omega_{nm} = E_n - E_m$. For short times and weak perturbations, when all states except $|m\rangle$ are empty initially (at $t = 0$),

$$c_n(t) = -\frac{i}{\hbar}\int_0^{t'} \langle n|H^{(1)}(t')|m\rangle \, e^{i\omega_{nm}t'} \, dt'$$

atom by mixing in p_z orbitals; if the field is applied in the x direction, it impresses 'x character' on the atom by mixing in p_x orbitals. The deep reason for this behaviour can be found in °group theory and the vanishing of integrals over the perturbation unless the wavefunctions match the shape of the perturbation.

The first-order correction of the energy of the system is the weighted average of the perturbation over the unperturbed wavefunction. The correction is of the order of the strength of the perturbation itself (see Box P.3). The second-order correction to the energy takes into account the distortion of the wavefunction: the wavefunction is distorted by the perturbation, and then the average value of the perturbation energy is calculated over this distorted wavefunction. The name 'second-order' indicates that the perturbation is involved twice. The second-order correction is of the order of the strength of the perturbation multiplied by the distortion the perturbation has caused in the wavefunction. Since the latter is of the order of λ,

$$\text{second-order energy correction} = \frac{(\text{strength of perturbation})^2}{\text{separation of energy levels}}$$

Once again, we see that a stiff system has an energy that differs only slightly from that of the unperturbed system. A theorem of perturbation theory shows that to calculate the nth order energy it is not necessary to know the wavefunction to better than order† $n-2$.

A helpful classical analogy that illustrates all the points made so far is the modification of the vibration of a violin string by suspending from it a number of small weights. The weights hanging from the nodes affect neither its motion nor its energy, but those hanging from the antinodes (the points of maximum displacement) may have a pronounced effect on its vibrational energy and waveform. The first-order energy correction is found by averaging the effect of the weights (the perturbation) over the wavefunction (in this case the displacement) of the string. The weights also distort the waveform of the string: the nodes are slightly shifted and the pure sinusoidal shape is lost. This distorted mode can be reproduced by taking a suitable superposition of the harmonics of the unladen string. If the weights are not too big only a few of the harmonics need be incorporated to give a good representation of the distortion. The second-order energy correction to the vibrational energy of the string is found by averaging the effect of the weights over the distorted waveform.

Time-dependent perturbation theory is used when the perturbation varies with time. It is used to calculate the °transition probability, the probability that the system will be found in a state other than the initial one. The distortions change with time and the admixture of the excited states of the system may be interpreted as transitions from the ground state to another state. For instance, the probability that the two-level system will be in the state with wavefunction ψ' is equal to λ^2, and in time-dependent perturbation theory λ changes with time. The **transition rate** is the rate of change of probability of being found in the final state, and is equal to the derivative $\mathrm{d}(\lambda^2)/\mathrm{d}t$. An important application is the effect of an electromagnetic field on the system, for that leads to the discussion of the intensities of spectral transitions (see °transition probability).

---

† "Really, if the lower orders don't set us a good example, what on earth is the use of them?" (Oscar Wilde, *The Importance of Being Earnest*, Act 1.)

Further information

See Chapter 8 of MQM for a description of perturbation theory and other techniques of approximation, such as the °variation principle. The mathematics of modern perturbation theory are described in some detail by Hirschfelder, Byes Brown, and Epstein (1964) and very thoroughly in the volume edited by Wilcox (1966). Both these references explain the differences between 'Rayleigh–Schrödinger perturbation theory' (the scheme set out in Box P.3) and the 'Wigner–Brillouin perturbation theory'. The convergence (or not) of a perturbation series to the exact energy is a frisky problem, and can be explored in Wilcox (1966). See Langhoff, Epstein, and Karplus (1972) for a survey of time-dependent perturbation theory. Bransden and Joachain (1989) present perturbation in a clear, straightforward manner, with plenty of illustrations. The utilisation of many-body perturbation theory (often called 'Møller-Plesset perturbation theory') to improve SCF calculations of wavefunctions is currently of considerable interest: see Wilson (1984) for a survey and Møller and Plesset (1934) for the origin. For the diagrammatic perturbation theory of the electrodynamic properties of molecules see Craig and Thirunamachandran (1984).

Phonon

The energy of a collective vibrational mode of frequency v of a solid lattice is an integral multiple of hv above its °zero-point energy. The excitation energy nhv can be interpreted as signifying the presence of n particles, each of energy hv. These hypothetical particles are the **phonons** of the lattice. A phonon is the mechanical analogue of a °photon.

Phonons of different frequency exist in a solid (just as in the electromagnetic field there are photons of different frequencies). As in the case of electromagnetic radiation, the phonons in a solid may be polarized; as well as **transverse phonons**, in which the lattice atoms are displaced perpendicular to the propagation direction, it is also possible to have **longitudinal phonons**, in which the displacement is along the direction of propagation, as in a sound wave.

A further classification of phonons applies in an ionic lattice where the collective motions of the cation and anion sublattices may be in-phase or out-of-phase with each other. If all the components of the unit cell do not move relative to each other, the vibration is an **acoustic mode**. All other vibrations are **optical modes**. The latter name arises from the modulation of the lattice dipole moment during the out-of-phase movements of the two sublattices, and the fact that electromagnetic radiation can interact with this oscillating dipole and so stimulate that particular mode of lattice vibration.

Further information

For a simple introduction to phonons see Jennings and Morris (1974), Hall (1974), and Anderson (1984). For more information see Cochran (1973) and

Reissland (1973). Stiffer descriptions are given by Kittel (1986, 1987) and Kittel and Kroemer (1980): these accounts describe the role of phonons in the electrical, thermal, and optical properties of solids. The authoritative monograph in this field is that by Maradudin (1971).

Phosphorescence

When a **phosphorescent material** is illuminated, it emits radiation that may persist for an appreciable time after the stimulating illumination has been removed. This persistence distinguishes phosphorescence from °fluorescence, for in the latter the emission ceases as soon as the exciting radiation is removed. The fundamental distinction between phosphorescence and fluorescence, however, is that the former involves a change in the multiplicity of the excited state but the latter does not.

The mechanism of phosphorescence can be understood by considering the excitation of a molecule from its singlet (spin paired) ground state S_0 to an upper singlet state S_1. The electronic excitation is accompanied by vibrational excitation (see Fig. P.13), and the excess vibrational energy is lost non-radiatively to the surrounding molecules. If the vibrational deactivation is not too fast and there is a nearby triplet (spin parallel) state, the molecule may convert to that state under the influence of the spin–orbit coupling (see °intersystem crossing). After the intersystem crossing has occurred, the vibrational deactivation continues down the ladder of levels belonging to the triplet state.

At the foot of the ladder the molecule is trapped. It cannot radiate its electronic energy and drop to the ground state because that involves a singlet–triplet transition (which is forbidden by the °selection rules). It cannot climb back to the crossing point and then step down the singlet ladder because the collisions with the lattice cannot supply enough energy. It cannot give up its electronic energy to the surrounding molecules by a radiationless transition because we have supposed that even the vibrational deactivation is weak, and that involved a smaller energy (which is easier to remove).

However, the °singlet–triplet transition is not strictly forbidden. The fact that intersystem crossing has occurred implies that there is enough °spin–orbit coupling present to break down the singlet–triplet selection rule, and so this transition becomes weakly allowed. As the $T_1 \rightarrow S_0$ transition is only weakly allowed it is slow, and it may persist even after the illumination has ceased. We can predict from Fig. P.13 that the wavelength of the emitted light should be longer (further into the red) than fluorescent emission because the lowest vibrational level of the triplet lies below that of the lowest excited singlet.

Phosphorescence may occur if there is a suitable triplet state in the

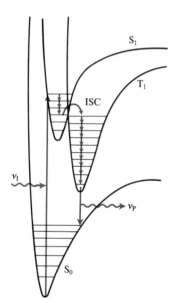

Fig. P.13 The mechanism of phosphorescence. Absorption take the molecule to the excited singlet state; as its vibrational excitation is discarded it undergoes an intersystem crossing into a triplet state. Nonradiative vibrational relaxation continues, to leave the molecule trapped in the triplet state. However, it can undergo a weak radiative transition to the ground singlet state because the selection rules are broken by the spin–orbit coupling.

vicinity of the excited singlet states of the molecule and if there is a sufficiently strong spin–orbit coupling to induce intersystem crossing. In the **heavy atom effect** the intersystem crossing is enhanced by that atom's strong spin–orbit coupling. There must also be enough time for the molecule to cross from one curve to the other, which means that the vibrational deactivation must not proceed so fast that the molecule is quenched and taken below the point where the curves intersect before intersystem crossing has time to occur. It is for this reason that many molecules that fluoresce in a fluid solution phosphoresce when they are trapped in a solid lattice.

It is not impossible for some molecules to clamber back into the singlet state S_1, and to fluoresce into the ground singlet state. In this process of **delayed fluorescence** the triplet state acts merely as a reservoir. Similarly, triplet–triplet collisions can repopulate the excited singlet state of one of the collision pair at the expense of the other, and the newly revived singlet state can decay radiatively.

Further information

See Chapter 12 of MQM for an introduction to the quantum mechanical description of the mechanism, and McGlynn, Azumi, and Kinoshita (1969) for details. Kasha (1984) has given a very interesting historical account of the emergence of the view that a triplet state is involved. Phosphorescence pathways are often displayed on a °Jablonski diagram. For chemical applications see Wayne (1988) and Barltrop and Coyle (1975, 1988).

Photoelectric effect

The **photoelectric effect** is the ejection of electrons when electromagnetic radiation falls on a solid surface. The ejected electrons are called **photoelectrons** and their ejection is called **photoejection**. Three observations are very important:

- Photoejection occurs only if the frequency of the radiation exceeds a certain minimum which is characteristic of the solid.

- Instantaneous photoejection occurs even at very low intensities.

- The kinetic energy of the photoelectrons is linearly proportional to the frequency of the incident radiation.

The quantum theory and its concept of °photons can explain all the features of the effect simply and directly. It recognizes that radiation of frequency v consists of photons of energy hv. When a photon strikes the surface it can eject an electron by imparting all its energy to it, but the ejection is successful only if the energy transferred hv is sufficient to overcome the **work function** Φ, the energy required to remove an electron from the solid (the analogue of the °ionization energy of a molecule). If $h\nu$ is less than the work function the electron will not be ejected, and the photon re-emerges from the surface as part of the reflected beam.

The intensity and time characteristics are explained on the same basis, for so long as the frequency exceeds the threshold value of Φ/h, the photon is able to eject an electron: the effect depends on single photon–electron collision events rather than the accumulation of energy from a passing wavefront. At low intensities (few photons) only a few collisions occur, but each photon carries the same energy $h\nu$ as the photons in a heavily populated intense beam of the same frequency.

The third point can be explained on the grounds that a successful photon is annihilated in the collision that ejects the electron, and, by the conservation of energy, all the energy of the photon must be transferred to the electron. Because an energy Φ is expended in removing the electron from the solid, the remainder $h\nu - \Phi$ must appear as the kinetic energy of the electron. It follows that the kinetic energy of the electron is proportional to the frequency of the incident light.

Further information

For an instructive account of the photoelectric effect see Bohm (1951) and Jammer (1966), both of whom analyze the classical and quantum versions of the explanation. The implications of the effect for the view that photons exist are explored by Born and Beim (1968). Einstein's classic paper on the photoelectric effect, which ostensibly won him the Nobel Prize, is Einstein (1905); a translated version has been published by Arons and Peppard (1965).

Photoelectron spectroscopy

In **photoelectron spectroscopy** (PES), electrons are ejected from molecules by a high-energy photon (from a short-wavelength source). The kinetic energy of the emitted photoelectron is the energy of the incident photon, $h\nu$, less the °ionization energy of the photoelectron. The latter is identified with the one-electron orbital energy by °Koopmans' theorem. Hence, by measuring the kinetic energies of the photoelectrons, the orbital energies of the molecule may be inferred.

When the radiation source is in the X-ray region each photon carries sufficient energy to eject electrons from atomic cores. This X-ray technique is known as **electron spectroscopy for chemical analysis** (ESCA) or **X-PES**. The energies of inner electron orbitals are largely independent of the state of chemical bonding of the atom, so X-PES spectra can be used to identify the elements present. When ultraviolet radiation is used in **UV-PES**, the photons carry enough energy to eject electrons from the valence orbitals and from the molecular orbitals they form.

Hence, UV-PES gives information about the energies of orbitals responsible for chemical bonding in the molecule.

When photoejection occurs from a valence shell, the resulting cation is left in a non-equilibrium conformation. It responds by bursting into vibration. Since different energies are needed to excite different vibrational states of the ion, the photoelectrons appear with different kinetic energies (Fig. P.14). The vibrational fine structure in the spectrum gives a good indication of the role the ejected electron plays in determining the force field in the original molecule, for a long °progression of lines suggests that the electron had a strong influence on the location of the nuclei. The ejection of an electron from a nonbonding orbital can be expected to leave the force field largely unchanged and the vibrational progression is very brief.

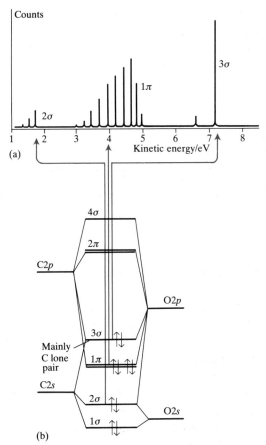

(a)

(b)

Fig. P.14 The uv-photoelectron spectrum of carbon monoxide and its interpretation in terms of the molecular orbital energy level diagram. Note the extensive vibrational structure when the photoelectron is ejected from the $2p\pi$ bonding orbital. The $2p\sigma$ orbital is largely nonbonding on the carbon atom, and loss of an electron from it has little effect on the force field; hence the vibrational structure of the band is very brief.

Further information

Introductions to photoelectron spectroscopy include those by Eland (1983), Ballard (1978), and Baker and Betteridge (1977) and the volumes edited by Brundle and Baker (1977). The chapters in Hollas (1983) and Ebsworth, Rankin, and Cradock (1987) are useful introductions to the technique.

Photon

A **photon** is a packet of electromagnetic radiation. Radiation of frequency v consists of photons of energy hv and linear momentum hc/v. (The latter conclusion stems from the °de Broglie relation $p = h/\lambda$, and $\lambda v = c$.)

It is important to keep distinct the relations between the energy and frequency and between intensity and number of photons. The greater the intensity of a beam of monochromatic radiation, the greater the number of photons it contains. The higher the frequency of the radiation (the shorter its wavelength), the greater the energy of the individual photons. The photon energies for some typical types of radiation are given in Table P.2.

A photon has zero rest mass, travels at the speed of light c, and has unit spin ($s = 1$). Its integral spin identifies a photon as a °boson, and a consequence of this classification is that there is no restriction on the numbers of photons that may occupy a given state (as specified by its frequency or energy, direction of propagation, and polarization).

Table P.2 Colour, frequency, and wavelength of electromagnetic radiation

Colour	Frequency $v/(10^{14}$ Hz)	Wavelength λ/nm	Energy per photon $hv/(10^{-19}$ J)	E/eV
X-rays and γ-rays	10^3 and above	3 and below	660 and above	420 and above
Ultraviolet radiation	10	300	6.6	4.1
Visible light				
Violet	7.1	420	4.7	2.9
Blue	6.4	470	4.2	2.6
Green	5.7	530	3.7	2.3
Yellow	5.2	580	3.4	2.1
Orange	4.8	620	3.2	2.0
Red	4.3	700	2.8	1.8
Infrared radiation	3.0	1000	2.0	1.3
Microwaves and radiowaves	3×10^{-3} and below	1 mm and above	2.0×10^{-3} and below	1.3×10^{-3} and below

Thus, intense, monochromatic, polarized, collimated beams of light may be prepared: this is done in lasers.

Since the photon is a relativistic particle, it has some peculiar properties. Thus, although $s = 1$, its spin can have only two (not three) projections along the direction of propagation, which correspond to the **helicity states** $\sigma = +1$ (clockwise around the direction of travel, Fig. P.15) and $\sigma = -1$ (counter-clockwise). These two helicity states correspond to left and right circularly polarized light respectively. (The electric field of right polarized light rotates clockwise as seen by an observer towards whom the light is travelling, and hence is rotating counter-clockwise around the direction of propagation, Fig. P.15.) Plane polarized light is a superposition of the two helicity states. Because a photon has an intrinsic spin angular momentum, it supplies that angular momentum to an atom or molecule that absorbs it, and removes it from the species that emits it. The conservation of angular momentum in this process is one of the origins of °selection rules.

Other quantum mechanical features of photons include the property that it is impossible to confine the wavefunction of a photon to a region with linear dimensions smaller than about one wavelength. Furthermore, in the sense of the °uncertainty principle, the number of photons in a beam is °complementary to the phase of the wave. That is, if the phase of a light wave is known exactly, nothing can be said about the number of photons present. This restriction is an aspect of the °dual character of electromagnetic radiation: the number of photons is an intrinsically particle property and the phase of the radiation is an intrinsically wavelike property; speaking precisely in the language of one precludes speaking precisely in the language of the other.

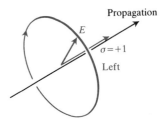

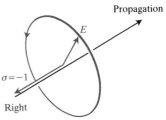

Fig. P.15 The helicity states of a photon are distinguished by the quantum number σ, the projection of the spin on the propagation direction. A photon with $\sigma = +1$ corresponds to left circular polarization and that with $\sigma = -1$ corresponds to right circular polarization. The different helicity states correspond to the different directions of rotation of the electric field vector around the propagation direction.

Further information

For an account of the emergence of the concept of photons, see Kidd, Ardini, and Anton (1989). The quantum mechanical description of electromagnetic radiation can be of virtually boundless complexity, but see Loudon (1979), Louisell (1973), and Goldin (1982). See Jauch and Rohrlich (1955) for a thorough account. The uncertainty relation for the phase of radiation and the number of photons is discussed correctly by Carruthers and Nieto (1968). The study of photons is widely used in analyses of the interpretation of quantum mechanics: see Dirac (1958), Feynman, Leighton, and Sands (1963), and Rae (1986). Are photons particles? See Born and Beim (1968) for a view.

Polar coordinates

The **cartesian coordinates** x, y, z are inappropriate in a system that has cylindrical or spherical symmetry. For these systems it is sensible to use **cylindrical polar coordinates** r, ϕ, z (Fig. P.16a) or **spherical polar**

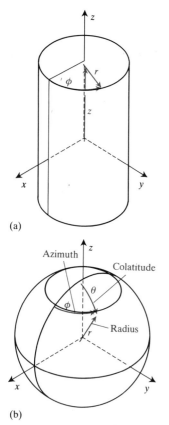

(a)

(b)

Fig. P.16 (a) Cylindrical coordinates and (b) spherical polar coordinates.

coordinates r, θ, ϕ (Fig. P.16b) respectively. The angle ϕ is the **azimuth** of a point and θ is the **colatitude** (the complement of the latitude); r is the **radius**. The relation between the spherical polar, cylindrical polar, and cartesian coordinates are specified in Box P.4.

Further information

Polar coordinates are described in most books that treat the mathematics of physics and chemistry. See, for example, Stephenson (1973), Riley (1974), and Arfken (1985). The book by Kyrala (1967) is particularly helpful, with some excellent diagrams of all kinds of curvilinear coordinates, including elliptic and parabolic cylindrical coordinates, prolate and oblate spherical coordinates, bispherical coordinates, and toroidal coordinates, as well as less abstruse coordinates.

Polarizability

When an electric field is applied to an individual atom or molecule the electron distribution is modified and the molecular geometry is distorted. Atoms and molecules respond to electric fields by acquiring an electric dipole moment (in addition to the one they may already possess) as the centroids of positive and negative charge are displaced. The **polarizability**, α, is the constant of proportionality between the induced dipole moment, μ', and the strength of the electric field, E: $\mu' = \alpha E$. If the applied field is very strong the induced dipole also depends on E^2 and higher powers; the coefficients of the higher powers of E are known as **hyperpolarizabilities**.

The total polarizability of a system can be divided into several contributions. The **atomic polarizability** is the contribution of the geometrical distortion. It is usually significantly smaller than the **electronic polarizability**, which is the contribution from the displacement of the electrons. The explicit formula for this contribution (which is obtained from °perturbation theory) is given in Box P.5. The **orientation polarizability** is the modification of the thermal averaging over the orientations

Box P.5 Polarizability

The **induced dipole moment** μ' in an electric field of strength $\mathscr{E}$ is written

$$\mu' = \mu + \alpha\mathscr{E} + \tfrac{1}{2}\beta\mathscr{E}^2 + \dots$$

where μ is the **permanent electric dipole moment**, α is the **polarizability**, and β is the **first hyperpolarizability**. The **polarizability volume**, α', is equal to $\alpha/4\pi\varepsilon_0$. The polarizability is calculated from

$$\alpha = \frac{2}{3}\sum_n{}' \frac{|\langle 0|\mu|n\rangle|^2}{E_n - E_0}$$

and the **dynamic polarizability**, $\alpha(\omega)$, far from an absorption band is calculated from

$$\alpha(\omega) = \frac{2}{3\hbar}\sum_n{}' \frac{\omega_{n0}|\langle 0|\mu|n\rangle|^2}{\omega_{n0}^2 - \omega^2}$$

The **molar polarization**, P_m, of a fluid medium is

$$P_m = \frac{N_A}{3\varepsilon_0}\{\alpha\mathscr{E} + \mu\mathscr{L}(x)\} \qquad x = \mu\mathscr{E}/kT$$

where $\mathscr{L}(x)$ is the **Langevin function**:

$$\mathscr{L}(x) = \frac{e^x + e^{-x}}{e^x - e^{-x}} - \frac{1}{x}$$

For $x \ll 1$, $\mathscr{L}(x) \approx \tfrac{1}{3}x$.

adopted by a polar molecule in the applied field. Its magnitude is calculated by applying the Boltzmann distribution to determine the mean dipole moment of the sample, and the result is given by the **Langevin function** (Box P.5 and Fig. P.17).

The electronic polarizability of an atom increases with the atom's radius and with the number of electrons it contains. This can be understood in terms of it being easier for a field to distort the electronic

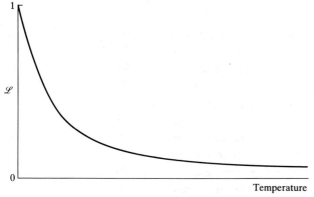

Fig. P.17 The Langevin function, which gives the variation of the orientation polarization of a medium with temperature. At high temperatures the polarization approaches zero because the thermal motion overcomes the orienting effect of the applied electric field.

Table P.3 Polarizabilities and polarizability volumes

	$\alpha'/(10^{-24}\,\text{cm}^3)$	$\alpha/(10^{-40}\,\text{J}^{-1}\,\text{C}^2\,\text{m}^2)$
Ar	1.66	1.85
C_6H_6	10.4	11.6
CCl_4	10.5	11.7
CH_2Cl_2	6.80	7.57
CH_3Cl	4.53	5.04
CH_3OH	3.23	3.59
CH_4	2.60	2.89
$CHCl_3$	8.50	9.46
CO	1.98	2.20
CO_2	2.63	2.93
H_2	0.819	0.911
H_2O	1.48	1.65
HBr	3.61	4.01
HCl	2.63	2.93
He	0.20	0.22
HF	0.51	0.57
HI	5.45	6.06
N_2	1.77	1.97
NH_3	2.22	2.47

Data: Böttcher and Bordewijk (1978).

distribution when the electrons are far from the nucleus or well shielded from its charge. The expression given in Box P.5 for this contribution reflects the usual °perturbation theoretical result that the closer the excited states of a molecule to the ground state the 'softer' and more responsive the system to distorting influences.

The numerator in the expression for α given in Box P.5 is proportional to the square of the °electric transition dipole moment from the ground to the excited states. Since the intensity of an °electric dipole transition is also proportional to this quantity, we can conclude that the greater the intensity of a transition to an excited state, the greater the contribution of the upper state to the mixture that describes the distortion. Likewise, the lower the energy of the state above the ground state, the greater its contribution. Hence, we can expect molecules that have intense transitions in the visible or lower frequency region of the spectrum to be highly polarizable whereas those that absorb only in the ultraviolet can be expected to be stiff and unpolarizable. Thus aliphatic hydrocarbons, which have weak optical transitions in the ultraviolet, are only weakly polarizable.

The electronic polarizability can also be interpreted in terms of the fluctuations in the instantaneous electric dipole moment of the molecule. It is possible to imagine local transient shifts of charge in the molecule that give rise to a dipole moment, and the greater these fluctuations the greater the polarizability. This interpretation is related to the first, because the fluctuations are greater in large, weakly bound systems.

The polarizability depends on the frequency of the distorting electric field. At low frequencies (below about 10^{12} Hz, of the order of the rotation rate of the molecules) polar molecules can follow the changing direction of the applied field and all the contributions to the polarizability are present. At higher frequencies (above 10^{12} Hz) the molecules cannot reorientate sufficiently quickly, and so the orientation polarization ceases to contribute. At still higher frequencies (above 10^{14} Hz, the rate at which molecules vibrate) the atomic nuclei respond too sluggishly to follow the oscillating field, and the atomic contribution disappears. Hence, at high frequencies (in the visible range) the only contribution to the polarizability is the electronic contribution, but even this ceases at very high frequencies. The frequency dependence (the °dispersion) of the polarizability is a helpful property for distinguishing the various contributions.

The polarizabilities of molecules contribute to the relative permittivity (dielectric constant) of the medium they compose (see Box P.5) and to its °refractive index. Since the refractive index is normally measured at optical frequencies it is related to the electronic polarizability.

Further information

See MQM Chapter 13 for an introduction to the calculation of the electric properties of molecules using perturbation theory. Atkins (1990) indicates how some of the properties are measured. Other discussions will be found in van Vleck (1932), Davies (1967), Hinchcliffe amd Munn (19850, and Craig and Thirunamachandran (1984); for a thorough treatment, see Böttcher and Bordewijk (1978). Hyperpolarizabilities and their measurement are described by Buckingham and Orr (1967). The variations of polarizability and refractive index with the frequency of the incident radiation are described in the entry on °dispersion.

Polaron

A **polaron** is a defect in an ionic crystal that is formed when an excess of charge at a point polarizes the lattice in its vicinity. Thus, if an electron is captured by a halide ion in an alkali halide crystal the metal ions move towards it and the other negative ions shrink away. As the electron moves through the lattice it is accompanied by this distortion. Dragging this distortion around effectively makes the electron into a more massive particle.

Further information

An elementary discussion is given in Cox (1987). See Anderson (1984), Kittel (1986, 1987), and Mott and Davies (1971) for more detailed accounts.

Potential energy surface

A **potential energy surface** is a generalization of the concept of molecular potential energy curve which shows how the energy of an assembly of atoms varies with their relative locations. In accord with the °Born–Oppenheimer approximation, the nuclei are stationary. The potential energy surface for two atoms is the same as the molecular potential energy curve because there is only one separation. The potential energy surface for a cluster of three atoms is three-dimensional because there are three interatomic separations to consider. If the atoms are constrained to be collinear, the surface reduces to two dimensions. A fragment of such a surface is depicted in Fig. P.18 together with a representation in terms of contours of equal potential energy. The surface shows a region in which ABC is stable and a channel for its fragmentation into $AB + C$.

In principle, a potential surface may be calculated by solving the Schrödinger equation for a large number of different arrangements of the atoms. However, this is a strikingly difficult procedure, and in many cases the surface is represented instead by a number of empirical functions that describe the changes in energy that occur for various distortions of the 'molecule'.

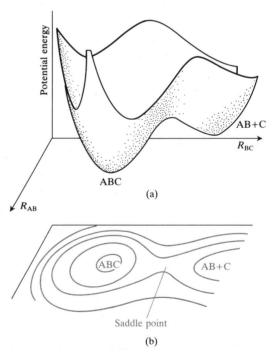

Fig. P.18 (a) The potential energy surface for a linear species ABC that can undergo fragmentation into $AB + C$. The atoms are constrained to be collinear. (b) The corresponding contour plot of the surface: the contours are of equal potential energy.

Molecular potential energy surfaces can be determined, to a certain extent, experimentally by the appropriate treatment of spectroscopic data. One approach is to fit the observed vibration–rotation transition frequencies to eigenstates calculated on the basis of a parametrized potential. Another, which is suitable for diatomic molecules, is to use the **Rydberg–Klein–Rees method** to locate the turning points of the vibrational motion.

Further information

Thorough discussions of the topics touched on in this entry have been given by Hirst (1985, 1990) and Murrell *et al.* (1984). For the use made of potential energy surfaces in the theory of unimolecular reactions see Gilbert and Smith (1990) and Levine and Bernstein (1987).

Precession

The **precession** of the axis of a spinning body (a top or a gyroscope) is a motion of the axis that sweeps out a conical surface (Fig. P.19a). The concept is adopted in the °vector model of the atom to indicate the energy of a state of angular momentum, with a high precession rate signifying a high energy.

In the vector model of the atom and in the general theory of °angular momentum, an angular momentum with quantum numbers j and m_j is represented by a vector of length $\{j(j+1)\}^{1/2}$ that makes a projection of length m_j on a z axis. According to the °uncertainty principle, nothing can be said about the x and y components of the angular momentum if the z component is known, and so the vector may lie anywhere on a cone (Fig. P.19b). In the absence of magnetic fields the vector is at rest at an indeterminate position on the conical surface.

When a magnetic field is applied along the z axis, the states with different projections have different energies by virtue of the magnetic moment associated with the angular momentum. The energy of the state m_j, which is $m_j\mu_B B$ for orbital angular momentum (and twice that value for °spin), can be expressed as a frequency—called the **Larmor frequency**, ω_L—by dividing by $\hbar$; thus $\omega_L = |m_j|\mu_B B/\hbar$. The Larmor frequency of each state (and hence the energy, $\hbar\omega_L$) is then represented on the vector diagram by supposing that each precesses on its cone at its Larmor frequency (Fig. P.19c). The name commemorates the Irish mathematician and physicist Sir Joseph Larmor (1857–1942) who proposed (in 1897) that an electron should undergo a type of precessional motion.

As the field is made weaker the precession frequency slows, and in the limit of zero field the static, indeterminate distribution of vectors is regained. The state with the greatest value of $|m_j|$ precesses most

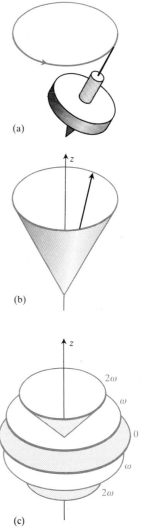

Fig. P.19 (a) The classical precessional motion of a spinning top: the axis sweeps out a cone with the tip of the top at its apex. (b) When there is no field present, the angular momentum vector lies at an indeterminate location on a cone surrounding the z axis. (c) When a field is present, the angular momentum vector precesses at the Larmor frequency: the greater the magnitude of the energy, the greater the rate of precession.

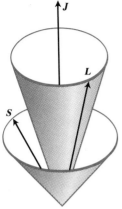

Fig. P.20 The orbital angular momentum and spin angular momentum vectors precess about their resultant J at a rate that is proportional to the energy of their interaction by spin–orbit coupling.

rapidly, and that with $m_j = 0$ does not precess at all. The rate of precession about an axis represents the strength of coupling to a field that defines the axis.

The concept of precession may be extended to cases in which the energy arises from sources other than external magnetic fields. If the spin–orbit coupling is strong, s and l precess rapidly around their resultant j (Fig. P.20). When the coupling is weak, the precession is slow and the coupling can easily be broken by other influences.

Further information

See the entry on the °vector model of the atom for more information and further references. As remarked in that entry, the vector model is a representation of the spin density matrix, so precession is actually a representation of the time-evolution of density matrices: see Munowitz (1988). Applications of the concept in atomic spectroscopy are described in Herzberg (1944), Candler (1964), and Kuhn (1962). See the °Hund coupling cases and references there for applications in linear molecules. The language of precession is widely used in magnetic resonance, and it may be pursued through the entries on °electron spin resonance and °nuclear magnetic resonance and the references therein. An example of the decoupling of two precessing vectors is provided by the °Paschen–Back effect.

Predissociation

Dissociation occurs when a molecule is excited to a state that possesses more energy than the separated fragments (Fig. P.21a). Above the dissociation limit of a spectrum the line structure is lost because the upper states are unquantized translational states. **Predissociation** is dissociation that occurs in a transition before the dissociation limit is attained, hence its name. It is detected by a blurring of the absorption lines (Fig. P.21b) followed by the resumption of sharp lines at higher frequency before the onset of the true dissociation limit.

Predissociation can be interpreted in terms of an °internal conversion from the upper electronic state A to a dissociative state B that crosses it (Fig. P.21b). The vibrational states of A in the vicinity of the intersection of curves in Fig. P.21b are a mixture, and they acquire some of the translational character of the states of B at that energy. Therefore a state near the intersection can dissociate by virtue of its B character even though its energy may be far below the dissociation energy for the state A. When the energy of the incident radiation is great enough to excite A to a vibrational state above the intersection, the lines are again observed to be sharp, for now the molecule is unable to switch into the state B.

Induced predissociation is predissociation that is induced by some external influence, such as collisions ('collision-induced predissoci-

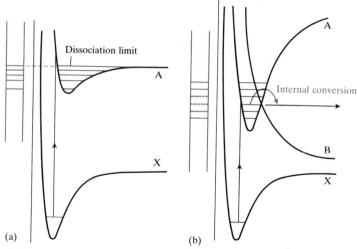

(a)   (b)

Fig. P.21 (a) Dissociation occurs when electron excitation raises the molecule to an energy above that of the uppermost bound state. It is detected by the cessation of vibrational structure in the spectrum and the onset of a continuous absorption. (b) Predissociation occurs when excitation to a vibrational state below the dissociation limit leads to dissociation because the bound electronic state intersects an unbound state, and internal conversion to the latter can occur. It is detected by a broadening of the absorption lines, followed by the resumption of sharp lines at higher energies.

ation') or an applied field. Collisions are able to knock the excited molecule from A to B, and an applied field may relax some of the selection rules that govern the internal conversion from A to B.

Further information

See Chapter 12 of MQM. Herzberg (1950, 1966) gives, as always, a very thorough discussion of predissociation, with numerous examples. See also Hollas (1983) and the entry on °internal conversion.

Progression

An °electronic excitation of a molecule is accompanied by excitation of °vibration (see the °Franck–Condon principle), and instead of a single line in the spectrum there may be an extensive band of transitions. A **progression** is a series of lines that arise from transitions from the same vibrational level of one of the states (the ground electronic state for absorption) to successive vibrational levels of the other state. For instance, the $v''=0$ progression is the series of transitions starting in the $v''=0$ level of the ground electronic state of the molecule and terminating in the $v'=0$, 1, 2,... levels of the upper electronic state (Fig. P.22). The $v''=1$ progression is a series starting in the $v''=1$ level

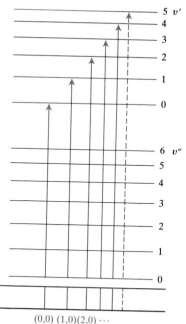

Fig. P.22 The formation of a progression. The $v''=0$ progression is the series of transitions starting in $v''=0$ and terminating in the $v'=0$, 1, 2, ...levels of the upper electronic state.

of the ground electronic state. The lines in a progression are labelled (v',v''); therefore the $v'' = 0$ progression consists of the transitions $(0,0)$, $(1,0)$, $(2,0)$, and so on.

Further information

Detailed information on the analysis, appearance, and formation of progressions will be found in Hollas (1983, 1987) and Herzberg (1950).

Quadrupoles and other multipoles

An electric **multipole** is an array of positive and negative point electric charges. An electric **monopole** is a single point charge. An °electric **dipole** consists of two equal and opposite point charges (Fig. Q.1) with zero net charge. An electric **quadrupole** consists of four point charges with overall zero charge and zero dipole moment. It may be pictured as two dipoles arranged so that their dipole moments cancel (Fig. Q.1). Higher multipoles may be constructed in an analogous way. Thus, a 2^n-pole consists of 2^n point electric charges in an array that has no lower multipole moment (see below). Some multipoles are shown in Fig. Q.1, but alternative arrangements may also be envisaged. Any arbitrary distribution of charge may be expressed as a sum of multipoles.

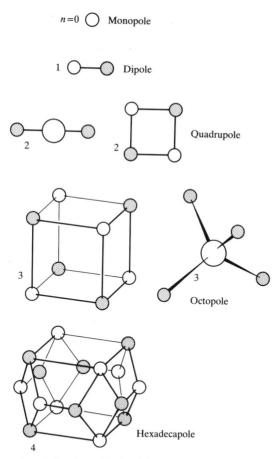

Fig. Q.1 Examples of electric multipoles. The tinted and untinted spheres represent equal and opposite charges. Each multipole has zero lower multipole moments (so that, for instance, the quadrupole has a quadrupole moment but no dipole or monopole moment).

The multipoles that characterize a molecule's charge distribution are determined by the molecule's symmetry. For example, the linear CO_2 molecule has neither a net charge (no electric monopole moment) nor an electric dipole moment. However, since there is a partial negative charge on each O atom and a compensating partial positive charge on the C atom, it has a non-vanishing quadrupole moment.

The electric potential arising from a 2^n-pole decreases with distance as $1/r^{n+1}$ (Fig. Q.2). The electric field (the negative gradient of the potential, $\boldsymbol{E} = -\nabla\phi$) varies with distance as $1/r^{n+2}$. The steeper decrease in the potentials of higher multipoles may be understood in terms of how a cluster of electric charges, when viewed from a great distance, is hardly distinguishable from a zero point charge. The apparent cancellation at large distances occurs more effectively the higher the rank of the multipole because there are more (cancelling) charges in the cluster.

The potential arising from a general charge distribution is the sum of potentials of the multipoles into which the distribution may be divided. Since the potential decays more rapidly the greater the value of n, the potential of a given distribution is dominated by its lowest moments at great distances, and the lower moments contribute more strongly as the molecule is approached.

An electric monopole interacts with the electric potential itself (see Box Q.1). An electric dipole interacts with the gradient of the potential, since its two charges must experience different potentials if there is to be a net interaction. In other words, an electric dipole interacts with the electric field (which is the negative gradient of the potential). An electric quadrupole interacts with the second derivative of the potential (with the gradient of the electric field). This may be understood from

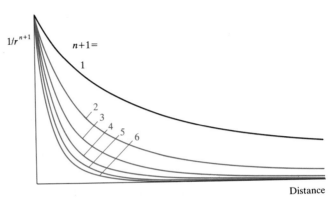

Fig. Q.2 The distance dependence of the electric potentials of multipole moments. Note how the higher the order n of the multipole, the sharper the decrease in potential.

Box Q.1 Multipole fields

Electric field at r arising from a point n-pole, $\mathscr{E} \propto 1/r^{n+2}$.

The field is related to the potential by $\boldsymbol{E} = -\nabla \phi$.

Potential at r arising from a point n-pole, $\phi \propto 1/r^{n+1}$

$$\text{Dipole: } \phi = \frac{\mu}{4\pi\varepsilon_0} \frac{\cos\theta}{r^2}$$

$$\text{Quadrupole: } \phi = \frac{\tfrac{1}{2}eQ}{4\pi\varepsilon_0} \frac{3\cos^2\theta - 1}{r^3}$$

Energy of point multipole:

$$\text{Monopole: } E = q\phi$$

$$\text{Dipole: } E = -\mu\mathscr{E}\cos\theta = \mu\left(\frac{\partial\phi}{\partial z}\right)\cos\theta$$

$$\text{Quadrupole: } E = -\tfrac{1}{8}eQ\left(\frac{\partial\mathscr{E}}{\partial z}\right)(3\cos^2\theta - 1)$$

the cases depicted in Fig. Q.3. When the gradient of the potential (the field) is constant over the quadrupole, the energy of interaction is independent of orientation and there is no net interaction. When the potential has curvature (and hence where the field has a gradient), the energy depends on the orientation and hence there is a net interaction.

The **moment** of a 2^n-pole is the sum of each charge multiplied by the nth power of its distance from the charge centroid. The 'monopole moment' (q) is the charge itself, the dipole moment (μ) is proportional to the separation of the charges, and the quadrupole moment (Q) is proportional to the squares of the distances of the charges (Box Q.1). It follows that the magnitude of a molecular electric dipole moment is of the order of eR, where R is a measure of the extent of the molecule, which is of the same order of size as its diameter. Hence, for $R \approx 100$ pm, $\mu \approx 10^{-29}$ C m (see °dipole moment). Likewise, the order of magnitude of a molecular electric quadrupole moment is $eR^2 \approx 10^{-39}$ C m².

Further information

A good introduction to multipoles is given by Wangsness (1979). See also Jackson (1962) and Rose (1955, 1957). The use of multipoles to express the distribution of charge in a molecule is described by Stone and Alderton (1985). The role of multipole moments in spectroscopy is described by Flygare (1978) and Hollas (1983, 1987). The calculation of molecular quadrupole moments is described by Davies (1967) and Hinchcliffe and Munn (1985), and their measurement is described by Buckingham (1965), who also considers their role in °intermolecular forces (see that entry too). For the role of changing multipoles in the absorption and emission of photons see °electric dipole transitions.

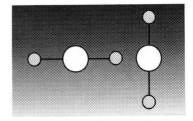

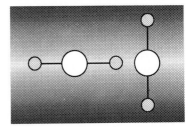

Fig. Q.3 The density of the tint represents the magnitude of the electric potential. (a) The energy of the quadrupole is the same in the two orientations because the adverse change when one positive charge moves into a region of high potential is offset by the other positive charge moving into a region of low potential. (b) The energy of the quadrupole does depend on its orientation if the potential has curvature, for now *both* positive charges move into a region of high potential when the orientation of the quadrupole is changed.

Quantum

A **quantum** of energy is the smallest amount of energy that a system can accept. Similarly, a quantum of angular momentum is the smallest amount of angular momentum that a system may accept. Such amounts may be arbitrarily small according to classical physics. In quantum mechanics, the size of a quantum is determined by the characteristics of the system and the state it is in. The word 'quantum' is Latin for 'amount'.

A °harmonic oscillator of frequency v provides an example. According to classical mechanics, the oscillator may swing at its natural frequency with a continuously variable energy. Quantum mechanics, however, shows that the energy is confined to the values $E = (v + \frac{1}{2})hv$, where $v = 0, 1, 2,\ldots$, with the result that the minimum change in the oscillator's energy is hv, where h is Planck's constant. The harmonic oscillator is unusual in that the same quantum of energy is required to excite it whatever its initial state. That is not true in general. For example, the size of the quantum needed to excite molecular °rotation in three dimensions increases linearly with the degree of excitation, so that successive steps require quanta with sizes 1, 2, 3,... units. The precise size of the quantum required to excite a system from a specific state is obtained by solving the appropriate °Schrödinger equation.

The energy quanta of macroscopic objects are so small that changes in their energy appear to be continuous. That near continuity is what misled the classical physicists. However, quantization cannot be ignored for microscopic objects. For example, the quantum of energy of a pendulum of period 1 s is only 6×10^{-34} J but the oscillation of a bond in a molecule occurs at a frequency about 10^{14} times higher; the quantum of energy required is that much larger too, and the non-continuous character of the excitation cannot be neglected.

The °angular momentum of a rotating body may be changed only in steps of the order of $\hbar$, where $\hbar = h/2\pi$. Thus, although we appear to be able to accelerate a wheel smoothly to any angular momentum, in fact we can do so only in steps of the order of $\hbar$. The angular momentum quanta are so minute for macroscopic bodies that we can treat the stepwise acceleration as virtually continuous. However, the quantization of the °angular momentum of atoms and molecules is of profound importance.

The energy of the °electromagnetic field is quantized in the sense that radiation of a given frequency v can be excited only in steps of magnitude hv. This stepwise excitation gives rise to the concept of the °photon. However, since there is no limitation on the value that v may take (the frequency of electromagnetic radiation is continuously variable over an enormous range), the *total* energy of an electromagnetic field is continuously variable (see °black body radiation).

Further information

See the early chapters of MQM for an introduction to the quantum mechanics of simple systems. The books by Heisenberg (1989) and Hey (1987) and the introductory text by Eisberg and Resnik (1985) all convey the flavour of the role of the quantum in physics.

Quantum chemistry

The term **quantum chemistry** denotes the application of the principles of quantum mechanics to topics of chemical interest. It therefore includes the calculation of molecular structures and properties, the analysis of spectroscopic data, and the description of chemical reactions in terms of individual molecular events. The term 'theoretical chemistry' is widely used as a synonym for quantum chemistry, but strictly it is a broader term since it includes other activities, particularly statistical thermodynamics.

The conventional division of the field into **computational quantum chemistry** and **non-computational quantum chemistry** is currently being blurred by the ubiquitous use of computers in all kinds of problems. However, the general sense of the division is that computational quantum chemistry is primarily concerned with the numerical computation of molecular electronic structures by °*ab initio* and °semi-empirical techniques whereas non-computational quantum chemistry is largely the formulation of analytical expressions for the properties of molecules and their reactions.

The great bulk of quantum chemistry is computational, and quantum chemists are among the heaviest users of computers and supercomputers. Typical problems include the calculation of the electron distributions in molecules, the prediction of molecular geometries, the computation of °potential surfaces for reactive systems, and the screening of molecules for potential pharmacological activity. A great deal of computational chemistry is also concerned with the discovery of efficient methods of solving molecular electronic structures.

Typical problems in non-computational quantum chemistry include the elucidation of the electromagnetic properties of molecules using °perturbation theory, the formulation and manipulation of °hamiltonians that represent the °hyperfine interactions in molecules, the analysis of the symmetry properties of fluxional molecules, and the analysis of features of potential energy surfaces for their consequences for rate constants and other aspects of molecular collisions. A particularly thriving field, particularly for the analysis of reactive molecular collisions, is that of **semiclassical mechanics**, in which quantum mechanically intractable problems are analyzed in terms of theories that are a blend of quantum and classical mechanics.

Further information

The scope of MQM gives a reasonable impression of the scope of quantum chemistry, as do the introductory texts by McQuarrie (1983) and Levine (1983); if self-reference is admissible, the whole of this book gives an indication of the scope of the subject. The more advanced aspects of non-computational quantum chemistry can be encountered in Hinchcliffe and Munn (1985) and Craig and Thirunamachandran (1984) for the electromagnetic properties of molecules, in Child (1974, 1991) for the application of semiclassical mechanics to problems of chemical interest, and in books on group theory, such as Bunker (1979) and Cotton (1990), for a taste of how symmetry is deployed. As for computational quantum chemistry, see Hirst (1990), Clark (1985), McWeeny (1989), Szabo and Ostlund (1982), and Hehre *et al.* (1986) for a thorough bird's-eye view of the enormously broad landscape. Richards (1983,1989) gives an impression of how quantum chemistry is used in pharmacological studies.

Quantum defect

The spectrum of atomic °hydrogen (Fig. H.18) consists of several °series of lines with wavenumbers that can be expressed as the difference of two °terms of the form R/n^2, where R is the °Rydberg constant and n is an integer. The spectra of alkali metal atoms may also be grouped into series that can be expressed similarly. However, for these atoms the **effective principal quantum number**, n, is not an integer but is equal to $n' - \delta$, where n' is an integer and δ is a correction called the **quantum defect**.

The quantum defect arises from the effect of the other electrons in the atom on the electron of interest. It decreases as the principal quantum number of the electron increases, for as n increases the electron is progressively further away from the nucleus and its surrounding core electrons increasingly resemble a single point charge. The greatest defects occur for s orbitals, which °penetrate most closely to the nucleus. The quantum defect is a guide to the extent of penetration, but it has little other theoretical significance or importance.

Further information

The quantum defect occupies more of the ancient literature than it does of the modern. Mention of it will be found in §3.6 of King (1964), §1.5 of Herzberg (1944), and Chapter 11 of Kuhn (1962). Kuhn lists some values and discusses why they were important.

Quantum number

A **quantum number** is an integer that labels the state (and wavefunction) of a system and determines the value of a physical observable of the system in that state. Half-integral quantum numbers may occur if the system includes °spin.

As an example, the state of a °particle in a box is labelled by the quantum number $n = 1, 2, ...$, and its energy in that state is equal to $n^2(h^2/8mL^2)$. In general, more than one quantum number is needed to

---

Box Q.2 Quantum numbers

Upper-case letters are used for many-particle systems.

F **Total angular momentum quantum number**, including nuclear spin. Significance as for j. Positive values only.

I **Nuclear spin quantum number**. Significance as for j. Each nuclide has a fixed value of I, which may be integral or half integral. See Table S.1. Positive values only.

j, J **Total angular momentum quantum number** (excluding nuclear spin):

         Magnitude of angular momentum $= \{j(j+1)\}^{1/2}\hbar$.

         Number of projections: $2j + 1$ (see m_j).

     Permitted values of j given by the °Clebsch–Gordan series. Positive values only.

J **Rotational quantum number**. Significance as for j; may have $2J + 1$ projections on external axis and (except for linear molecules) $2J + 1$ projections on figure axis of molecule (see K). Positive values only.

K **Angular momentum projection quantum number** (no official name). Component of rotational angular momentum about the figure axis of a molecule is $K\hbar$, with $K = J, J - 1, ..., -J$.

l, L **Orbital angular momentum quantum number** (l once called the 'azimuthal quantum number'). Interpretation as for j, but l and L are restricted to integral values. Positive values only.

m, M **Magnetic quantum number** (and specifically m_l, M_L, m_s, M_S, m_j, M_J, and so on). Gives component of angular momentum on z axis as $m\hbar$. Restricted to the $2j + 1$ values $m = j, j + 1, ..., -j$.

n **Principal quantum number** for hydrogenic atoms, where it determines the energy (through $-hcR/n^2$) and the number of states (n^2) of a given energy. Restricted to $n = 1, 2, ...$ (positive values only). **General quantum number** when a system is quantized, as in a °particle in a box, for which $n = 1, 2, ...$

s, S **Spin quantum number**. Denotes spin of elementary particle (specifically an electron or many-electron system). Significance and interpretation of s, S as for j, J. The quantum number $s = \frac{1}{2}$ for an electron. Positive values only.

v **Vibrational quantum number**. Determines energy of harmonic oscillator through $(v + \frac{1}{2})\hbar\omega$. Restricted to $v = 0, 1, 2, ...$ (positive values only).

α, β **Spin states** of spin $-\frac{1}{2}$ particle (specifically electron or proton): α corresponds to $m_s = +\frac{1}{2}$ and β corresponds to $m_s = -\frac{1}{2}$.

Λ **Orbital angular momentum projection quantum number** in a linear molecule. The component of angular momentum about the internuclear axis is $\Lambda\hbar$. Positive or negative integral values are permitted.

Σ **Spin projection quantum number** in a linear molecule. The component of spin angular momentum about the internuclear axis is $\Sigma\hbar$. May be integral or half-integral, and positive or negative.

Ω **Total electronic angular momentum projection quantum number** in a linear molecule. The component of electronic (spin and orbital) angular momentum about the internuclear axis is $\Omega\hbar$.

specify a state fully. For instance, the state of an electron in a °hydrogen atom is fully determined by the numerical values of the four quantum numbers n, l, m_l, and m_s, where:

- n determines the energy of the state through the expression $-hcR/n^2$,

- l determines the angular momentum through the expression $\{l(l+1)\}^{1/2}\hbar$,

- m_l determines the orientation of that angular momentum in terms of the projection $m_l\hbar$ it has on an arbitrary axis,

- m_s likewise determines the orientation of the electron's spin.

Each spatial °orbital of the atom is uniquely specified by the values of n, l, and m_l.

Box Q.2 lists some common quantum numbers and the properties they determine: for more information, consult the appropriate entry.

Further information

See throughout MQM, and any text of quantum mechanics, for examples of quantum numbers and the roles they play in labelling states and determining observables.

Quantum theory

The view that energy can be transferred between systems only in discrete amounts (see °quantum) arose from observations on the interaction of matter and radiation and the measurement of the properties of solids at low temperatures. The first evidence that classical physics was wrong came from the study of °black body radiation, the °photoelectric effect, the °Compton effect, °atomic spectra (especially the spectrum of atomic °hydrogen), and the °heat capacities of solids.

The concept of quantization was introduced by the German physicist Max Planck (1858–1947), who in 1900 deduced the distribution law for black body radiation. The first calculation in which the quantum ideas were applied to a dynamical system was °Bohr's calculation (in 1913) of the energy levels of atomic hydrogen. The ingenious cobbling together of a rag-bag of concepts drawn from classical physics and the newly emerging quantum physics is now called the **old quantum theory**.

The old quantum theory was displaced by our current quantum theory in 1926, when °Schrödinger proposed his equation and Heisenberg his °matrix mechanics. These initially apparently quite different but actually mathematically equivalent descriptions of the world entailed a wholesale revision of classical physics. That was particularly true of the introduction of the °uncertainty principle, the probabilistic interpretation of the °wavefunction, and the recognition of the °duality of matter.

Special relativity was combined with quantum mechanics by °Dirac in 1928. Since then, quantum theory has been firmly established, and however bizarre its predictions (for those who are classically conditioned to certain expectations), no exceptions to the theory have been found. Experiments were carried out during the 1980s by the French physicist A. Aspect and his collaborators under the stimulation of a theoretical analysis; this analysis, which culminated in the formulation of **Bell's inequalities** by J.S. Bell, has also eliminated the suspicion that there may be local **hidden variables** that determine the outcome of observations and hence reduce the perceived probabilistic interpretation of the theory.

Apart from the perennial problem of interpretation, the only outstanding problem (which, like other black clouds, might bring about the downfall of quantum theory) is the synthesis of quantum theory and gravitation.

Further information

Jammer (1966) gives an excellent survey of the emergence of quantum mechanics, as does Kuhn (1978) for the earlier stages of its emergence. For a perceptive account of Schrödinger's role, which shows something of the ebb and flow of ideas, see Moore (1989); a similar insight into Einstein's mind will be found in Pais (1982). A collection of the significant early papers (in translation) has been edited by van der Waerden (1967). Heisenberg has reflected in Heisenberg (1930, 1989). For surveys of the interpretation of quantum mechanics see Jammer (1974), Hughes (1989), Bacry (1988), d'Espagnat (1976), and Redhead (1987). A simple introduction to Bell's theorem will be found in Bransden and Joachain (1989). See also Bell (1987), a collection of papers that includes two that have revolutionized the discussion of the foundations of quantum theory.

Quenching

The orbital angular momentum is said to be **quenched** when it is eliminated by an electric field that destroys the spherical or cylindrical symmetry of a system.

The potential energy of an electron in an atom is independent of the angular coordinates, and its orbital angular motion can occur smoothly and without hindrance (Fig. Q.4a). In such a case the °orbital angular momentum is well defined. When the atom is surrounded by ligands, the potential energy of the electron depends on its angular coordinates, and hence it experiences a force that accelerates it in a complicated

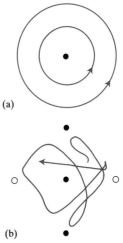

Fig. Q.4 (a) The smooth classical orbits of a particle in a centrosymmetric field: the orbital angular momentum is well-defined. (b) In a non-centrosymmetric field, the path of the particle is much more complex and there is no net orbital angular momentum: the angular momentum has been quenched by the additional field.

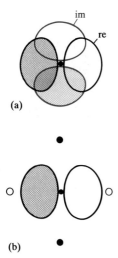

(a)

(b)

Fig. Q.5 In a centrosymmetric (or axially symmetric) field, the real and imaginary components of a *p* orbital are degenerate, and the orbital angular momentum is nonzero. (b) In a less symmetrical field, the degeneracy is removed, and the electron occupies either the real or the imaginary component: the orbital angular momentum is zero.

manner (Fig. Q.4b). In classical terms, the acceleration continuously changes the direction of the electron's motion and the average angular momentum is zero.

In quantum mechanical terms, quenching occurs when the degeneracy of the real and imaginary components of a °complex wavefunction is removed by the potential (Fig. Q.5). Thus, if the wave corresponding to $m_l = +1$ is placed in a ligand field of low symmetry, its sine and cosine components have a different energy and the electron occupies the one of lower energy. In this way a travelling wave is replaced by a standing wave (a p_x or a p_y orbital), and the electron no longer travels around the nucleus.

The °magnetic properties of *d*-metal complexes can often be interpreted on the basis that the orbital angular momentum has been quenched by the °ligand field, and hence that only the spin and not the orbital magnetic moment makes a contribution to the total magnetic moment.

Further information

The quenching of angular momentum is described in Chapter 14 of MQM. For further information see the entry on °magnetic properties and the references therein.

Racah parameters

The energy of an atomic °term depends on the kinetic energies of the electrons, their attractive Coulombic interactions with the positively charged nucleus, and their mutual Coulombic repulsions. The last are very difficult to take into account quantitatively, yet they make an important contribution to the energy of the atom and to the relative energies of its °terms. In particular, the repulsions between the valence electrons of incomplete shells depend on the orbitals the electrons occupy, and hence terms with different total orbital angular momentum have different energies.

The interaction energies between electrons are expressed as complicated integrals over the wavefunctions of the electrons. However, for electron configurations of the form p^n or d^n the integrals group together in characteristic ways, and the total repulsion energy of a term depends on only three combinations of the integrals. These three combinations of integrals, denoted A, B, and C, are the **Racah parameters**. The coefficients were introduced by the Italian theoretical physicist Giulio Racah in 1942.

As an example, the configuration d^2 gives rise to two triplet terms (^3P and ^3F) and three singlet terms (^1S, ^1D, and ^1G). These terms differ in total orbital angular momentum, and hence in the details of the occupation of the five d orbitals that are available in the valence shell. The total repulsion energy of each term can be expressed in terms of the Racah parameters as follows:

$E(^1S) = A + 14B + 7C$
$E(^1G) = A + 4B + 2C$
$E(^1D) = A - 3B + 2C$
$E(^3P) = A + 7B$
$E(^3F) = A - 8B$

If we are interested only in the relative order of the terms, we can ignore the common contribution represented by A. Moreover, if we are interested only in the relative order of the triplet terms we can also ignore C.

Since the Racah parameters are all positive (they represent repulsions), we can predict that the ^3F term will always lie lowest in energy (as °Hund's rule predicts), but that the ^3P term lies below the ^1D term only if $C > 5B$ (Fig. R.1). When $C < 5B$ it is better to ensure that the electrons stay well apart, which is achieved if they occupy orbitals that result in a high orbital angular momentum.

Further information

The application of Racah parameters is described briefly in Shriver, Atkins, and Langford (1990), who give the Tanabe–Sugano diagrams for a range of d^n

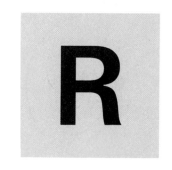

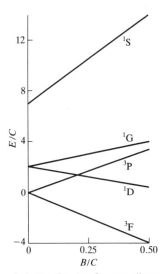

Fig. R.1 The Tanabe–Sugano diagram for a d^2 electron configuration in an octahedral ligand field.

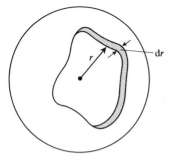

Fig. R.2 The radial distribution function gives the total probability that an electron will be found anywhere between the walls of a spherical shell of thickness d*r* at a distance *r* from a nucleus. Note that the volume of the shell is proportional to r^2, and so is zero when $r = 0$.

configurations in octahedral symmetry. See the references in the entry on °ligand field theory, such as Ballhausen (1979), for almost all of them deal with the parameters in their discussion of *d* metal complexes. Griffith (1964) is a characteristically thorough source. Racah's original paper, Racah (1942), is in the collection edited by Biedenharn and van Dam (1965). The entries on °Hund's rules and °Tanabe–Sugano diagrams are relevant.

Radial distribution function

The **radial distribution function**, *P*, gives the probability that a particle will be found on a spherical shell of thickness d*r* (Fig. R.2). The total probability of finding a particle at a radius *r* taking into account all angular positions is the value of *P*d*r*.

Consider a spherically symmetrical charge distribution described by the wavefunction ψ (which could be an °*s* orbital of a hydrogenic atom). The probability of finding a particle in the volume element dτ at a point **r** is equal to ψ^2dτ. Since the total volume enclosed between the walls of a spherical shell of thickness d*r* and radius *r* is its surface area times its thickness, or $4\pi r^2 \times$ d*r*, the *total* probability of finding the particle between its walls is its total volume times the value of ψ^2 at that radius, or $4\pi r^2 \psi^2$d*r*. Hence, for a spherical distribution of charge, $P = 4\pi r^2 \psi^2$. For a non-spherical distribution (such as a *p* orbital), the integration over the angular variation of ψ^2 must be done explicitly.

Since the surface area of the shell increases with radius as r^2, its 'catchment' volume increases in the same manner. Hence, *P* is zero at the nucleus, follows the oscillations of the orbital and passes through one or more maxima, and then decreases to zero as the wavefunction decays exponentially at large distances (Fig. R.3). The radius at which

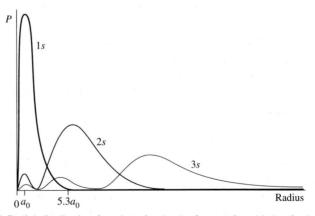

Fig. R.3 Radial distribution functions for the 1*s*, 2*s*, and 3*s* orbitals of a hydrogen atom. Note how the maximum moves away from the nucleus as the principal quantum number increases.

P passes through a maximum is the radius at which the electron will most probably be found. For a hydrogenic atom in its ground state this most probable radius is equal to a_0/Z, where a_0 is the °Bohr radius.

Further information

Analytical expressions and references to the radial distribution functions of numerous atoms are given by McGlynn *et al.* (1972), particularly in their Appendix B. See also Herman and Skillman (1963) for numerical tables.

Raman spectra

The **Raman process** is the inelastic scattering of a photon by a molecule. An inelastic process is one in which energy is transferred between the two colliding systems. In Raman scattering the incident photon may lose energy to the molecule by exciting its rotation or vibration, in which case it emerges from the collision with a lower frequency. Alternatively, the photon may acquire energy from the molecule if a mode is already excited and hence emerge with a higher frequency. Since molecular rotation and vibration are °quantized, the energy transfer can occur only in packets, and so the scattered light contains frequency components that are shifted from the incident frequency by discrete amounts. The frequency composition of the scattered radiation is the **Raman spectrum** of the molecule. The effect was discovered in 1928 by the Indian physicists Sir Chandrasekhara Raman (1888–1970) and K.S. Krishnan but had been predicted some years previously by Werner Heisenberg in Germany and A. Smekal in the Soviet Union. There is an continuing dispute over priority, for the Soviet physicists G. Landsberg and L. Mandelstam observed the effect at about the same time as Raman.

In practice, an intense, monochromatic beam of incident radiation from a laser passes through the sample and the radiation scattered perpendicular to the propagation direction is detected and analyzed. The spectrum consists of a strong **Rayleigh component** at the incident frequency, which arises from elastic collisions between the photons and the sample, and a series of lines to high and low frequency of that component. The lines to low frequency are the **Stokes lines**. They arise from collisions in which photons lose energy to the molecule (**1**). The lines to high frequency are the **anti-Stokes lines**. They arise from collisions in which photons gain energy from the molecule (**2**). The Stokes lines are generally more intense than the anti-Stokes lines because the latter require the presence of a pre-existing population of excited molecules.

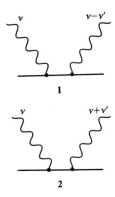

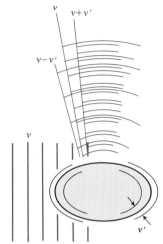

Fig. R.4 The classical interpretation of the Raman effect. The polarizable molecule interacts with a monochromatic incident wave, and the dipole induced in the vibrating or rotating molecule emits radiation at frequencies shifted from that of the incident radiation. In a sense, the frequency of the internal mode v' beats with the incident frequency v, and the scattered radiation has components that are the sum and difference of the incident and internal frequencies.

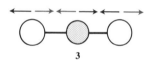

3

That the Raman effect depends on the °polarizability of the molecule can be understood in terms of a picture of the process in which the incoming radiation induces a °dipole moment in the molecule which then acts as the source of the scattered radiation. The efficiency of the process depends on the ease with which the molecule can be distorted by the incident radiation, and hence it depends on its polarizability.

The classical description of the Raman effect focuses on the frequency at which the induced dipole of the molecule oscillates when it is exposed to an oscillating electromagnetic field of frequency v. If the polarizability is unvarying, the induced dipole oscillates at the same frequency as the incident radiation. However, if the motion of the molecule results in an oscillation of its polarizability at a frequency v', this oscillation is impressed on the motion of the induced moment, which then oscillates at the sum and difference of the two frequencies (Fig. R.4) as well as at the frequency v. Hence, the molecule is a source of radiation of frequencies $v + v'$ and $v - v'$ as well as the unshifted component.

For a molecule to show a **rotational Raman spectrum**, in which energy transfer is to or from the rotational motion of the molecule, its polarizability must be anisotropic and change as the molecule rotates. Thus, a rotating H_2 molecule is Raman active because it has different polarizabilities parallel and perpendicular to the bond. A CH_4 molecule is rotationally Raman inactive because its polarizability is independent of its orientation.

For a molecule to have a **vibrational Raman spectrum**, in which energy transfer is to or from the vibrations of the molecule, its polarizability must change as it vibrates. An H_2 molecule is vibrationally Raman active because its polarizability depends on how greatly the bond is stretched. On the other hand, the antisymmetrical vibration of CO_2 (**3**) does not affect the polarizability of the molecule (it leaves the molecule roughly the same size) and so this vibration is Raman inactive.

The **exclusion rule** states that:

If a molecule has a centre of inversion, no mode of motion can be both infrared and Raman active.

(A mode may be inactive in both.) It follows that the Raman effect is useful in the study of vibrations and rotations that are inaccessible to normal absorption spectroscopy.

The Raman °selection rules are $\Delta v = \pm 1$ for vibrational transitions and $\Delta J = 0, \pm 1$, or ± 2 for rotations. The possibility that J can change by 2 but v only by 1 is connected to the fact that whereas the polarizability of a molecule returns to its original value only after a full period of vibration (Fig. R.5), it returns after only half a period of rotation. Hence, the polarizability appears to change at the same rate as a vibration but at twice the rate of the rotation. The quantum mechanical

explanation in terms of the unit spin of the °photon is more complicated, but is based on the fact that *two* photons are involved in the
scattering (one in, one out), and their directions allow the angular
momentum of the molecule to change by two units.

Vibrational Raman transitions are accompanied by rotational structure, just as in the case of the °branches of a rotation-vibration spectrum. In addition to the P, Q, and R branches of the latter, a Raman
spectrum also has O ($\Delta J = -2$) and S ($\Delta J = +2$) branches.

Further information

For a thorough survey of Raman spectroscopy, including non-linear adapta-
tions of the technique, see Long (1977). A lot of information will also be found
in Barron (1983), Hollas (1983, 1987), and Herzberg (1945). The original papers
on the effect are Raman and Krishnan (1928) and Landsberg and Mandelstam
(1928). For an analysis of Raman's contribution see Miller and Kauffman
(1989).

Reduced and effective mass

The **reduced mass**, μ, of two particles of mass m and m' is calculated
from the expression $1/\mu = 1/m + 1/m'$. The **effective mass** is a measure
of the total amount of matter that moves when a particular mode of
motion occurs. The effective mass and the reduced mass are synonym-
ous in systems composed of two particles. For example, in a hydrogen
atom the electron and proton move jointly about their centre of mass,
and the effective mass in motion is equal to the reduced mass of the
two particles.

The reduced mass is always smaller than the mass of the lighter
particle. It is also always closer to the mass of the lighter particle, since
that particle contributes more to the motion of the system. For instance,
if one particle is a ping-pong ball and the other is a cannon ball, the
reduced mass is effectively that of the ping-pong ball, because the
cannon ball barely participates in the joint motion about their centre
of mass. If the two particles are identical (as in positronium, a bound
state of a positron and an electron), the reduced mass is one-half the
mass of one of the particles.

When only two atoms participate in a vibration, the effective mass
of the mode is equal to the reduced mass of the two atoms. Thus, the
stretching motion of a C—H bond is determined by the reduced mass,
which is given by $1/\mu = 1/m(C) + 1/m(H)$: both atoms move as the bond
vibrates, and so the masses of both atoms determine its frequency.
With $m(C) = 12$ u (the cannon ball) and $m(H) = 1$ u (the ping-pong ball),
we find $\mu = 0.92$ u, which is close to the mass of the lighter atom, the
atom that undergoes most motion. When the bond is deuterated its

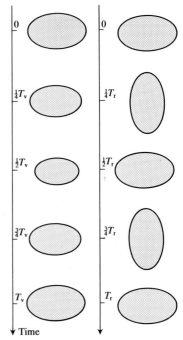

Fig. R.5 Whereas the polarizability of
a molecule oscillates at the same
frequency as the molecule vibrates (left
series, top to bottom), it returns to its
original value twice as rapidly as the
molecule rotates (right series).

vibrational frequency is changed only to the extent that the lighter atom contributes to the motion. With $m(D) = 2$ u we find that $\mu = 1.7$ u, a change of less than a factor of 2.

In general, the effective mass of a °normal mode of vibration of a polyatomic molecule is a combination of the masses of all the atoms participating in the motion weighted according to the extent to which each of them moves. For example, the mass of the central C atom of CO_2 does not contribute to the reduced mass of the symmetric stretch (**4**) because in that mode the atom is stationary.

The formal derivation of the expression for the reduced mass is the technique of **separation of variables**, in which the Schrödinger equation for a composite system is factorized into contributions from distinct types of motion. In the case of the H atom, the motion of the two particles is expressed in terms of their motion around the centre of mass (which introduces μ) and the motion of the centre of mass, which represents the motion of the atom as a whole (and is determined by the total mass of the atom). Likewise, the separation of the internal motions of a polyatomic molecule on the assumption that the atoms move in a parabolic potential results in a separation into dynamically independent motions; the latter may be expressed as the harmonic vibration of a single particle of a certain effective mass along a composite displacement of the atoms (such as the symmetric stretch of CO_2).

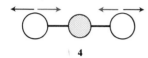

4

Further information

The separation of variables technique is described in MQM in a number of places. Appendix 6 is particularly relevant, for it shows how the hydrogen atom hamiltonian is separated into the motion of its centre of mass and the relative motion of its components. Similarly, the effective mass of normal modes is dealt with in an example in Appendix 16. See Bunker (1979), Allen and Cross (1963), and Wilson, Decius, and Cross (1955). For simpler introductions, see Graybeal (1988), Struve (1989), and Hollas (1983, 1987). In general, good sources on classical mechanics are Goldstein (1982) and Kibble (1985).

Refractive index

The **refractive index**, n_r, of a medium is the ratio of the speed of light in a vacuum, c, to its speed in the medium, v: $n_r = c/v$. The refractive index depends on the strength of the interaction of the electromagnetic field with the medium, and hence on the polarizability of the molecules or atoms from which it is made.

The equations relating refractive index and polarizability are set out in Box R.1, which also gives the quantum mechanical expression for the refractive index obtained from °perturbation theory (see °polarizability). The formula expresses the °dispersion of the refractive index,

Box R.1 Refractive index

The **refractive index**, n_r, is defined as

$$n_r = c/v$$

where c is the speed of light in a vacuum and v its speed in the medium of interest. It is related to the relative permittivity, ε_r, by

$$n_r = \varepsilon_r^{1/2}$$

The **molar refractivity**, R_m, is defined as

$$R_m = \frac{M}{\rho} \times \frac{n_r^2 - 1}{n_r^2 + 2}$$

where M is the molar mass and ρ is the density of the medium. Then, in terms of the dynamic °polarizability (Box P.5), the refractive index at any frequency may be determined from

$$3\varepsilon_0 R_m = N_A \alpha(\omega)$$

its variation with frequency. At optical frequencies the increase in refractive index with frequency (and the greater refraction by a prism of blue light than red) is due to the energetic blue photons being more able than red photons to excite the molecules.

The refractive index varies strongly at frequencies very close to absorption bands because the molecule and the electromagnetic field are effectively degenerate (see the °Bohr frequency condition for an elaboration of this remark). If the refractive index is known throughout the frequency range it is possible to calculate the absorption spectrum of the molecule (and vice versa) by evaluating an integral over the curve using the **Kramers–Krönig dispersion relation**.

Further information

The calculation of the refractive index using perturbation theory is described in some detail in Chapter 13 of MQM. A very thorough account will be found in Caldwell and Eyring (1971); see also Hinchliffe and Munn (1985) and Böttcher and Bordewijk (1978). An elementary introduction to the far from elementary Kramers–Krönig relation is given by Slichter (1988); for more advanced treatments see Roman (1965, 1969), and for the derivation in two lines see Hu (1989).

Relativistic effects

See °Dirac equation.

Relaxation

The term **relaxation** signifies the return to thermal equilibrium of a non-equilibrium population of states. The return to equilibrium is

t $t+T_1$

(a)   (b)

characterized by one or more time-constants known as **relaxation times**. The term 'relaxation' also has other meanings in physics, such as the displacement of ions surrounding a defect in a solid or at a surface or grain boundary.

A particular case of relaxation to thermal equilibrium (the only aspect we shall consider), which is important for the interpretation of magnetic resonance spectra, is that of a collection of spins in a magnetic field. A collection of spins at thermal equilibrium in a magnetic field has a Boltzmann distribution of populations and, in classical terms, a random distribution of orientations in the xy plane. There are therefore two features of the relaxation: the return to a Boltzmann distribution of populations (Fig. R.6) and the randomization of orientations around the z axis, the direction of the applied field (Fig. R.7). The former is known as **longitudinal relaxation** since it is accompanied by a change in the net magnetic moment of the system parallel to the applied field. The latter, the loss of 'phase coherence', is known as **transverse relaxation** since it results in a change in the net magnetization in the plane perpendicular to the applied field. The two relaxation processes in general occur with different relaxation times, but under some circum-

Fig. R.6 An illustration of the process involved in longitudinal relaxation. (a) Initially, there are more β (down) spins than α spins, but the Boltzmann distribution requires there to be more α spins than β spins (we are supposing that the spins are those of protons and that there is a magnetic field present in the z-direction), as depicted in (b). The time constant for the return to equilibrium is the spin-lattice relaxation time T_1.

t $t+T_2$

(a)   (b)

Fig. R.7 An illustration of the process involved in transverse relaxation. (a) Initially the spins are aligned at the same angle around the z-axis, but (b) at thermal equilibrium they have random phases relative to this axis. The time constant for the achievement of random phases is the transverse relaxation time.

stances the relaxation rates are equal. The longitudinal relaxation time is denoted T_1 and the transverse relaxation time T_2.

Longitudinal relaxation is also called **spin–lattice relaxation** since it involves the transfer of energy from the spin system to the surroundings, the 'lattice'. This transfer is driven by fluctuating magnetic fields which induce a transition in the spin state. The coupling between the spin and the lattice is most efficient if the fluctuations oscillate at the transition frequency.

Transverse relaxation is also called **spin–spin relaxation** since it corresponds to the relative randomization of spins. It arises .from two processes. One is the transition between spin states that causes spin–lattice relaxation, for these transitions also help to randomize the orientations of the spins in the xy plane. The other process is a result of the spins experiencing slightly different local magnetic fields parallel to the applied field, and hence having slightly different °precession rates. Thus, their relative angles are randomized as they move out of step (Fig. R.8).

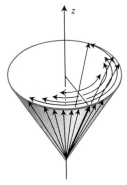

Fig. R.8 The transverse relaxation time is determined by the range of Larmor precession frequencies in the sample: the wider the range, the more rapidly the initial state of the system becomes a collection of spins with random phases around the z-axis.

Further information

A sensible place to start with the intricate subject of spin relaxation is one of the books on magnetic resonance mentioned in the entries on °nuclear magnetic resonance and °electron spin resonance. See particularly Harris (1986) and Sanders and Hunter (1987). The book by Slichter (1988) is an excellent introduction to the field.

Resonance

If two pendulums are weakly linked the motion of one may be transferred to the other, and the ebb and flow of energy continues indefinitely in the absence of damping forces. The exchange of energy is most effective when the pendulums have the same natural frequency, and they are then said to **resonate**. In classical physics, **resonance** is the strong coupling of systems with the same natural frequency, as in the coupling of a radio receiver to the weak electromagnetic field of a distant transmitter.

In spectroscopy, resonance denotes the strong coupling that occurs when the frequency of an electromagnetic field matches the transition frequency of a sample. The strong coupling results in strong absorption, the rapid transfer of energy from the field to the sample. This is the basis of °nuclear magnetic resonance and °electron spin resonance.

In °valence bond theory, resonance denotes the formation of a °superposition of wavefunctions to produce a better approximation to the actual wavefunction of a molecule. If the wavefunction of a molecule is described by the superposition $\psi + \psi'$ of two structures, it is said to be a **resonance hybrid** of them, or to **resonate** between them. A resonance hybrid is a blend of the contributing structures. The formation of a superposition results in a state with an overall lower energy (by the °variation principle), and the reduction in energy is greater the closer the energies of the contributing structures. Thus, two degenerate states 'resonate' most strongly, and the energy of their superposition may be substantially lower than that of the individual components. The most famous example of resonance in valence bond theory is °benzene, where the contributing Kekulé structures resonate to give a superposition that has a substantially lower energy than either structure alone (see °resonance energy).

Ionic–covalent resonance, in which the wavefunction of a molecule is expressed as a superposition of ionic and covalent structures, also achieves a lowering of energy because the overall wavefunction is a better description of the molecule than either structure alone.

Further information

For the considerable field of magnetic resonance see the entries on °nuclear magnetic resonance and °electron spin resonance. The concept of resonance in °valence bond theory can be pursued through that entry but see Pauling (1960) for a virtuoso deployment of the concept and Klein and Trinajstić (1990) for the concept in a more modern context.

Resonance energy

The **resonance energy** is the difference between the true energy of a molecule and a selected reference state. The choice of reference is problematic, and several suggestions have been made.

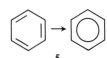

5

The **Hückel definition** of the resonance energy of a conjugated hydrocarbon takes the reference state as the hypothetical molecule with localized double bonds but with the same geometry as the actual molecule (**5**). Thus, if the π-electron energy of °benzene is found to be $3\alpha + 4\beta$, and the energy of each of the three ethene bonds is $\alpha + \beta$, the resonance energy is β. In the **Mulliken–Parr definition** the reference state is the localized bonding version of the molecule, but with the geometry optimized to the bond orders. Thus, for benzene the reference state would be the hypothetical cyclohexatriene molecule with alternating short and long bonds (**6**).

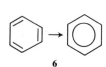

6

The revival of interest in °valence bond techniques, and in particular the formulation of the spin-coupled version of the theory in which it is no longer necessary to invoke hundreds of contributing structures, has led to a more precisely formulated and less arbitrary definition of the resonance energy. One definition is that it is the stabilization of the total spin-coupled solution relative to the solution based on the single most important spin-coupled structure.

Further information

See Coulson (1979) and Yates (1978) for discussions of conventions; for modern valence bond approaches, see Cooper *et al.* (1989), Cooper, Gerratt, and Raimondi (1990), and articles in the volume edited by Klein and Trinajstić (1990).

Rotational motion

The rotational energy of a molecule arises from its °angular momentum. Because the latter is quantized, so too is the rotational energy. The separation of adjacent rotational states is determined by the moment of inertia of the molecule, with large, heavy molecules having closely spaced energy levels. The **moment of inertia** of a molecule is the sum of the masses of the atoms multiplied by the squares of their perpendicular distances from the axis about which the rotation occurs (Fig. R.9).

Molecules are classified into four groups according to their moments of inertia around three perpendicular axes (Fig. R.10). These three axes are the **principal axes** of the molecule in the sense that the total rotational kinetic energy of the molecule is the sum of three quadratic terms of the form $\frac{1}{2}I\omega^2$, with one term for each of the three axes. For a cylinder, one principal axis is parallel to the axis and the other two

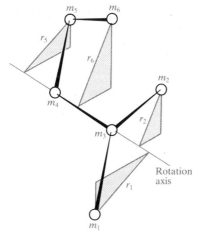

Fig. R.9 The moment of inertia of a collection of particles about a given axis is the sum of each mass multiplied by the square of its distance from the axis of rotation.

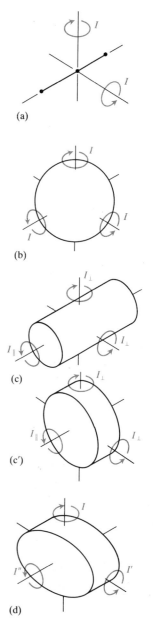

Fig. R.10 The classification of rigid rotors. (a) Linear rotor, (b) spherical rotor, (c) prolate symmetric rotor, (c') oblate symmetric rotor, (d) asymmetric rotor.

Box R.2 Moments of inertia†

1. *Diatomics*

$$I = \frac{m_A m_B}{m} R^2 = \mu R^2$$

3. *Symmetric rotors*

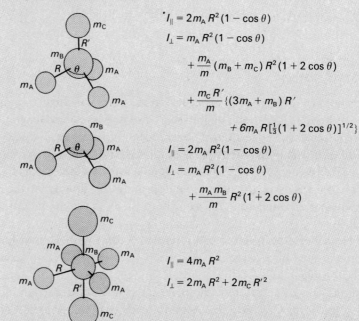

$$I_{\parallel} = 2m_A R^2 (1 - \cos\theta)$$

$$I_{\perp} = m_A R^2 (1 - \cos\theta)$$

$$+ \frac{m_A}{m} (m_B + m_C) R^2 (1 + 2\cos\theta)$$

$$+ \frac{m_C R'}{m} \{(3m_A + m_B) R'$$

$$+ 6m_A R [\tfrac{1}{3}(1 + 2\cos\theta)]^{1/2}\}$$

$$I_{\parallel} = 2m_A R^2 (1 - \cos\theta)$$

$$I_{\perp} = m_A R^2 (1 - \cos\theta)$$

$$+ \frac{m_A m_B}{m} R^2 (1 + 2\cos\theta)$$

$$I_{\parallel} = 4m_A R^2$$

$$I_{\perp} = 2m_A R^2 + 2m_C R'^2$$

2. *Linear rotors*

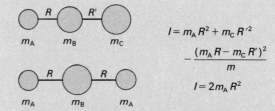

$$I = m_A R^2 + m_C R'^2$$

$$- \frac{(m_A R - m_C R')^2}{m}$$

$$I = 2m_A R^2$$

4. *Spherical rotors*

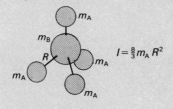

$$I = \tfrac{8}{3} m_A R^2$$

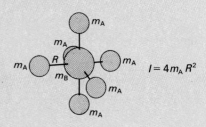

$$I = 4m_A R^2$$

† In each case *m* is the total mass of the molecule.

are perpendicular to it. The **principal moments of inertia** are the moments of inertia about the principal axes. The classification is as follows:

- **Linear rotor**: a body for which one principal moment of inertia (around the line of atoms) is zero.

- **Spherical rotor**: a body with all three principal moments of inertia equal.

- **Symmetric rotor**: a body with two principal moments of inertia equal but different from the third.

- **Asymmetric rotor**: a body with three different principal moments of inertia.

Symmetric rotors have at least one three-fold axis of symmetry. A **prolate** molecule is a cigar-shaped symmetric rotor; an **oblate** molecule is a pancake-shaped rotor. Spherical rotors have more than one three-fold axis (and hence are tetrahedral, octahedral, or icosahedral). The relation between molecular geometry and moment of inertia is summarized for some common cases in Box R.2, and the energy levels of each type of rotor are specified in Box R.3.

Box R.3 Rotational energies

Energies are expressed in terms of equivalent wavenumbers, using $E = hc\tilde{F}$.

Symmetric rotor:

$$F = BJ(J+1) + (A-B)K^2$$

$$J = 0, 1, 2, \ldots, \qquad K = J, J-1, \ldots, -J,$$

$$M_J = J, J-1, \ldots, -J \text{ (does not affect energy)}$$

$$A = \frac{\hbar}{4\pi c I_\parallel} \qquad B = \frac{\hbar}{4\pi c I_\perp}$$

Spherical rotor:

$$F = BJ(J+1) \qquad J = 0, 1, 2, \ldots$$

$$K = J, J-1, \ldots, -J, \qquad M_J = J, J-1, \ldots, -J \text{ (do not affect energy)}$$

$$B = \frac{\hbar}{4\pi c I}$$

Linear rotor:

$$F = BJ(J+1) \qquad J = 0, 1, 2, \ldots$$

$$M_J = J, J-1, \ldots, -J \text{ (does not affect energy)}$$

$$B = \frac{\hbar}{4\pi c I}$$

For the allowed rotational states of molecules, see °nuclear statistics.

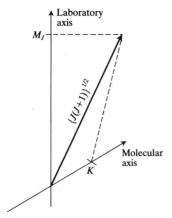

Fig. R.11 The vector model of the angular momentum of a rotor. The quantum number J gives the magnitude of the vector representing the rotational angular momentum of the body, and K and M_J give the projections of the angular momentum on to the molecular axis and the laboratory-fixed axis respectively. The greater the magnitude of K, the more rapid the rotation of the molecule about its axis.

Each rotational state is specified by three quantum numbers (Fig. R.11):

- J, which determines the overall angular momentum of the molecule and (for linear and spherical rotors) its rotational energy.

J can take any of the values 0, 1, 2,... without limit.

- K, which gives the projection of the angular momentum on the molecular axis as $K\hbar$. The energy of a symmetric rotor depends on K because it is important to know how the motion is apportioned between the different axes.

$K = 0, \pm 1,... \pm J$ for symmetric and spherical rotors ($2J + 1$ values for each value of J), but $K = 0$ for linear rotors because the angular momentum must be perpendicular to the line of atoms. Equal and opposite values of K correspond to opposite senses of rotation around the molecular axis.

- M_J, which gives the projection of the angular momentum on an axis fixed in space. The energy of a rotor is independent of M_J unless an electric field is present and the molecule is polar.

$M_J = 0, \pm 1,... \pm J$ for all types of rotor, giving $2J + 1$ values for each value of J.

The degeneracy of a linear rotor in a state with quantum number J is $2J + 1$ (because there are $2J + 1$ values of M_J but only one value of K). The degeneracy of a spherical rotor, however, is $(2J + 1)^2$ since for each value of J and M_J there are also $2J + 1$ values of K (Fig. R.12). One macroscopic consequence of the much greater density of states of spherical rotors compared with linear rotors is their greater heat capacity. At high temperatures the rotational heat capacity of a spherical rotor is 50 per cent greater than that of a linear rotor (see the °equipartition theorem).

Not all the rotational states of a molecule may be occupied if the rotational motion of the molecule interchanges identical nuclei (such as the two hydrogen nuclei in H_2O under a 180° rotation). This feature is a consequence of the °Pauli principle and is described in the entry on °nuclear statistics.

°Electric dipole transitions between the rotational states of molecules can occur only if the molecule has a permanent dipole moment. The dipole acts as a kind of lever for the interaction, and through it the field accelerates the molecule by exerting a torque. The °selection rules for electric dipole transitions of polar molecules are $\Delta J = \pm 1$ and $\Delta K = 0$. The former reflects the conservation of angular momentum and the fact that a photon carries unit spin angular momentum. The fact that K cannot change stems from the fact that the molecule has no component of electric dipole moment perpendicular to its symmetry axis, so the field cannot exert a torque about that direction. It follows that a rotational spectrum is a set of lines spaced by $2B$ with an intensity distribution governed largely by the initial (Boltzmann) populations of the rotational states.

Further information

The rotation of molecules is dealt with in Chapter 11 of MQM, where the expressions given in Box R.3 are derived. See also Hollas (1983, 1987) and Flygare (1978). A standard work on molecular rotation is the book by Townes and Schawlow (1955). See also Gordy and Cook (1984), Allen and Cross (1963) and, for spectroscopic applications, Herzberg (1945, 1950, 1966). The classical mechanics of rotational motion is described in Goldstein (1982) and Kibble (1985); for a thorough treatment of the angular momentum of diatomic molecules see Judd (1975). For a good general introduction to the angular momentum of rotating molecules see Zare (1987).

Russell–Saunders coupling

See °Vector model of the atom.

Rydberg constant

The **Rydberg constant R_∞** is

$$R_\infty = \frac{m_e e^4}{8h^3 c\varepsilon_0^2}$$

and its numerical value is $1.097 \times 10^5 \text{ cm}^{-1}$. The constant appears in the expression for the energies of hydrogenic atoms, which are equal to $E = -hcZ^2 R/n^2$, with R a modification of the Rydberg constant that takes into account the joint motion of the electron and the nucleus around their common centre of mass. Specifically, $R = R_\infty \times \mu/m_e$, where μ is the °reduced mass of the atom.

The constant was first introduced empirically by the Swedish spectroscopist Johannes Rydberg (1854–1919) in his summary of the spectrum of atomic °hydrogen. Its theoretical value was first obtained by °Bohr in 1913 in his treatment of the hydrogen atom, and then by Schrödinger in his solution of his equation for the atom in 1926.

Further information

See the entries on °atomic spectra, the °hydrogen atom, and the °Bohr model of the atom for more information.

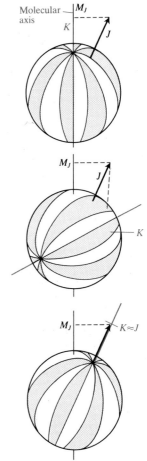

Fig. R.12 A spherical rotor has a very high degeneracy. These diagrams show the wavefunction for a state with $J=4$ and a fixed value of K. However, the axis on which K is defined may have $2J+1$ orientations with respect to a laboratory axis, and all the states have the same energy.

Rydberg level

An electronic transition in a molecule might excite an electron into a molecular orbital that is best regarded as formed from atomic orbitals belonging to shells outside the valence shells of the atoms. The state so formed is a **Rydberg state** of the molecule and the electron occupies a **Rydberg level**. An example of the formation of a Rydberg state is the excitation of a $2p$ electron of an F atom in an F_2 molecule into a molecular orbital composed of $3s$ orbitals.

Electrons in Rydberg levels are characterized by small °quantum defects. They are so diffuse that they hardly interact with the inner electrons, and even the nuclei of a diatomic molecule appear to them as a single point positive charge. Hence their wavefunctions resemble those of the °hydrogen atom.

Further information

Rydberg levels are discussed in some detail in §10.3 of King (1964) and §VI.5 of Herzberg (1950). For thorough discussions see Duncan (1971) and Stebbings and Dunning (1983).

s orbital

An **s orbital** is an °atomic orbital with $l = 0$. An electron in an s orbital is called an **s electron**. There is one s orbital in each shell of an atom. All s orbitals are spherically symmetrical (with no angular nodes, Fig. S.1) but as the number of radial nodes is equal to $n - 1$, higher shells have progressively more radial nodes (Fig. S.2).

All s orbitals have non-zero amplitude at the nucleus; therefore an s electron has a non-zero probability of being found very close to the nucleus. This difference in behaviour from all other atomic orbitals (which have zero amplitude at the nucleus) is a consequence of the s orbital having zero orbital angular momentum, with the result that there is no centrifugal force acting on the electron however closely it approaches the nucleus. That an s electron can °penetrate through inner shells and approach the nucleus closely is of profound importance for the periodicity of the elements and the structure of the periodic table.

Further information

See the entry on °atomic orbitals for a description of s orbitals in the context of atomic structure, and the entry on the °hydrogen atom for an insight into their form. The explicit radial dependences of s orbitals are set out in Box S.1.

σ orbital

A **σ orbital** is a °molecular orbital with locally cylindrical symmetry around the internuclear axis (Fig. S.3). That is, a σ orbital looks like an s orbital when viewed along the bond. An electron in a σ orbital is called a **σ electron**, and a pair of electrons in a σ orbital is called a **σ bond**. A σ orbital (which may be either bonding or antibonding) may be formed from the overlap of any pair of orbitals with cylindrical

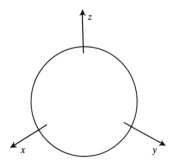

Fig. S.1 The boundary surfaces of all s orbitals are spherical, with the nucleus at the centre.

(a)
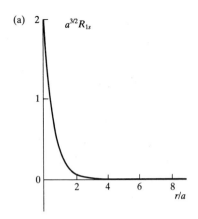

(b)
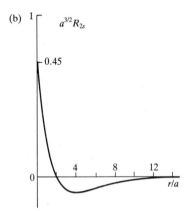

(c)
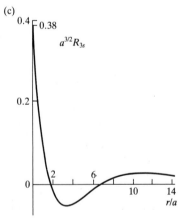

Fig. S.2 The radial component of three *s* orbitals of a hydrogen atom (a) 1*s*, (b) 2*s*, (c) 3*s*. Note that the number of radial nodes is 0, 1, and 2 respectively. In each case the orbital has non-zero amplitude at the nucleus (there is a change in scale between the diagrams).

Box S.1 *s* orbitals

$$\psi = R(r)\, Y, \qquad Y = \left(\frac{1}{4\pi}\right)^{1/2}$$

Radial wavefunctions:

$$n = 1 \qquad R = 2\left\{\frac{Z}{a_0}\right\}^{3/2} e^{-\rho/2} \qquad \rho = 2Zr/a_0$$

$$n = 2 \qquad R = \frac{1}{2\sqrt{2}}\left\{\frac{Z}{a_0}\right\}^{3/2} (2-\rho)\, e^{-\rho/2} \qquad \rho = Zr/a_0$$

$$n = 3 \qquad R = \frac{1}{9\sqrt{3}}\left\{\frac{Z}{a_0}\right\}^{3/2} (6 - 6\rho + \rho^2)\, e^{-\rho/2} \quad \rho = 2Zr/3a_0$$

Average values:

$$\langle r \rangle = \frac{3n^2 a_0}{2Z} \qquad \langle 1/r \rangle = \frac{Z}{a_0\, n^2}$$

Probability density at the nucleus

$$n = 1 \qquad \psi^2(0) = \frac{1}{\pi}\left\{\frac{Z}{a_0}\right\}^3$$

$$n = 2 \qquad \psi^2(0) = \frac{1}{8\pi}\left\{\frac{Z}{a_0}\right\}^3$$

$$n = 3 \qquad \psi^2(0) = \frac{1}{27\pi}\left\{\frac{Z}{a_0}\right\}^3$$

See also Box H.7 for general expressions.

symmetry around the internuclear axis. Common examples are orbitals formed from (s,s), (s,p_z), and (p_z,p_z) overlap (Fig. S.3).

The classification of an orbital as σ is strictly applicable only to linear molecules. However, it is common practice to extend the classification to orbitals that have local cylindrical symmetry with respect to a specific individual internuclear axis. For instance, it is legitimate to speak of the σ bonds in benzene.

Further information

See the entry on °molecular orbital theory for a discussion of the formulation of σ orbitals, along with a more general discussion and further references. The surveys by Baird (1986) and DeKocK (1987) are worth looking at in this context. For a general reference see DeKock and Gray (1980) and Coulson (1979).

Schrödinger equation

The **Schrödinger equation** is the equation which, when solved, gives the °wavefunction of a system and thence, through the wavefunction, all the properties of the system. It was proposed by the Austrian theoretical physicist Erwin Schrödinger (1887–1961) in 1926, and its formulation marked the joint foundation (with Heisenberg and his °matrix mechanics) of quantum mechanics.†

The Schrödinger equation is a second-order partial differential equation in space (it contains terms such as d^2/dx^2) and a first-order partial differential equation in time. Various forms of it are shown in Box S.2. The **time-dependent Schrödinger equation** is used to calculate the time evolution of a wavefunction. In many cases it is possible to separate the time dependence of a wavefunction from its spatial variation and to write the total wavefunction, Ψ, as a product of the spatial wavefunction, ψ, and a °complex oscillating function of the form $e^{-iEt/\hbar}$, where E is the energy of the state. The spatial component ψ satisfies the **time-independent Schrödinger equation**, which is a second-order differential equation in the spatial coordinates. The latter equation has the same form as a classical equation for a standing wave. This property gives rise to the name 'wave mechanics', which is used to distinguish Schrödinger's approach to quantum mechanics from Heisenberg's °matrix mechanics.

The time-independent Schrödinger equation may be interpreted as an equation for the *curvature* of the wavefunction, in the sense that it shows how the second derivative of the wavefunction, ψ'', is determined by the value of the total energy and the potential and amplitude of the wavefunction at each point. The essential feature to note is that the sharpness with which ψ is curved is proportional to the difference $E - V$, and hence to the °kinetic energy of the particle (see the °de Broglie relation). The sign of the curvature at each point depends on the signs of $E - V$ and ψ at the point in question, as displayed in Fig. S.4.

How the Schrödinger equation unfurls a wave from a specified amplitude and slope at a given point can be seen by considering a free particle with $V=0$ everywhere. Suppose $\psi > 0$ at the point of interest, and that $\psi' = 0$ there. Then the curvature of ψ is negative, and the function droops down towards zero (Fig. S.5) with ever decreasing curvature (because ψ is decreasing). The droop causes the wavefunction to pass through zero (with zero curvature when $\psi = 0$) and become negative. As soon as it does so, its curvature changes sign and it begins to curl up towards zero. It crosses zero, and begins to droop again.

† 'While visiting Paris he (Victor Henri) received from Langevin a copy of 'the very remarkable thesis of de Broglie'; back in Zurich and having not very well understood what it was all about, he gave it to Schrödinger who after two weeks returned it to him with the words: "That's rubbish". When visiting Langevin again, Henri reported what Schrödinger had said. Whereupon Langevin replied: "I think Schrödinger is wrong; he must look at it again". Henri, having returned to Zurich, told Schrödinger "You ought to read de Broglie's thesis again, Langevin thinks this is a very good work"; Schrödinger did so and "began his work".' Jammer (1966, p.258).

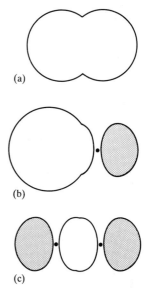

Fig. S.3 A σ orbital can be formed from the overlap of any kind of orbital with cylindrical symmetry around the internuclear axis. This diagram shows the result of overlap of (a) two *s* orbitals, (b) an *s* orbital and a *p* orbital, and (c) two *p* orbitals.

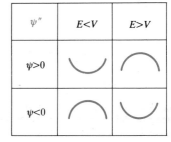

Fig. S.4 This chart shows how the curvature of the wavefunction is related to the sign of the wavefunction (left column) and the relative sizes of the total energy E and the potential energy at the point (top row).

Box S.2 The Schrödinger equation

The **time-dependent Schrödinger equation** *is*

$$H\Psi = i\hbar\frac{\partial\Psi}{\partial t}$$

where H is the °hamiltonian for the system. If H does not change with time, Ψ may be expressed as $\Psi = \psi\, e^{-iEt/h}$, where ψ satisfies the **time-independent Schrödinger equation**:

$$H\psi = E\psi$$

Some typical forms of the time-independent Schrödinger equation are as follows:

One-dimensional system:

$$-\frac{\hbar^2}{2m}\frac{d^2\psi}{dx^2} + V(x)\psi = E\psi$$

For example, $V(x) = \tfrac{1}{2}kx^2$ for the °harmonic oscillator.

Three-dimensional system:

$$-\frac{\hbar^2}{2m}\nabla^2\psi + V(r)\psi = E\psi$$

For example, $V = -Ze^2/4\pi\varepsilon_0 r$ for a °hydrogenic atom.

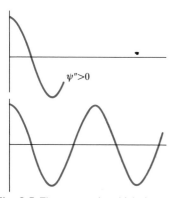

Fig. S.5 The manner in which the Schrödinger equation traces out a wavefunction from a given amplitude and slope is shown in this sequence of diagrams. At each stage, the curvature of the function can be worked out from the information in Fig. S.4.

The process continues indefinitely, and the wavefunction unfolds as a wave of constant amplitude and wavelength. The distance between the crossing points decreases as E is made larger, and hence the wavelength decreases as the particle's energy—its °kinetic energy in this instance—is increased. If the particle is in a region of varying potential, the shape of the wavefunction is more complex, but in each case it is governed by the relation between $E - V$, the amplitude, and the curvature.

A second-order differential equation has an infinite number of solutions. However, only some of the mathematically possible solutions satisfy the stringent requirements of the Born interpretation and only those which do are physically acceptable. In particular the wavefunctions must (Fig. S.6):

● be finite everywhere (except possibly over an infinitesimal region),

● be continuous everywhere,

● have a continuous slope everywhere (except at locations where the potential becomes infinite),

● be single valued everywhere (unless °spin is involved).

These conditions imply that the only acceptable wavefunctions are solutions that satisfy certain **boundary conditions**, or values at certain locations in space. For example, the only acceptable wavefunctions for a °particle in a box have zero amplitude at each wall, for all other mathematically possible solutions will conflict with one or more of the physical requirements stated above. The wavefunctions for a particle on a ring are acceptable only if they satisfy **cyclic boundary conditions**, that the wavefunction has the same value at the angles separated by 2π. Solutions that satisfy the boundary conditions can be found only for certain values of E; hence, only those values of E are possible for the system. Thus the imposition of boundary conditions on the wavefunction implies the °quantization of energy.

Some of the solutions of the Schrödinger equation are discussed under the appropriate headings (see °particle in a box, °harmonic oscillator, °angular momentum, and °hydrogen atom). When analytical solutions cannot be obtained the equation may be solved numerically or by one of the available approximation procedures, such as °perturbation theory, °variation theory, or the method of °self-consistent fields.

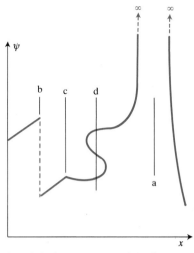

Fig. S.6 Transgressions of the four criteria that a wavefunction must obey if it is to be acceptable. This candidate wavefunction is multiply unacceptable because it (a) is infinite over a non-infinitesimal range, (b) has a discontinuous amplitude and (c) slope, and (d) is not single-valued.

Further information

See Chapter 2 of MQM for a justification of the form of the Schrödinger equation in terms of an action principle and the analogy between the propagation of light and the propagation of particles. All introductory books on quantum mechanics deal at length with the equation and its solutions, so see, for example, MQM, Davydov (1976), Das and Melissinos (1986), and Bransden and Joachain (1989). For Schrödinger's original papers, see Schrödinger (1926) and van der Waerden (1967), and for the originality of much of Schrödinger's life, see the fascinating biography by Moore (1989). As usual, Jammer (1966, 1974) gives an authoritative and interesting account of the intellectual development of the equation. For a survey of exactly solvable model systems, see Albeverio (1988), and for some pictures of its solutions, see Brandt and Dahmen (1985). For a problem book, see Johnson and Pedersen (1973).

Second quantization

Ordinary, common or garden **first quantization** is the replacement of physical observables by °operators that obey a certain °commutation relation. The properties of a system are then calculated by allowing these operators to act on the wavefunction ψ obtained by solving the °Schrödinger equation. In **second quantization** these wavefunctions are interpreted as operators acting on an abstract space. First quantization introduces a new fundamental constant into physics (Planck's constant); second quantization does not introduce a new constant, and hence it is best regarded as a mathematical manipulation rather than a fundamental realignment of our understanding of the world.

In a second-quantized description, the wavefunction ψ becomes the **annihilation operator** ψ. When it acts, it annihilates the state that the

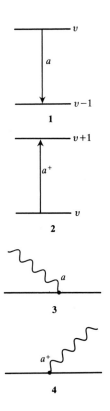

wavefunction ψ represents. Likewise, the complex conjugate wavefunction ψ^* becomes the **creation operator** ψ^+. When it acts, it creates the state represented by the wavefunction ψ.

The operators ψ and ψ^+ are normally expressed in terms of two operators a and a^+ that generate states with quantum numbers differing by ± 1 from the initial state. Thus, when the annihilation operator a acts on a harmonic oscillator in a state with quantum number v it generates the state with quantum number $v-1$ (**1**). When the creation operator a^+ acts on the state v it generates the state with quantum number $v+1$ (**2**). In the entries on °photons and °phonons it is explained that successive excitations of an oscillator may be interpreted as the presence of additional particles. Thus, when a and a^+ are interpreted as operators for the electromagnetic field, a annihilates photons (**3**) and a^+ creates them (**4**).

Further information

See MQM Appendix 4 for an introduction to annihilation and creation operators for the harmonic oscillator. A much more complete introduction to the application of the techniques in chemistry is Jorgensen (1981). See also standard texts, such as Bransden and Joachain (1989), for straightforward introductions. The connection between second quantization and angular momentum is established in Schwinger (1965).

Secular determinant

A °molecular orbital constructed from a °linear combination of N atomic orbitals is written $c\psi + c'\psi' + \ldots$, where the ψs are the atomic orbitals. The N coefficients c are chosen by searching for the set of values that gives the lowest energy. This optimum set of coefficients is obtained by solving a set of N simultaneous equations (see Box S.3) called the **secular equations**, in which the factors multiplying the unknowns (the coefficients c) include the energy E of the system, which at this stage is also unknown.

The N possible values of E are found by noting that a set of N simultaneous equations has a non-trivial solution only if the $N \times N$ determinant of the factors of the unknown quantities is zero. This determinant is the **secular determinant** of the system. It follows that if we can find the N roots of the determinant (the values of E that ensure it is zero), then we can use each of these values in turn in the secular equations, and solve those equations for the N coefficients c. The lowest root of the secular determinant is then identified with the lowest energy of the molecular orbitals attainable with the given basis set of atomic orbitals, and the corresponding coefficients give the composition of this most bonding orbital. The remaining $N-1$ roots and the

Box S.3 Secular equations and the secular determinant

For a linear combination of N atomic orbitals of the form

$$\psi = \sum_n c_n \phi_n$$

the Schrödinger equation $H\psi = E\psi$ becomes

$$\sum_n c_n \{H\phi_n - E\phi_n\} = 0$$

Multiplication from the left by ϕ_k and integration gives

$$\sum_n c_n \{H_{kn} - ES_{kn}\} = 0$$

which is a set of N simultaneous equations, the **secular equations**, for the c_n. The equations have a solution if the determinant of the coefficients of the unknowns c_n vanishes. That is, if

$$|H_{kn} - ES_{kn}| = 0$$

where $|H_{kn} - ES_{kn}|$ is the $N \times N$ **secular determinant**.

In the special case of a 2×2 determinant with $H_{12} = H_{21}$, $S_{11} = S_{22} = 1$ and $S_{12} = S_{21} = 0$, the secular equations have solutions for

$$E = H_{11} - H_{12}\cot\theta$$

$$E = H_{22} + H_{12}\cot\theta$$

and the normalized linear combinations are respectively

$$\psi = \phi_1 \sin\theta + \phi_2 \cos\theta$$

$$\psi = \phi_1 \cos\theta - \phi_2 \sin\theta$$

where

$$\theta = \tfrac{1}{2}\arctan\frac{2H_{12}}{H_{11} - H_{22}}$$

coefficients c obtained from the secular equations in each case are then identified with molecular orbitals of successively higher energy.

A secular determinant is encountered wherever a linear combination of wavefunctions is optimized. If there is a °perturbation that can mix one state with another, the true ground state of the system is expressed as a linear combination of the two unperturbed states. The energy of the new system is given by the lowest root of the secular determinant, and the set of coefficients corresponding to that root gives the optimum modification of the wavefunction. Two examples of this type of calculation are °configuration interaction and the °Hückel method for conjugated molecules.

The origin of the name 'secular' is found in classical mechanics, and especially in celestial mechanics. A secular variation in the motion of a body, in particular the orbit of a planet, is one that gradually develops over a long period of time (*saeculum* is Latin for age or generation). If we imagine the interaction between atomic orbitals in a molecule to grow to its final value as the atoms are brought together, the perturbation gradually accumulates and the different linear combinations of atomic orbitals diverge until they attain the separations characteristic of the molecule. In the LCAO method we are actually considering a strong secular perturbation on the atomic orbitals, and the resulting levels are found from the secular determinant.

Further information

See MQM Chapters 8 and 10. The former deals with the occurrence of the secular determinant in degenerate state perturbation theory, and the latter with its occurrence in the treatment of a number of problems related to molecular structure. Further information will be found in Levine (1983), McGlynn *et al.* (1972), and Das and Melissinos (1986). For the secular determinant in classical mechanics see Goldstein (1982) and Kibble (1985).

Selection rule

A **selection rule** is the statement of the changes of state that may occur in a specific type of transition. For example, there are °electric dipole selection rules that govern the changes of state during an electric dipole transition. Changes of state that may occur are called **allowed transitions**; changes that may not occur by the specified interaction are called **forbidden transitions**.

A **gross selection rule** is a statement about the overall property that a molecule must possess in order for it to undergo a specific class of transitions. For example, a molecule must be polar if it is to be able to undergo electric dipole transitions between its rotational states. A **specific selection rule** is a statement about the changes that may occur in a °quantum number that characterizes the system. For example, the quantum number of a harmonic oscillator can change only by $+1$ or -1 (that is, $\Delta v = \pm 1$) in an electric dipole transition. Some selection rules are summarized in Box S.4.

A gross selection rule expresses the requirement that a molecule has some way of interacting with the electromagnetic field. The possession of a permanent dipole moment, for instance, means that the electromagnetic field can accelerate the molecule rotationally by exerting a torque on it. The specific selection rules can generally be understood in terms of the spin angular momentum of the °photon and the conservation of angular momentum, but some are characteristics of the symmetry of the system. Thus, when a photon is absorbed it transfers its angular momentum to the atom or molecule and either increases or decreases the latter's angular momentum (Fig. S.7) depending on the photon's helicity (its direction of spin relative to its motion). Selection rules such

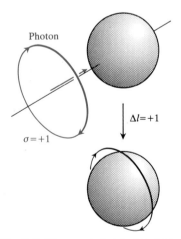

Photon

$\Delta l = +1$

$\sigma = +1$

Fig. S.7 When a photon is absorbed by an atom in an electric dipole transition, its angular momentum is transferred to an electron. In this example, a photon with $\sigma = +1$ (and left circularly polarized) is absorbed by an atom, and the latter's angular momentum increases by one unit.

Box S.4 Selection rules

Atoms

Electric dipole transitions:

$\Delta J = 0, \pm 1$ but $J = 0 \nrightarrow 0$

$\Delta L = 0, \pm 1$ but $L = 0 \nrightarrow 0$

(Russell–Saunders coupling)

$\Delta S = 0$ (Russell–Saunders coupling)

$\Delta l = \pm 1$

Laporte selection rule: $g \rightarrow u$, $u \rightarrow g$

Magnetic dipole transitions:

$\Delta J = 0, \pm 1$ but $J = 0 \nrightarrow 0$

$\Delta L = 0, \pm 2$

$g \rightarrow g$, $u \rightarrow u$

Electric quadrupole transitions:

$\Delta J = 0, \pm 1, \pm 2$ but $J = 0 \nrightarrow 0$

$\Delta L = 0, \pm 1, \pm 2$ but $L = 0 \nrightarrow 0$

$g \rightarrow g$, $u \rightarrow u$

Molecules

Electric dipole electronic transitions:

$\Delta J = 0, \pm 1$ but $J = 0 \nrightarrow 0$

$+ \rightarrow -, \; - \rightarrow +, \; g \rightarrow u, \; u \rightarrow g$

$\Delta \Lambda = 0, \pm 1$

but for $\Lambda = 0 \rightarrow 0$: $\Sigma^+ \rightarrow \Sigma^+$, $\Sigma^- \rightarrow \Sigma^-$

$\Delta S = 0, \Delta \Sigma = 0$

$\Delta \Omega = 0, \pm 1$ but $\Omega = 0 \nrightarrow 0$ if $\Delta J = 0$

Vibrational transitions:
Infrared: Dipole moment must change for displacement along °normal coordinate; then $\Delta v = \pm 1$.
Raman: Polarizability must change with vibration; then $\Delta v = \pm 1$.

Rotational transitions:
Microwave: Molecule must have a permanent electric dipole moment; then $\Delta J = \pm 1$, $\Delta K = 0$.
Raman: Polarizability must be anisotropic; then $\Delta J = 0, \pm 1, \pm 2$.

Vibration–rotation transitions:
Infrared: $\Delta v = \pm 1$ and $\Delta J = 0, \pm 1$ but $\Delta J \neq 0$ for linear molecules (no Q branch).
Raman: $\Delta v = \pm 1$ and $\Delta J = 0, \pm 1, \pm 2$.

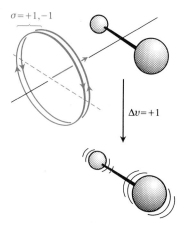

$\sigma = +1, -1$

$\Delta v = +1$

Fig. S.8 A plane polarized photon is a superposition of two counter-rotating circularly polarized states. When it is absorbed by a molecule the angular momentum is conserved in the sense that *both* components are absorbed. The outcome is that the molecule starts to vibrate harmonically (for that motion may also be expressed as the superposition of two counter-rotating motions of the molecule).

as $\Delta l = \pm 1$ and $\Delta J = \pm 1$ reflect this conservation of angular momentum, and are discussed further under °electric dipole transitions and in the sections on the different types of spectra. (See also °magnetic dipole and electric quadrupole transition for a variant of this rule.)

The conservation of angular momentum also accounts for there being a $\Delta v = \pm 1$ selection rule for one-dimensional harmonic oscillators even though the latter can possess no angular momentum. In this case, the incoming plane polarized photon is a superposition of photon states of opposite helicity, and that superposition is transferred to the oscillator (Fig. S.8). The oscillator acquires the *superposition* of angular momentum states (not just one angular momentum state, as in rotational transitions), which we observe as mechanical oscillation in the same plane as the polarization of the absorbed photon.

All selection rules may be deduced by establishing the conditions that enable the transition moment to be non-zero (see °electric dipole transition). One very important way of establishing the criteria is to use °group theoretical arguments based on the symmetry of the initial and final states and the perturbation that connects them. Specifically, the transition moment must vanish unless the product of the irreducible representations of the initial state, the final state, and the transition moment operator spans the totally symmetric irreducible representation.

Further information

See MQM Chapter 7 for a discussion of the group theoretical basis of selection rules. Individual selection rules for the various classes of spectroscopic transition are discussed in Hollas (1983, 1987), Herzberg (1944, 1945, 1950, 1966), Graybeal (1988), and specific sources mentioned in the appropriate entries.

Self-consistent field

The **self-consistent field** (SCF) method of calculating atomic and molecular structures was originated by the Cambridge mathematician Douglas Hartree, improved by the Russian physicist Vladimir Fock, and used by almost everyone. The procedure begins with a set of approximate °orbitals for all the electrons in the system. One electron is then selected, and the potential in which it moves is calculated by freezing the distribution of all the other electrons and treating their averaged distribution as a centrosymmetric source of potential. The Schrödinger equation for the electron is solved for this potential, which gives a new orbital for it. The procedure is repeated for all the other electrons in the system, using the electrons in the frozen orbitals as the source of the potential. At the end of one cycle we have a set of orbitals which in general differs from the original set. Now the cycle is repeated, but the improved wavefunctions generated by the first cycle are used as the starting point. A complete cycle generates a new set of improved functions. This sequence is continued until passing a set of orbitals through the cycle leaves them unchanged. The orbitals are then self-consistent.

The original **Hartree method** expresses the total wavefunction of the system as a product of one-electron orbitals (see the °orbital approximation). In the **Hartree–Fock method** the wavefunction is an antisymmetrized determinantal product of one-electron orbitals (see °Slater determinant), and so it allows for the effects of electron °exchange. In order to implement the method, the Schrödinger equation is converted into a set of **Hartree–Fock equations** (Box S.5). The **unrestricted Hartree–Fock method** (UHF) allows more freedom to the form of the orbitals by permitting the spatial form of the orbital to depend on whether the electron has an α or a β °spin.

Since the Hartree–Fock (HF) method replaces the instantaneous location of all the electrons other than the one considered by an averaged charge located at the nucleus, it neglects the °correlation of electrons. A major component of modern SCF calculations looks for ways of taking into account correlations between the locations of electrons.

Box S.5 The Hartree–Fock equations

$$\left\{ H\,(\text{core}) + 2\sum_j J_j(1) - \sum_j K_j(1) \right\} \psi_i(1) = \varepsilon_i \psi_i(1)$$

where the **Coulomb operator** is

$$J_j(1) = \kappa \int d\tau_2\, \psi_j^*(2)\, \psi_j(2)\, \frac{1}{r_{12}}$$

with $\kappa = 1/4\pi\varepsilon_0$, and the **exchange operator** is defined through

$$K_j(1)\,\psi_i(1) = \kappa \int d\tau_2\, \psi_j^*(2)\, \psi_i(2)\, \frac{1}{r_{12}}\psi_j(1)$$

Then, the energy is given by

$$E = \sum_{i,j} \{2\varepsilon_i - 2J_{ij} + K_{ij}\}$$

in terms of the **Coulomb integral:**

$$J_{ij} = \int \psi_i^*(1)\, J_j(1)\, \psi_i(1)\, d\tau_1$$

and the **exchange integral:**

$$K_{ij} = \int \psi_i^*(1)\, K_j(1)\, \psi_i(1)\, d\tau_1$$

In the application of the SCF procedure to molecules, the molecular orbitals are expressed as °linear combinations of atomic orbitals, and the self-consistency procedure is applied to the combination coefficients, not to the atomic orbitals themselves. The equations employed in the optimization procedure are called the **Roothaan equations**. In one choice of a **basis set** the initial orbitals are the °Slater atomic orbitals, but in practice it is possible to use these orbitals only for atoms and linear molecules (on account of difficulties with integrations). The evaluation of molecular integrals is considerably simplified if °gaussian orbitals are used, but more of these must be used if the atomic wavefunctions are to be accurately represented.

Further information

A very good discussion of the Hartree–Fock equations and how they are set up and solved by analytic or numerical procedures (where by 'analytic' is meant expressing the orbitals as a linear combination of °Slater type orbitals) will be found in Condon and Odabaşi (1980), who provide tables of coefficients and exponents and many references to the literature.

Semiconductor

A **semiconductor** is a substance with an electrical conductivity that increases as the temperature is raised. In terms of °band theory, a semiconductor is a solid in which an empty band is thermally accessible from a filled band (Fig. S.9). Hence, as the temperature is raised, the number of electrons in the upper band increases (approximately exponentially, see the °Fermi–Dirac distribution) and more holes are left in the valence band. The increase in the number of carriers is responsible for the increase in conductivity.

An **intrinsic semiconductor** is one in which semiconduction is a property of the band structure of the pure material. An **extrinsic semiconductor** is one in which carriers are present as a result of the replacement of some atoms (to the extent of about 1 in 10^9) by **dopant** atoms, the atoms of another element. If the dopant increases the number of electrons in the material, and hence partially populates the conduction band (Fig. S.10a), it is called an **n-type semiconductor** (where the n denotes the presence of negative electrons). An example is silicon (Group IV/14) doped with arsenic (Group V/15). We can picture the semiconduction as arising from the transfer of an electron from an As atom to a neighbouring Si atom, when it becomes able to travel through the solid in the silicon conduction band. If the dopant decreases the number of electrons it introduces holes in the previously full valence band (Fig. S.10b) and gives rise to a **p-type semiconductor**

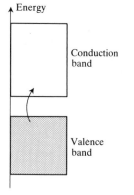

Fig. S.9 In an intrinsic semiconductor, there is only a small gap between the valence and conduction bands, and the latter is thermally populated by electrons. The higher the temperature, the greater the population in the conduction band.

(where the p denotes the relatively positive holes). An example is silicon doped with indium (Group III/13). Now we can picture the semiconduction as arising from the transfer of an electron from a Si atom to a neighbouring In atom. The electrons at the top of the silicon valence band are now mobile, and carry current through the solid.

A **compound semiconductor** is an intrinsic semiconductor that is a combination of different elements (as distinct from minuscule doping). Examples are the **III-V compounds** formed from elements of Groups III/13 and V/15 (such as GaN), and the **II-VI compounds** formed from elements of Groups II/2 and VI/16 (such as CdS). Compound semiconductors are not all covalent compounds, for the class also includes many *d*-block oxides.

A further classification is into **direct-gap semiconductors** and **indirect-gap semiconductors**. As explained in the entry on °band theory, these differ in the relative linear momenta of electrons at the top of the valence band and the bottom of the conduction band.

Further information

An introduction to semiconductors will be found in Chapters 3 and 18 of Shriver, Atkins, and Langford (1990). For chemical aspects see Cox (1987), Cheetham and Day (1987), Rao and Gopalakrishnan (1986), and Duffy (1990). For more physical aspects see Anderson (1984), Kittel (1986, 1987), and Ridley (1988).

Semiempirical methods

When the labour of a °self-consistent field molecular orbital calculation appears, or actually is, too great, approximations are introduced in a more or less rational fashion and integrals are estimated from empirical data. Such methods give rise to the numerous **semiempirical techniques** of molecular orbital theory. The programs that are available typically require very little knowledge of quantum mechanics, and are widely used as a resource in industry, particularly the pharmaceutical industry. Until recently, semiempirical methods have been the only way to study molecules of more than about 30 atoms, but new developments in °*ab initio* techniques and bigger, faster computers now mean that full SCF calculations can be carried out on much larger systems.

The grandfather of all semiempirical techniques is °Hückel theory, in which all overlap integrals are neglected and the remaining integrals are expressed as two parameters α and β. In the **zero differential overlap** (ZDF) methods, a less severe approximation is made. In these procedures, the product of two atomic orbitals is set equal to zero everywhere if they belong to separate atoms. The fact that the functions themselves

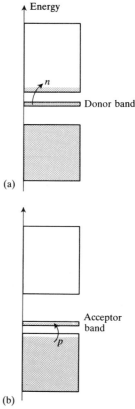

Fig. S.10 (a) In an n-type semiconductor, an electron-rich dopant supplies electrons to the conduction band (so long as the temperature is above zero), and they act as carriers. (b) In a p-type semiconductor, an electron-poor dopant makes available a band of states to which electrons may be thermally excited. The holes so introduced in the valence band act as carriers.

are set equal to zero is the source of the term 'differential', which distinguishes the approximation from the assumption that the overlap *integral* is zero (the differential of an integral is the function itself). The ZDF approximation eliminates a large number of electron–electron repulsion integrals, and those remaining are estimated by a variety of procedures.

Early ZDF procedures included the **CNDO** technique, in which there is complete neglect of differential overlap (much as in Hückel theory there is complete neglect of integral overlap). The CNDO method is still quite widely used, perhaps more than is appropriate given that better procedures are readily available. The ZDF techniques were followed by the **MINDO** family of techniques, where the initials stand for modified intermediate neglect of differential overlap. Intermediate neglect is less draconian than complete neglect, for differential overlap between orbitals on the *same* atom is not neglected in one-centre electron–electron repulsion integrals but it is neglected in two-centre electron–electron repulsion integrals. The remaining integrals are estimated. In **MINDO-1**, the parameters were chosen to give the best fit to the enthalpy of formation of the compound. In **MINDO-2** and **MINDO-3**, which are later developments, the treatment of certain integrals is modified, and the technique then produces good enthalpies of formation and reasonable molecular geometries.

The MINDO family served its purpose well, but has been superseded by other techniques, particularly **AM1** (for 'Austin method'), a third generation procedure devised by Dewar (at the University of Texas, Austin). AM1 is based on the **MNDO** method, also devised by Dewar, which is less restrictive than the MINDO family.

Further information

The field is vast, and there are numerous books, reviews, and papers. Useful surveys of semiempirical techniques have been compiled by Murrell and Harget (1972) and Hirst (1990). See also Levine (1983), McGlynn *et al.* (1972), Szabo and Ostlund (1982), Naray-Szabo, Surjan, and Angyan (1987), and Clark (1985) for introductions and further references. A helpful survey of techniques and applications will be found in the two volumes edited by Segal (1977). More specific references are Bingham, Dewar, and Lo (1975) for MINDO/3, Dewar and Thiel (1977) for MNDO, and Dewar *et al.* (1985) for AM1. A special issue of the *Journal of computer aided molecular design* in early 1990 is also relevant. For techniques outside the Dewar family see Stewart (1989) for PM3, and Brown, Gerloch, and McMeaking (1988) for modern AOM.

Singlet and triplet states

A **singlet state** is one in which the net °spin of a many-electron system is zero ($S = 0$). A **triplet state** is a state in which $S = 1$. The names 'singlet' and 'triplet' denote the number of values that the quantum number M_S (the projection of the spin on the z axis) may take in each case. For $S=0$, M_S is allowed the single value of 0; for $S=1$, M_S may take any of the three values $+1$, 0, and -1. In a singlet state the electrons are °paired in the sense of having antiparallel spins. In a triplet state of two electrons the spins are parallel.

The expressions 'parallel' and 'paired' merit further elucidation. We shall show that two electron spins are strictly antiparallel if they are paired, but that two electrons are not strictly parallel in a triplet state. We shall also show that two electrons may be 'parallel' even if one is α and the other is β.

The combined state of two electrons may be any of the four states $\alpha\alpha$, $\alpha\beta$, $\beta\alpha$, and $\beta\beta$. None of these states corresponds to a resultant spin vector of fixed length because the individual vectors can make any azimuthal angle to each other on their °precessional cones. However, we can form a vector of zero length by arranging an α electron at $180°$ to a β electron (Fig. S.11a). In fact, as we cannot know which electron is α and which is β, we should use one of the superpositions $\alpha\beta + \beta\alpha$ or $\alpha\beta - \beta\alpha$ to denote this state. The fact that the two spins must be out of phase (at $180°$) implies that we should use the combination $\alpha\beta - \beta\alpha$.

The remaining three states, $\alpha\alpha$, $\alpha\beta + \beta\alpha$, and $\beta\beta$ correspond to the three states of the triplet. The first has a total projection of $+1$, the second of 0, and the third -1 on the z axis. Our task is to show that even though the state $\alpha\beta + \beta\alpha$ is composed of α and β spins, all three states corresponds to $S=1$ and hence have 'parallel' spins.

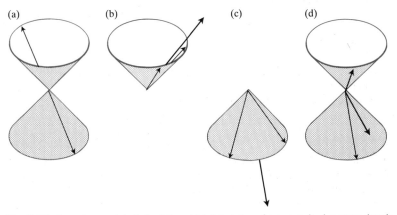

Fig. S.11 A vector model of singlet and triplet states of a two spin-$\frac{1}{2}$ system (such as two electrons). (a) $S=0$ and $M_S=0$. The spins are perfectly antiparallel and the net spin is zero. (b to c) The three states of a spin triplet $S=1$, $M_S=+1$, -1, 0 respectively. Note that the angle is the same between the two spins (and that two α spins make an angle to each other on their cone, as do two β spins on theirs), but that their nonzero resultant is in a different orientation in each case. Two 'parallel' spins are only partially parallel.

According to the °vector model of angular momentum, a spin $s = \frac{1}{2}$ is represented by a vector of length $3^{1/2}/2$ and a spin $S = 1$ by a vector of length $2^{1/2}$. Clearly, even two α spins cannot be strictly parallel because their resultant would then be of length $3^{1/2}$. In fact, they must make an angle of about $71°$ (specifically, arccos 1/3) to each other in order to have the appropriate resultant (Fig. S.11b). The same is true of the $\beta\beta$ state (Fig. S.11c). Finally, since the angle between the α and β cones is also $71°$, it follows that we can obtain the same resultant with an α and a β spin in the arrangement shown in Fig. S.11d. The two spins are now in phase (have the same azimuth), as is signified by the positive sign in $\alpha\beta + \beta\alpha$. We conclude that the three states of the triplet all correspond to $S = 1$, and that by 'parallel' spins we actually mean two spins that make $71°$ to each other. Note that electrons with α and β spin are just as parallel in Fig. S.12d as they are in the $\alpha\alpha$ and $\beta\beta$ states: in all three cases the angle between the spins is $71°$.

A singlet arrangement of two spins can be converted into a triplet arrangement (and vice versa) by causing one electron spin to precess faster than the other. See the entry on °intersystem crossing.

Further information

See Chapters 6 and 12 of MQM for a more detailed discussion of singlets and triplets and their interconversion. The role of the triplet state in chemistry is described in McGlynn, Azumi, and Kinoshita (1969), who give many references to the early literature. The photochemical consequences of excited singlets and triplets are discussed by Wayne (1988) and Barltrop and Coyle (1975, 1988). One of the consequences of °intersystem crossing is °phosphorescence, and those two entries should be consulted for more information.

Slater determinant

According to the °Pauli principle, a many-electron wavefunction must change sign whenever the labels of any two electrons are interchanged. It follows that a simple product of the form $A^{\alpha}(1)A^{\beta}(2)B^{\alpha}(3)...Z^{\beta}(N)$, where electron 1 occupies orbital A with spin α, and so on, is not an acceptable wavefunction. It is possible to ensure that a product of one-electron wavefunctions does satisfy the Pauli principle by writing it as a determinant:

$$\psi = (1/N!)^{1/2} \begin{vmatrix} A^{\alpha}(1) & A^{\alpha}(2) & A^{\alpha}(3) & ... & A^{\alpha}(N) \\ A^{\beta}(1) & A^{\beta}(2) & A^{\beta}(3) & ... & A^{\beta}(N) \\ B^{\alpha}(1) & B^{\alpha}(2) & B^{\alpha}(3) & ... & B^{\alpha}(N) \\ \vdots & \vdots & \vdots & & \vdots \\ Z^{\beta}(1) & Z^{\beta}(2) & Z^{\alpha}(3) & ... & Z^{\beta}(N) \end{vmatrix}$$

Expansion according to the rules of manipulating determinants leads to $N!$ terms, half with a positive sign and half with a negative sign. The factor $(1/N!)^{1/2}$ ensures that the wavefunction remains °normalized. That this **determinantal wavefunction** now satisfies the Pauli principle follows automatically from the property that interchange of any pair of rows or columns of a determinant reverses its sign. Suppose that we interchange electrons 1 and 2. The effect on the determinant is to interchange the first and second columns, and so the sign changes.

°Dirac appears to have been the first to suggest that wavefunctions should be written in this way, but the procedure was developed by the American physicist J.C. Slater, and determinantal wavefunctions are normally associated with his name. A wavefunction can be written as a single **Slater determinant** only for closed-shell species. When the shell is incomplete a linear combination of determinants must be used (unless the spin has its maximum value for the configuration, in which case a single determinant is adequate).

A word on notation: the orbitals with their accompanying spin are known as **spin-orbitals**. Instead of being written ψ^{α}, a spin-orbital is sometimes denoted ψ with the α spin understood; the β spin-orbital is denoted $\bar{\psi}$. Many trees would die if a determinantal wavefunction were always written in full; therefore it is normally denoted by listing only the terms on the diagonal and ignoring (but remembering) the normalization constant. The determinant above then becomes $\psi = |A(1)\bar{A}(2)B(3)...\bar{Z}(N)|$ in this notation.

Further information

See Richards and Cooper (1983) for a simple introduction to Slater determinants and the way they are manipulated. McGlynn *et al.* (1972), Szabo and Ostlund (1982), and McWeeny (1989) provide much more detail.

Slater type orbitals

One-electron orbitals in many-electron atoms (see the °orbital approximation) have a complicated dependence on position which can be represented accurately only by listing their amplitude numerically. For many purposes, however, it is desirable to have an analytical function rather than a table of numbers, and the **Slater type orbitals** (STO) are analytical functions that have been fitted to the numerical SCF solutions (Box S.6). The orbitals were introduced by J.C. Slater (see preceding entry).

The radial part of each orbital has the form shown in Box S.6. The term ζ is called the **orbital exponent**. The angular component of the orbital is a °spherical harmonic corresponding to the values of the quantum numbers l and m_l. The rules set out in Box S.6 were devised

Box S.6 Slater type orbitals

Each STO has the form

$$\psi = N r^{n^* - 1} e^{-\zeta r} Y_{l,m_l}$$

where Y is a 'spherical harmonic. The values of n^* and ζ are obtained using the following rules or (for ζ) the values in Table P.1. The tabulated values are more reliable. n^* is related to the principal number n by the correspondence

n	1	2	3	4	5	6
n^*	1	2	3	3.7	4	4.2

ζ is related to the **effective nuclear charge** by $\zeta = Z_{eff}/n^*$, with $Z_{eff} = Z - \sigma$. The **screening constant**, σ, is calculated as follows:

- First, classify the orbitals into the sets

  $(1s), (2s, 2p), (3s, 3p), (3d),$

  $(4s, 4p), (4d), (4f),$

  $(5s, 5p), (5d), \ldots$

- Then add together the following contributions:

  From the electrons in orbitals following the set containing X: 0

  From electrons in orbitals belong to the same set as X: 0.30 if the electron is $1s$ but 0.35 otherwise.

  If the electron of interest is ns or np, for each electron with principal quantum number $n - 1$: 0.85, and 1.00 otherwise.

  If the electron of interest is nd or nf, for each electron in a group preceding X in the list: 1.00.

to estimate ζ for any atom, but they have been largely superseded by tabulated values (Table P.1).

A major defect of the Slater orbitals is that they have no radial nodes and hence are not mutually °orthogonal. They may be made mutually orthogonal by the Schmidt orthogonalization procedure (Box O.2), and so this defect can be overcome.

Further information

A useful discussion of Slater type orbitals is given in McGlynn *et al.* (1972), who in their Appendix B give many references to the expression of SCF orbitals in terms of STOs and also give a table of orbitals. Other useful sources, with similar information, are Schaefer (1972) and Condon and Odabaşi (1980). For a computationally more efficient alternative to Slater type orbitals see °gaussian orbitals.

Spherical harmonic

A **spherical harmonic**, $Y_{l,m}$, is a wavefunction for a specific state of orbital angular momentum. More formally, the spherical harmonics are °orthogonal functions of the angular coordinates θ and ϕ (as defined in Fig. P.16) that satisfy the differential equation $\Lambda^2 Y = -l(l+1)Y$, where Λ^2 is the legendrian operator (see °laplacian). The functions are polynomials in $\sin\theta$ and $\cos\theta$ (Box S.7). Spherical harmonics are the angular factors in centrosymmetric °atomic orbitals, and their shapes will be familiar from those of the °s ($l=0$), °p ($l=1$), °d ($l=2$), and °f ($l=3$) orbital entries (Fig. S.12). They are particularly

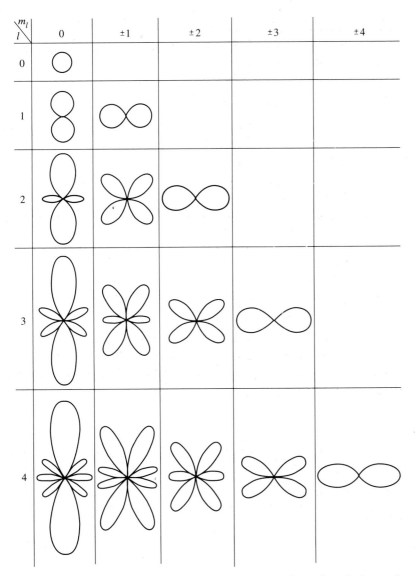

Fig. S.12 The θ dependence of the spherical harmonics for $l=0$ to 4; the z-axis points up the page. Note how the angular variation becomes more pronounced as l increases, and how the maximum amplitude shifts from the polar regions to the equatorial as $|m_l|$ increases.

Box S.7 Spherical harmonics, $Y = (1/4\pi)^{1/2} y$

l	m_l	y
0	0	1
1	0	$\sqrt{3} \cos\theta$
	± 1	$\mp\sqrt{\left(\frac{3}{2}\right)} \sin\theta \, e^{\pm i\phi}$
2	0	$\frac{1}{2}\sqrt{5}\,(3\cos^2\theta - 1)$
	± 1	$\mp\sqrt{\left(\frac{15}{2}\right)} \cos\theta \sin\theta \, e^{\pm i\phi}$
	± 2	$\sqrt{\left(\frac{15}{8}\right)} \sin^2\theta \, e^{\pm 2i\phi}$
3	0	$\frac{1}{2}\sqrt{7}\,(5\cos^3\theta - 3\cos\theta)$
	± 1	$\mp\frac{1}{4}\sqrt{(21)}\,(5\cos^2\theta - 1)\sin\theta \, e^{\pm i\phi}$
	± 2	$\sqrt{\left(\frac{105}{8}\right)} \sin^2\theta \cos\theta \, e^{\pm 2i\phi}$
	± 3	$\mp\frac{1}{4}\sqrt{35} \sin^3\theta \, e^{\pm 3i\phi}$

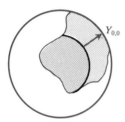

Fig. S.13 A representation of the amplitude of the spherical harmonic with $l = 0$ by the height above the surface of a sphere. This harmonic is a constant function.

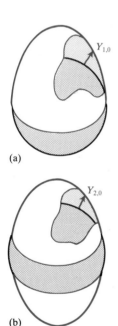

Fig. S.14 As in the preceding illustration, but for the spherical harmonics with $l =$ (a) 1 and (b) 2, with $m_l = 0$ in each case. The regions of negative amplitude are inside the sphere.

useful for expressing arbitrary angular distributions (of almost anything) as linear combinations of functions with simpler properties.

The spherical harmonic $Y_{0,0}$ is a constant. It can be represented as a surface of constant height above a sphere (Fig. S.13). Since the function is independent of θ and ϕ, it corresponds to a state of zero orbital angular momentum. It follows that a particle with $l = 0$ is uniformly distributed over all orientations.

The spherical harmonic $Y_{1,0}$ is positive in the northern hemisphere of the coordinate system, negative in the southern hemisphere, and has an equatorial °node (Fig. S.14a). The curvature (relative to the underlying spherical surface) arising from the node implies that a particle with this wavefunction has non-zero °orbital angular momentum. The other two spherical harmonics with $l = 1$ are °complex functions; both resemble $Y_{1,0}$ but with a circumpolar node in place of the equatorial node. The orientation of the node corresponds to the particle having orbital angular momentum around the z axis.

The shapes of spherical harmonics with larger values of l may be inferred from the shapes of °d orbitals (which have Y_2 functions as their angular factor, Fig. S.14b) and °f orbitals (which have Y_3 functions).

Further information

The mathematical properties of spherical harmonics, and their components, the associated Legendre functions, are tabulated in Abramowitz and Stegun (1965). For the properties of spherical harmonics see Arfken (1985) and Riley (1974). Since the spherical harmonics are °angular momentum wavefunctions, they are also discussed in that entry and in books on the subject; therefore see Rose (1957), Edmonds (1974), and Brink and Satchler (1968). Although it was written long ago, the book by Kauzmann (1957) is still worth looking at for its discussion of spherical harmonics in relation to waves on a flooded planet. For a portrayal of the complex spherical harmonics see Breneman (1988).

Spin

The **spin** of a particle is its intrinsic °angular momentum. Spin is a fixed characteristic of a particle, like its charge. To a very guarded extent spin has certain analogies with classical spinning motion about a point (which is the origin of its name); however, it is a purely quantum mechanical phenomenon without a true classical analogue, and picturing it as a spinning motion can be very misleading if the analogy is taken too far. The concept of electron spin was introduced in 1925 by the Dutch–American physicists George Uhlenbeck and Samuel Goudsmit to account for certain features of °atomic spectra.

The quantum mechanical description of the spin of a particle leads to the following conclusions:

- The spin angular momentum quantum number s (I is used for nuclei) has a positive, unique, integral or half-integral value which is characteristic of the particle.

- The magnitude of the spin angular momentum is $\{s(s+1)\}^{1/2}\hbar$

An electron has $s=\frac{1}{2}$, which implies that its spin angular momentum is $(3^{1/2}/2)\hbar$ whatever the other attributes of its state. See Table S.1 for a list of the spins of some nuclei.

- The z component of the spin angular momentum is quantized, and restricted to the values of $m_s\hbar$, where $m_s = s, s-1, \ldots, -s$.

Table S.1 Nuclear spins

Nuclide	Natural abundance %	Spin
^{1}n*		$\frac{1}{2}$
^{1}H	99.9844	$\frac{1}{2}$
^{2}H	0.0156	1
^{3}H*		$\frac{1}{2}$
^{13}C	1.108	$\frac{1}{2}$
^{14}N	99.635	1
^{17}O	0.037	$\frac{5}{2}$
^{19}F	100	$\frac{1}{2}$
^{31}P	100	$\frac{1}{2}$
^{33}S	0.74	$\frac{1}{2}$
^{35}Cl	75.4	$\frac{3}{2}$
^{37}Cl	24.6	$\frac{3}{2}$

* Radioactive.

For an electron $m_s = +\frac{1}{2}$ (which is variously called an α electron, spin-up, $\uparrow$) or $-\frac{1}{2}$ (a β electron, spin-down, $\downarrow$).

Spin is a non-classical property in the sense that if $\hbar$ were zero the spin angular momentum would vanish. This is in contrast to orbital angular momentum, for $\{l(l+1)\}^{1/2}\hbar$ can remain non-zero as $\hbar \to 0$ by allowing l to become infinite. There is no limit to the value of l, but for a given particle s is fixed.

Spin distinguishes particles into two major classes with fundamentally different behaviour. Particles with half-integral spin are °fermions and satisfy °Fermi–Dirac statistics; particles with integral spin (including zero) are °bosons and satisfy Bose–Einstein statistics. The two classes obey different forms of the °Pauli principle and consequently show profoundly different behaviour. The fundamental particles that we perceive as matter are all fermions. The fundamental particles that we identify as conveyors of force (such as the photon in electromagnetic interactions) are all bosons.

Further information

See MQM Chapter 6 for an introduction to the quantum mechanical description of spin. For a horse's mouth discussion see Dirac (1958). The complete treatment of spin as an aspect of °angular momentum can be pursued in that entry; see also Altmann (1986) for the group theoretical treatment of systems with spin. For an account of the chemical manifestations of spin see Wheatley (1970) and McWeeny (1970). The original papers in which spin was introduced are by Uhlenbeck and Goudsmit (1925, 1926). Their suggestion was developed into a consistent theory by Pauli and then put on a firm (relativistic) foundation by °Dirac. The historical development of the concept is described by Jammer (1966). Why spin is not necessarily a relativistic phenomenon is described by Galindo and Sanchez del Rio (1961).

Spin correlation

See °Fermi hole.

Spin–orbit coupling

The **spin–orbit coupling** of an electron is the magnetic interaction between its spin and orbital angular momenta.

An electron in an atom possesses a magnetic moment on account of its °spin and its °orbital magnetic momenta. The two magnetic moments interact (Fig. S.15), and the strength of the interaction depends on the relative orientation of the moments as well as their magnitudes. The energy of this interaction appears in atomic spectra

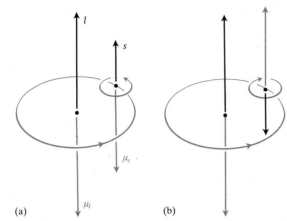

Fig. S.15 The spin–orbit coupling in an atom is a magnetic interaction between the spin and orbital magnetic moments of an electron. The energy of the interaction depends on the relative orientation of the magnetic moments and hence on the relative orientations of the angular momenta. (a) The state with high total angular momentum corresponds to high energy (the magnetic moments are parallel, N pole to N pole and S pole to S pole), and (b) the state with low angular momentum corresponds to low energy (the moments are antiparallel, N pole to S pole).

as a splitting of lines known as °fine structure. The strength of the couping is expressed by the **spin–orbit coupling constant** ζ (see Table S.2).

Spin–orbit coupling is stronger in heavy atoms than in light atoms, and in hydrogenic atoms it varies as Z^4. The dependence of the strength of the spin–orbit coupling on the atomic number of the element can be understood by imagining riding on an electron that is orbiting a nucleus of charge Ze. The perceived motion of the nucleus is equivalent to a positive circulating current and it gives rise to a magnetic field. For a given orbital angular momentum, the current, and hence the magnetic field, will be stronger the greater the charge on the nucleus. Hence, the spin–orbit interaction energy increases with the atomic number of the nucleus. That it increases as strongly as Z^4 stems from the fact that the orbital the electron occupies also becomes more compact as Z increases, and hence the electron experiences a stronger field both from the enhanced charge and because it is closer to the nucleus.

Table S.2 Spin–orbit coupling constants, ζ/cm^{-1}

B	C	N	O	F
10.7	13.4	(76)	−79.6	−265
Al	Si	P	S	Cl
60.5	64.0	(150)	−184	−545
Ga	Ge	As	Se	Br
460	399	(780)	−825	−2194

Data: Richards, Trivedi, and Cooper (1981); values in parentheses are estimates.

Further information

For a more detailed consideration of the topics mentioned here, see first MQM Chapter 9, then Moss (1973), and then Condon and Odabaşi (1980). For thorough treatments of the role of spin–orbit coupling in diatomic molecules see Richards, Trivedi, and Cooper (1981) and Judd (1975). A thorough treatment of spin–orbit coupling in polyatomic molecules is given by McGlynn, Azumi, and Kinoshita (1969). The paper by Fano (1970) is worth reading.

Spin–spin coupling

The interaction between nuclei which in °nuclear magnetic resonance gives rise to the °fine structure of the spectrum (Fig. N.10) is called **spin–spin coupling**. The coupling indicates that the energy of a system composed of two (or more) nuclear spins depends on their relative orientations. The strength of the interaction is generally denoted J and reported in hertz (Hz). If $J > 0$, the two nuclei have a lower energy when their spins are antiparallel, and if $J < 0$, they have a lower energy when their spins are parallel. The coupling between most protons lies in the range -20 Hz to $+40$ Hz, with the commonly observed range from 0 Hz to $+10$ Hz.

One possible mechanism for spin–spin interaction is a **direct °dipole–dipole interaction** between the two nuclear spin °magnetic moments. However, as the spherical average of such an interaction is zero, it cannot account for the fine structure of nuclei in a molecule that is freely rotating in a fluid medium. It does contribute in solid samples.

Another mechanism utilizes the electrons in the bonds as intermediaries in transmitting the interaction between the nuclei. This **polarization mechanism** is illustrated by the coupling of two H nuclei attached to the same C atom (Fig. S.16). The key to the mechanism is the °hyperfine interaction between the nuclei and the electrons.

When a spin-$\frac{1}{2}$ nucleus A has α spin, the °Fermi contact interaction results in it being energetically favourable for the β electron in the bond to be found nearby. The other electron in the bond must be α (by the °Pauli exclusion principle), and is most likely to be found far from the first electron (on account of the interaction of their charges) and therefore close to the C atom. According to °Hund's rule of maximum multiplicity, the lowest energy is achieved if electrons in *different* orbitals have parallel spins; hence the electron in the second C—H bond that is closer to the C atom is likely to have α spin too. The second electron in that bond must have β spin, and hence the most favourable state of the second proton is α (on account of the Fermi interaction, as before). Therefore, by the collaboration that we can represent Fermi|Pauli|Hund|Pauli|Fermi, it is energetically favourable for the two nuclei to be antiparallel. If the orientation of either

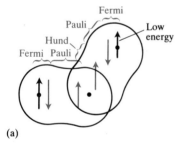

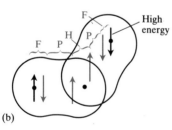

Fig. S.16 The polarization mechanism of spin–spin coupling. In each case, the proton on the left has α spin, and the lowest energy arrangement is as shown for the sequence Fermi, Pauli, Hund, Pauli. The final Fermi interaction implies that when the other proton spin is (a) α the overall system has a lower energy than when it is (b) β.

proton is changed, the energy rises slightly. Hence, the energy of the nuclei depends on their relative orientation, and we report that there is a spin–spin interaction.

The size of J depends on the strength of the hyperfine interactions and the extent to which the electrons in the bonds are found predominantly close to the most favourable nuclei. The latter is a kind of °polarization of the bond, for the orbital of each electron is distorted by the nuclear spins to a different extent. The ease of polarization depends on the closeness in energy of excited states. However, in order to polarize the electrons according to their spin orientation, it turns out that we need to use the energy of excitation to *triplet* configurations of the molecule (see °singlet and triplet states). This point can be understood in very broad terms as the outcome of the formation of the superposition $\alpha\beta + \lambda\alpha\alpha$, which tends to reduce the contribution of the β spin, or of $\alpha\beta - \lambda\alpha\alpha$, which tends to reduce the contribution of the α spin (Fig. S.17).

Although we have dismissed the direct dipole–dipole interaction in fluids, the electron–nucleus dipolar °hyperfine interaction can contribute to spin–spin coupling between nuclei. We have seen that the polarization mechanism involves two hyperfine interactions, one at each end of the chain; the rotational average of the product of two dipolar interactions does not disappear, and so two electron–nuclear dipole interactions, one at each proton, can contribute. This type of interaction is important in atoms other than hydrogen where p orbitals occur in the valence shell.

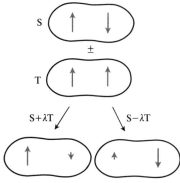

Fig. S.17 When some triplet state wavefunction is mixed into the ground state singlet wavefunction, the spin angular momenta partly cancel (depending on which combination is taken) and the resulting wavefunctions are spin polarized.

Further information

See MQM Chapter 14 for a more detailed introduction to the calculation of the spin–spin coupling constant. Further details will be found in Carrington and McLachlan (1967), Slichter (1988), and Memory (1968). A good, accessible introduction to the mechanism and consequences of spin–spin coupling is Harris (1986). For data on J values and their measurement, see Harris (1986), Sanders and Hunter (1987), and Derome (1987).

Stark effect

The **Stark effect** is the modification of the energy levels, and therefore of the spectra, of atoms and molecules by the application of an electric field. The effect was first observed by the German physicist Johannes Stark (1875–1957) in 1913.

The **first-order atomic Stark effect**, which is linearly proportional to the applied field, is large but rare. The effect depends on the presence of °degeneracy in an atom or molecule, which enables it to respond strongly to the perturbing influence of the applied field. The linear

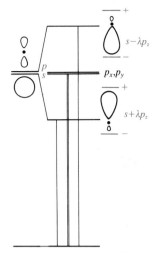

Fig. S.18 The linear Stark effect in the first excited level of a hydrogen atom. In the absence of an applied field, the 2s and 2p orbitals are degenerate. When a field is applied in the z-direction, the 2s and 2p_z orbitals mix and the two linear combinations move apart in energy. As a result, the single line in the spectrum is split into separate components.

effect occurs in the first excited level of atomic hydrogen because the 2s and 2p orbitals are degenerate, and so are easily mixed together by the applied field. The combination $2s + 2p_z$ gives a distribution strongly biased towards the low-potential region, and the combination $2s - 2p_z$ is concentrated on the high-potential side of the nucleus (Fig. S.18). Therefore the transitions to or from the $n = 2$ shell occur at different frequencies according to which of the mixed states is involved.

The Stark effect also causes lines to broaden slightly, and shifts the series limits to lower frequency. Both effects are related to the presence of the low potential on one side of the atom, for it enables an electron to escape (Fig. S.19). An electron in a state not far from the ionization limit may be able to °tunnel through the remaining potential energy barrier and emerge into a region where the applied field can pluck it from the atom. Tunnelling reduces its lifetime in the excited state, and the energy of that state is made imprecise (see °lifetime broadening). The electron need not be excited into so high an energy state for ionization to occur, and so the field also reduces the energy of the series limit.

In many-electron atoms (which do not have hydrogen's characteristic degeneracy), the first-order effect is replaced by the much weaker **second-order effect**. One power of the field is used to distort the atom from spherical symmetry (to °polarize it), and the second power is used in the interaction with the dipole moment of the distorted atom, so

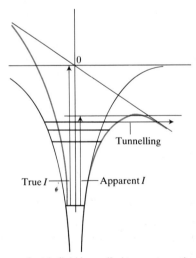

Fig. S.19 When a strong electric field is applied to an atom, the electrons experience the sum of the Coulombic field of the nucleus and the applied field. Not only are the energy levels shifted and, specifically, the series limit shifted to lower wave-number), but excited states may be able to tunnel through a potential barrier (upper right) and escape from the atom. The latter effect reduces the lifetimes of the upper states and hence broadens the spectral lines.

causing the energy separation. Since normal fields can polarize atoms only slightly, the induced dipole is small and its interaction with the field weak; therefore the second-order Stark energy shifts are small and huge fields ($100 \, \text{kV cm}^{-1}$) must be used. The **molecular Stark effect** may also be of first- or second-order. The first-order effect is shown by symmetric top molecules with permanent dipole moments; the applied field causes °rotational states of the same value of J but different values of M_J to have different energies (they are degenerate in the absence of the field), and the splitting is proportional to the permanent dipole moment of the molecule μ. Since fields of the order of $50 \, \text{kV cm}^{-1}$ give splittings of the order of 20 MHz, and this is easily and accurately detectable in a microwave spectrum, the method is a powerful way of determining dipole moments.

The molecular Stark effect is of considerable importance in the study of permanent dipole moments of the molecules that can be examined by microwave spectroscopy, and it is also important in the technology of microwave spectrometers, for an oscillating electric field (usually of several $10 \, \text{kV cm}^{-1}$) will modulate the position of spectral lines, and therefore the intensity of absorption or emission at a particular frequency. Spectrometers using detectors that make use of this oscillation of the intensity are called **Stark-modulation spectrometers**.

Further information

A good, accessible discussion of the atomic and molecular Stark effects is given in Bransden and Joachain (1989). For more information about the molecular Stark effect, see Hollas (1983, 1987). Townes and Schawlow (1955) also give an authoritative account. The most chemical version of the Stark effect is the early version of the crystal field theory of d-block complexes, in which the degeneracy of the d orbitals of the free atom is removed by the ligands: see °ligand field theory.

Stefan–Boltzmann law

The Austrian Josef Stefan (1835–1893) found experimentally that the **emittance**, M, of a °black body radiator, the power it emits per unit area summed over the entire spectral range, is proportional to the fourth power of the temperature (Fig. S.20). The addition of Ludwig Boltzmann's name reflects his theoretical contribution, for he derived the law from kinetic theory and thermodynamics, and showed that it applied only to black body radiators.

The constant σ in the expression is called the **Stefan–Boltzmann constant** and has the value $5.67 \times 10^{-8} \, \text{W m}^{-2} \, \text{K}^4$ (so, when multiplied by T^4, it results in a quantity with units watts per square metre). Thus, the total power radiated per square metre of the surface of a black

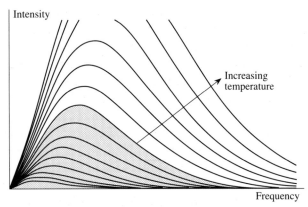

Fig. S.20 As the temperature is increased, the emittance from a black body increases as T^4. The curves are the Planck distributions for a series of temperatures: the total energy density inside the cavity is given by the area under each one and the diagram shows how strongly the area grows as the temperature is increased.

body at 1000°C is 150 kW m^{-2}. Stefan used his law to make the first estimate of the temperature of the sun's surface, which acts as a black-body emitter with a temperature close to 6000°C.

The Stefan–Boltzmann law is explained by the Planck distribution, for the integral of the Planck formula over all frequencies (which gives the total energy density in the black body) increases as T^4, and the power radiated is proportional to the total energy density. Physically, the strong T^4 temperature dependence arises both from the exponential increase in the degree of excitation of the oscillators that are already active and from the increase in the number of high frequency oscillators that are active. The joint effect is a very pronounced dependence on the temperature.

Further information

For the role of the Stefan–Boltzmann law in the early development of quantum theory see Jammer (1966) and Kuhn (1978).

Stern–Gerlach experiment

The **Stern–Gerlach experiment** demonstrated the existence of space quantization (see °angular momentum). It was carried out in 1920 by the German physicist Otto Stern (1888–1969) in collaboration with Walter Gerlach.

In the experiment, a collimated beam of silver atoms evaporated from hot metal was passed through slits into a vacuum and then through an inhomogeneous magnetic field (Fig. S.21). (An inhomogeneous field is needed because it provides a force in a specific direction

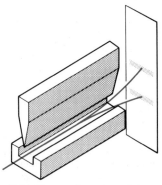

Fig. S.21 The configuration of the Stern–Gerlach experiment. The shaped magnets provide a strongly inhomogeneous field, and a beam of atoms with a single unpaired electron is split into two components.

across the beam.) Stern and Gerlach observed the distribution of the atoms deposited on a glass plate. If the atoms possessed a magnetic moment, the effect of the field would be to drive in one direction those that had one orientation, and in the opposite direction those that had the opposite orientation. According to classical physics, atoms with all intermediate orientations would also be observed and a continuous band of atoms would be detected.

In their first experiment, Stern and Gerlach observed the classically anticipated result. In their second, done with great care with a low pressure and a long exposure, they saw that the band of deposited atoms had two closely spaced components. This result is at variance with classical physics but in accord with quantum theory. Thus, each Ag atom possesses an unpaired *s* electron with °spin and therefore has a °magnetic moment that may take only two orientations in a magnetic field, not the full range of orientations permitted by classical physics.

The original explanation of the observation did not associate the magnetic moment with the °spin of the electron. That interpretation came later (in 1925), after Uhlenbeck and Goudsmit had introduced the concept of spin from their study of atomic spectra.

Further information

The Stern–Gerlach experiment is put into its historical context by Jammer (1966). The original papers are by Stern (1921) and Stern and Gerlach (1922). There is a good analysis of the experiment in Bohm (1951). For the philosophical issues to which the experiment has given rise (and which are still highly relevant to the theory of measurement and the interpretation of quantum mechanics) see Jammer (1966, 1974) and the abundance of papers collected by Wheeler and Zurek (1983).

Superconductor

A **superconductor** is a substance that has zero electrical resistance below a characteristic transition temperature. Until 1987, all known superconducting transitions occurred below about 20 K. However, in that year the first of a new breed of ceramic **high-temperature superconductors** (HTSC) were prepared that had transition temperatures close to 100 K (and possibly considerably higher). The central aspect of the **BCS theory** of low-temperature superconduction (where the initials are of the American physicists John Bardeen, Leon Cooper, and John Schrieffer) is the existence of a **Cooper pair**, a pair of electrons held together by their interaction with the nuclei of the atoms in the lattice. Thus, if one electron is in a particular region of a solid, the nuclei there move towards it to give a distorted local structure (Fig. S.22). Since that local distortion is rich in positive charge, it is favourable for a second

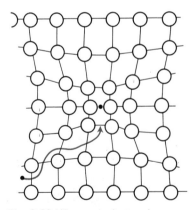

Fig. S.22 When one electron is present at some location (the central dot) it polarizes the surrounding lattice of cations, so providing a region of enhanced positive charge which attracts in another electron. As a result, the two electrons behave to some extent like a single particle. The effect is eliminated, however, if the temperature is raised and the thermal motion becomes violent.

electron to join the first. There is now a virtual attraction between the two electrons, and they move together as a pair. The local distortion is easily disrupted by thermal motion of the ions, so the virtual attraction occurs only at very low temperatures.

A Cooper pair undergoes less scattering than an individual electron as it travels through the solid, since the distortion caused by one electron can attract back the other electron should it be scattered out of its path in a collision. This has been likened to the difference between the motion of a herd of cattle, with members of the herd that are deflected from their path by boulders in their way, and a team of cattle yoked together, which will travel forward largely regardless of obstacles. Since the Cooper pair is stable against scattering, it can carry charge freely through the solid, and hence give rise to superconduction.

The mechanism of high-temperature superconduction is still obscure. The BCS theory does not appear to be applicable both on account of the high temperatures (which would disrupt Cooper pairs) and because there is no isotope effect. In low-temperature superconductivity the substitution of an atom by its isotope changes the transition temperature because the lattice vibrations mediate the electron–electron attraction. However, the mediation of the coupling by lattice phonons has not yet been ruled out, because pronounced isotope effects would be observed only if the modes responsible involved solely oxide ion motion. All the high-temperature superconductors are defective compounds: the corresponding pure oxides are not superconductors, but become so when doped (as with La_2CuO_4 doped to become $La_{2-x}Sr_xCuO_4$) or are made non-stoichiometric by oxidation or reduction.

Further information

For an elementary introduction to high-temperature superconductors see Ellis (1987) and Harris, Hills, and Hewston (1987). For an account of the theories of low-temperature superconduction, see Rose-Innes and Rhoderick (1976), Kittel (1986, 1987), and Tilley and Tilley (1986). A moderately accessible account of the theories, with a chemical bias, has been given by Matsen (1987).

Superoperator

An °operator operates on a function, and generates from it another function. A **superoperator** operates on an *operator*, and generates from it another operator. More specifically a superoperator $A*$ (or $A^\times$, notations vary) operates on an operator B and generates from it the value of the °commutator of A and B. That is, $A*B = [A,B]$. The seemingly trivial notational change from commutator to superoperator turns out to be very elegant. It greatly simplifies some expressions, and also brings out the similarities between otherwise apparently different equations. This is particularly clear in the case of the °density matrix, for its equation of motion turns out to be the superoperator analogue of the Schrödinger equation for the wavefunction (Box S.8).

Further information

Some insight into the deployment of superoperators can be found from Jeener (1982). They are used extensively, but obscurely, in a number of the articles in the collection edited by Muus and Atkins (1970).

Superposition principle

The **superposition principle** states that:

When a system is in any of a range of states, its wavefunction is a linear combination of the wavefunctions of the individual states.

That is, if the system may be in any of the states (of energy, angular momentum, and so on) with wavefunctions ψ, ψ',..., its total wavefunction is the weighted sum $c\psi + c'\psi' + \dots$. The probability that the system will be found in the state with wavefunction ψ is proportional to c^2 (for c real).

The superposition principle is one of the foundations of quantum mechanics. It results in the notable features of the differences between classical mechanics and quantum mechanics, for it leads to the concept of **interference** between different possible outcomes, as we shall now illustrate.

Suppose that a particle can travel from a point 1 to point 2 by the alternative paths A and B (Fig. S.23). Classical theory ascribes the probabilities P to one path and P' to the other, and goes on to say that the total probability of completing the journey is the sum of the individual probabilities, $P + P'$.

Quantum mechanics accepts that there are individual probabilities for each path, but it rejects the conclusion that the joint probability is their sum. According to the superposition principle, we first calculate the **probability amplitudes** (the °wavefunctions) for the individual paths (ψ and ψ'), then sum the individual wavefunctions (to obtain $c\psi + c'\psi'$), and from that composite wavefunction compute the overall probability (by taking its square).

Thus, the probability that the particle will complete its journey if only path A is open is c^2, and the probability that it will complete its journey if only path B is open is c'^2. However, the probability that it will complete its journey if *both* paths are open is calculated from $(c\psi + c'\psi')^2 = c^2\psi^2 + c'^2\psi'^2 + 2cc'\psi\psi'$. This total probability differs from the sum of the individual probabilities by the interference term $2cc'\psi\psi'$.

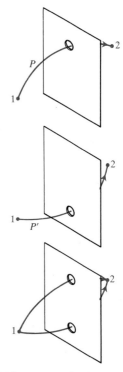

Fig. S.23 A schematic diagram of the superposition principle: to calculate the probability of the journey being completed when both paths can be used we must work with the sum of the probability amplitudes for the two paths, not with their probabilities directly.

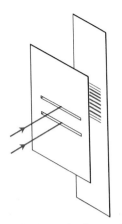

Fig. S.24 The Young's slit experiment for electrons. The pattern on the screen can be accounted for by interference between the wavefunctions for the electrons travelling by both paths.

In some locations the additional term will be positive, in which case there is a greater than classical probability of completing the journey. In other locations the interference term will be negative, and the probability will be reduced. The contribution of the interference term is the essence of the observation of the diffraction of particles (Fig. S.24), the analogue of Young's slit experiment on light.

The superposition principle applies to all the processes of quantum mechanics. Wherever a compound state is under consideration the calculations must be performed *on the wavefunction* and the observables calculated only at the end of the manipulation of ψ. In a sense, the difference between classical and quantum mechanics can be seen to be due to the fact that classical mechanics took too superficial a view of the world: it dealt with appearances. However, quantum mechanics accepts that appearances are the manifestation of a deeper structure (the wavefunction, the amplitude of the state, not the state itself), and that all calculations must be carried out on this substructure.

Further information

See Feynman, Leighton, and Sands (1963), §9.2 of Bohm (1951), and Chapter 1 of Dirac (1958). For a simple account of optical analogies see Chapter 2 of MQM. See Jammer (1966, 1974) for historical and philosophical attitudes; d'Espagnat (1976) and Hughes (1989) for the latter too. Feynman and Hibbs (1965) describe the formulation of quantum mechanics from the viewpoint of the superposition principle. Schlegel (1980) discusses superposition in physics from a more general point of view.

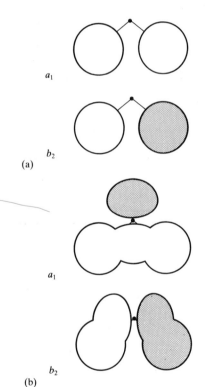

(a)

(b)

Fig. S.25 (a) Two symmetry adapted linear combinations of hydrogen $1s$ orbitals in H_2O and (b) the molecular orbitals that are formed by their overlap with oxygen p orbitals of the same symmetry type.

Symmetry-adapted linear combinations

A **symmetry-adapted linear combination** (SALC) of atomic orbitals is a linear combination of atomic orbitals on atoms that are related by symmetry. A given SALC has non-zero overlap with SALCs of the same symmetry on other atoms in the molecule, and hence may be combined with them to form molecular orbitals.

The formation of a SALC recognizes one of the principal features of °molecular orbital theory, that we should identify combinations of atomic orbitals with the same symmetry, and combines them together to build molecule-wide molecular orbitals. Note that a SALC is not necessarily a molecular orbital itself: it is a combination of atomic orbitals on equivalent atoms from which the orbital will in due course be built. Thus, SALCs are like a kit of parts from which molecular orbitals are to be constructed.

A simple example of a SALC is the combination of the two $H1s$ orbitals in the H_2O molecule (Fig. S.25). One combination (the addition of the two orbitals) has the same symmetry as the O2s and O2p_z

orbitals, and can overlap with them to form bonding and antibonding combinations. The other combination (the difference between the two orbitals) has the same symmetry as the $O2p_y$ orbital, and can overlap with that orbital to give bonding and antibonding combinations. Pictorial representations of SALCs for some common molecular shapes are shown in Table S.3.

Further information

Use is made of symmetry-adapted linear combinations throughout Shriver, Atkins, and Langford (1990), and there are more diagrams like those in

Table S.3 (a) Central atom orbitals

	$D_{\infty h}$	D_{2v}	D_{3h}	C_{3v}	D_{4h}	C_{4v}	D_{5h}	C_{5v}	D_{6h}	C_{6v}	T_d	O_h
s	σ	A_1	A_1'	A_1	A_{1g}	A_1	A_1'	A_1	A_{1g}	A_1	A_1	A_{1g}
p_x	π	B_1	E'	E	E_u	E	E_1'	E_1	E_{1u}	E_1	T_2	T_{1u}
p_y	π	B_2	E'	E	E_u	E	E_1'	E_1	E_{1u}	E_1	T_2	T_{1u}
p_z	σ	A_1	A_2''	A_1	A_{2u}	A_1	A_2''	A_1	A_{2u}	A_1	T_2	T_{1u}
d_{z^2}	σ	A_1	A_1'	A_1	A_{1g}	A_1	A_1'	A_1	A_{1g}	A_1	E	E_g
$d_{x^2-y^2}$	δ	A_1	E'	E	B_{1g}	B_1	E_2'	E_2	E_{2g}	E_2	E	E_g
d_{xy}	δ	A_2	E'	E	B_{2g}	B_2	E_2'	E_2	E_{2g}	E_2	T_2	T_{2g}
d_{yz}	π	B_2	E''	E	E_g	E	E_1''	E_1	E_{1g}	E_1	T_2	T_{2g}
d_{zx}	π	B_1	E''	E	E_g	E	E_1''	E_1	E_{1g}	E_1	T_2	T_{2g}

(b) Ligand orbitals

$D_{\infty h}$	C_{2v}
σ_g	A_1
π_g	A_2
π_u	B_1

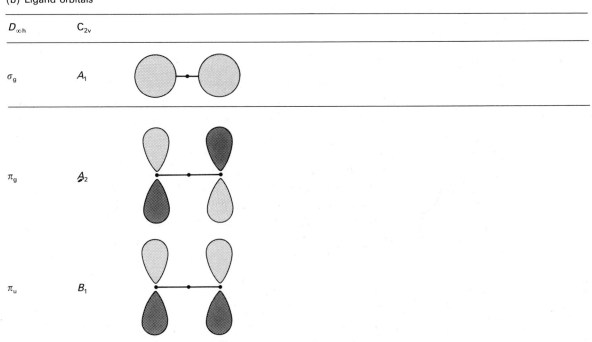

$D_{\infty h}$	C_{2v}		(*cont.*)

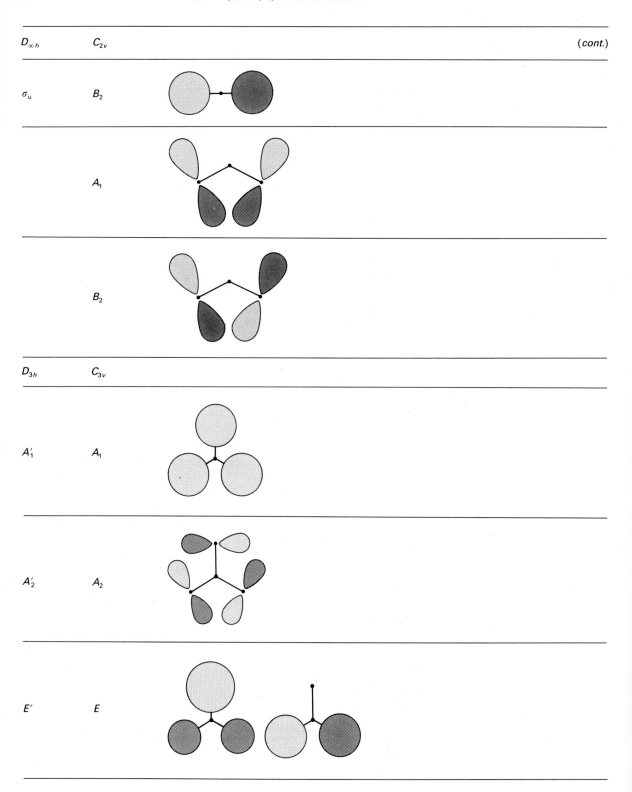

σ_u	B_2
	A_1
	B_2

D_{3h}	C_{3v}
A_1'	A_1
A_2'	A_2
E'	E

D_{3h}	C_{3v}	(cont.)

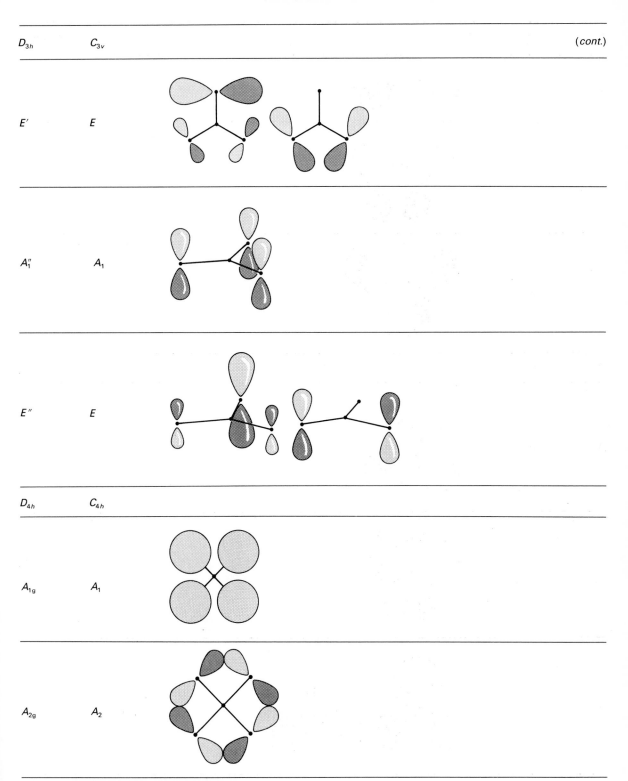

E'	E

A_1''	A_1

E''	E

D_{4h}	C_{4h}

A_{1g}	A_1

A_{2g}	A_2

| D_{4h} | C_{4v} | | *(cont.)* |

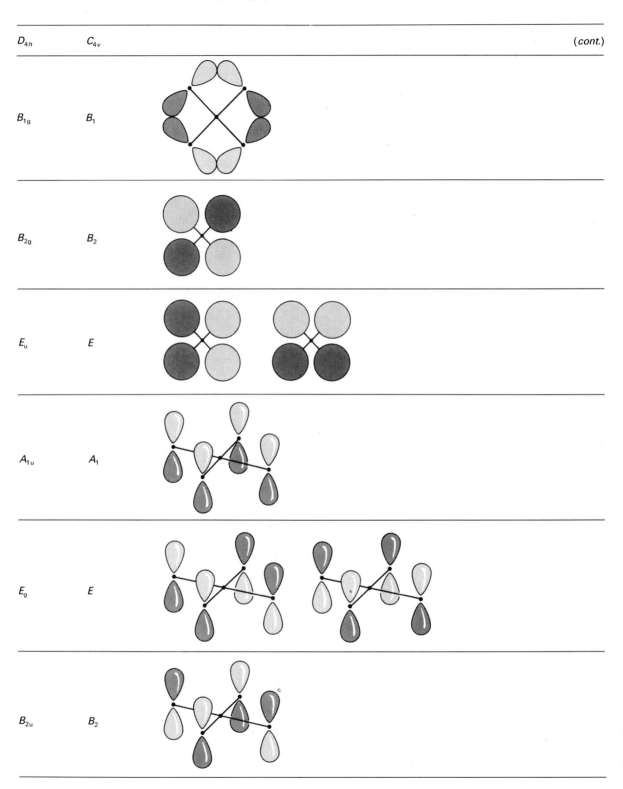

D_{5h}	C_{5v}		
A_1'	A_1		
A_2'	A_2		
E_1'	E_1		
E_2'	E_2		
A_2''	A_1		
E_1''	E_1		

| D_{5h} | C_{5v} | | | | | | *(cont.)* |

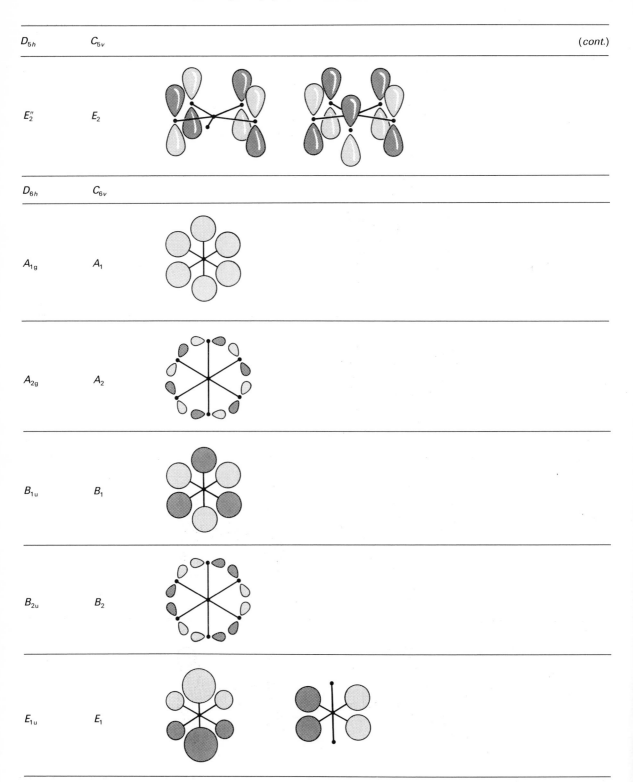

| E_2'' | E_2 |

| D_{6h} | C_{6v} |

| A_{1g} | A_1 |

| A_{2g} | A_2 |

| B_{1u} | B_1 |

| B_{2u} | B_2 |

| E_{1u} | E_1 |

D_{6h}	C_{6v}
E_{2g}	E_2
A_{2u}	A_1
B_{2g}	B_1
E_{1g}	E_1
E_{2u}	E_2

T_d

A_1

T_2

$O_h(\sigma)$

A_{1g}

E_g

T_{1u}

$O_h(\pi)$

T_{1u}

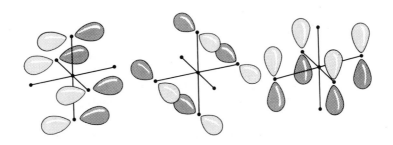

T_{2g}

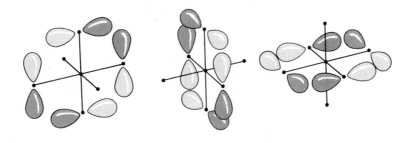

T_{1g}

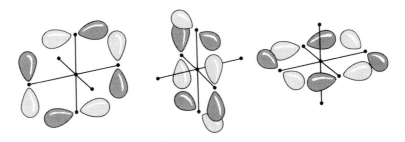

T_{2u}

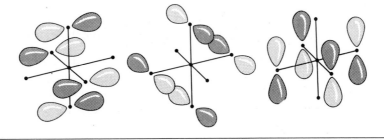

Table S.3 in their Appendix 6. Another very good source is Albright, Burdett, and Whangbo (1985). The construction of SALCs using group theory is explained in detail in Chapter 7 of MQM, Cotton (1971), and Kettle (1985).

Symmetry operations and elements

A **symmetry operation** is a transformation that, after its application, leaves an object indistinguishable from what it was initially. A **symmetry element** is a point, line, or plane with respect to which the symmetry operation is performed.

As an illustration, consider an undecorated square tile lying on a plane. If it is rotated through 90° about a vertical axis, the tile looks exactly the same as before (Fig. S.26). Hence a 90° rotation is a symmetry operation of the tile and the vertical axis is the symmetry element corresponding to that operation. A rotation through 180° is another symmetry operation of the tile, but a rotation through 30° is not. The study of the effect of symmetry operations is the basis of °group theory.

We shall confine attention to symmetry operations that leave at least one point unchanged. We shall not consider symmetry operations relating to translational motion. The five basic symmetry operations and their corresponding elements are then as follows (Fig. S.27):

- The **identity** E consists of doing nothing; the corresponding element is the entire object.

Since every molecule is indistinguishable from itself if nothing is done to it, every object possesses at least the identity element. One reason for including it is that some molecules (for example, CHClBrF) have only this symmetry element; another reason is technical and connected with the formulation of °group theory.

- An **n-fold rotation** (the operation) is a rotation through $360°/n$ about an **n-fold rotation axis** C_n (the corresponding element).

An H_2O molecule has a two-fold axis, C_2. An NH_3 molecule has a three-fold axis C_3 with which is associated two rotations, one by 120° clockwise and the other by 120° counter-clockwise. If a molecule possesses several rotation axes, the one (or more) with the greatest value of n is called the **principal axis**.

- In an **inversion** (the operation) through a **centre of symmetry** (the element), i, each point in a molecule is projected in a straight line through the centre to an equal distance on the other side.

Neither an H_2O molecule nor an NH_3 molecule has a centre of inversion, but both the sphere and the cube do have one.

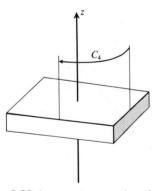

Fig. S.26 A symmetry operation of a square, undecorated tile is the 90° rotation designated C_4. The corresponding symmetry element, the four-fold axis, is the line around which the rotation takes place.

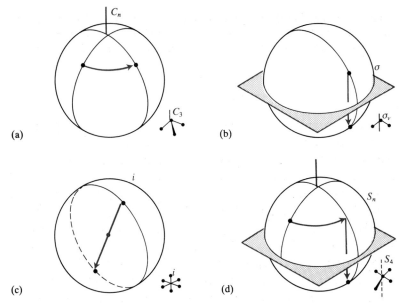

Fig. S.27 A representation of four symmetry operations and their associated elements. (a) An *n*-fold rotation about an axis of symmetry, (b) inversion through a centre of symmetry, (c) reflection in a plane of symmetry, and (d) a rotation-reflection about an axis of improper rotation. The small diagrams beside the spheres are illustration of the elements for some molecular shapes.

- A **reflection** (the operation) requires a **plane of symmetry** or a **mirror plane** σ (the element).

If the mirror plane is parallel to the principal axis, it is called **vertical** and denoted σ_v. An H_2O molecule has two vertical planes of symmetry and an NH_3 molecule has three. When the mirror plane is perpendicular to the principal axis it is called **horizontal** and denoted σ_h. The C_6H_6 molecule has a C_6 principal axis and a horizontal mirror plane (as well as several other elements). A vertical mirror plane that bisects the angle between two C_2 axes is called a **dihedral plane** and denoted σ_d.

- An **improper rotation** or a **rotary-reflection** (the operation) about an **axis of improper rotation** or a **rotary-reflection axis** S_n (the element) consists of an n-fold rotation followed by a horizontal reflection.

A CH_4 molecule has three S_4 axes, one of which is shown in (**5**).

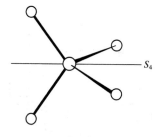

5 An S_4 axis

Further information

See Atkins (1990) and Chapter 7 of MQM for an account of symmetry operations and symmetry elements. Good detailed discussions are those of Cotton (1990) and Kettle (1985). For more physics orientated analyses, see Tinkham (1964) and Hamermesh (1962).

Tanabe–Sugano diagrams

A **Tanabe–Sugano diagram** shows how the energies of the °terms of an atom or ion change as the strength of a °ligand field increases. The diagrams provide a convenient way of analyzing the complicated electronic absorption spectra that are characteristic of d^n complexes.

The Tanabe–Sugano diagrams for d^n ions in an octahedral environment are shown in Fig. T.1. The terms of the free ion are given on the left of the diagram (compare Fig. R.1). Their positions are calculated from the °Racah parameters B and C and plotted as E/B. It is usual to make a plausible assumption about the relative values of B and C (typically, $C \approx 4B$), for then the energies can be expressed in terms of B alone. The horizontal axis depicts increasing ligand field strength (and is expressed as Δ/B, where Δ is the ligand field splitting parameter). The lines emerging from each free-atom term depict what terms they become in an octahedral environment. Thus, a ^3F term of a d^2 gas phase ion become the three sets of terms $^3A_2$, $^3T_1$, and $^3T_2$ in the complex.

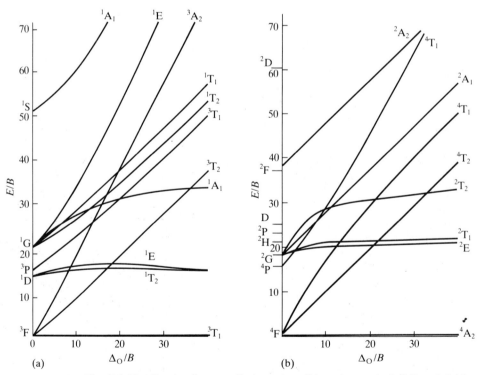

Fig. T.1 The Tanabe–Sugano diagrams for d^n ions in octahedral ligand fields. (a) d^2 with $C = 4.42B$, (b) d^3 with $C = 4.5B$, (c) d^4 with $C = 4.61B$, (d) d^5 with $C = 4.48B$, (e) d^6 with $C = 4.8B$, (f) d^7 with $C = 4.63B$, (g) d^8 with $C = 4.71B$.

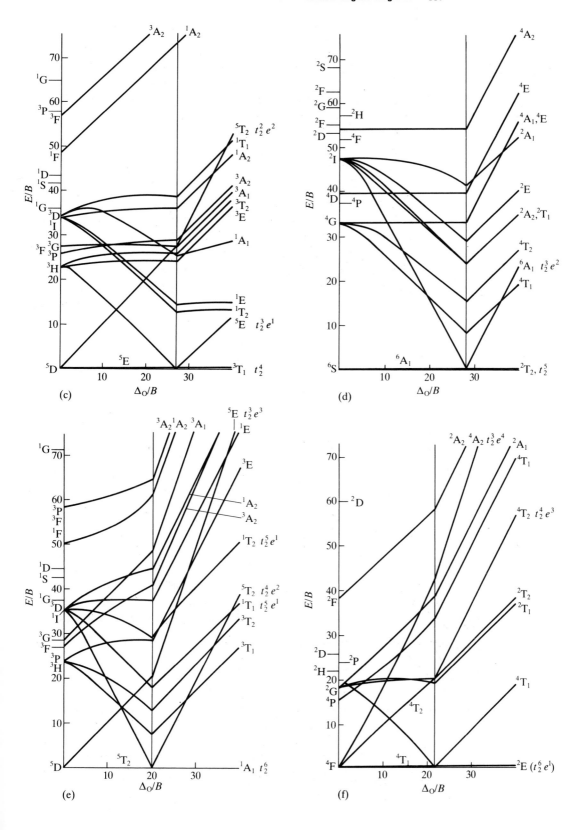

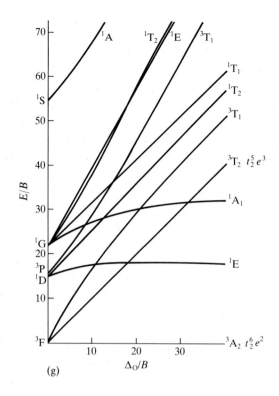

(g)

The energies of the terms diverge as the ligand field strength increases because they correspond to different occupations of the five d orbitals. Terms of the same symmetry obey the °non-crossing rule, and bend away from each other rather than cross. The zero of energy is taken as the energy of the lowest term. Hence, in some diagrams there are abrupt discontinuities where a new term takes over this role at a large enough ligand field strength.

A real complex corresponds to a ligand field strength somewhere along the horizontal axis: weak field ligands correspond to a location near the left, and strong field ligands correspond to locations progressively further to the right. It follows that the value of Δ/B may be identified by locating the position of a vertical line that cuts through the terms in such a way that their energies match the observed transitions. Since B is known from gas phase atomic spectra, the horizontal location of the transitions can then be interpreted in terms of Δ itself.

Further information

Tanabe–Sugano diagrams are introduced in Chapter 14 of Shriver, Atkins, and Langford (1990) and the illustrations shown here are from that text. Good accounts of how they are used will be found in Griffith (1964) and Ballhausen (1979). The diagrams were introduced by Tanabe and Sugano (1954).

Tensor

To a chemist, a **tensor** is a quantity that expresses the directional dependence of the response of molecules and solids to an external perturbation. For example, the **polarizability tensor** expresses the response of a molecule to an applied electric field. An **inertia tensor** expresses the response of the rotational state of a molecule to a torque. Other physical quantities that behave like tensors include those representing a stress/strain type of relation, and it is from this sense of 'tensing' an object that the name tensor arose.

As an example, consider the effect of an electric field E_z defining the z axis of a molecule. The field °polarizes the molecule and induces a dipole moment; however, in general, the field and the dipole it induces are not exactly parallel (Fig. T.2). The magnitudes of the x, y, and z components are proportional to the strength of the applied electric field and are written $\alpha_{xz}E_z$, $\alpha_{yz}E_z$, and $\alpha_{zz}E_z$. Similar expressions describe the induced dipole when the field is applied along the x and y axes. The nine quantities of the form $\alpha_{qq'}$ form the nine components of a second-rank polarizability tensor α. Knowing α we may predict the polarization in any direction when a field is applied in any direction.

There are three directions which, if the field is applied along them, result in a dipole moment that is exactly parallel to the applied field. These three directions are the **principal axes** of the tensor. The three components α_{qq} along these directions are the **principal components** of the polarizability tensor. They define the radii of the **polarizability ellipsoid** (Fig. T.3) which summarizes how the polarizability changes as the molecule rotates. The radius of the ellipsoid in any direction gives the magnitude of the dipole moment when the field is applied in that direction.

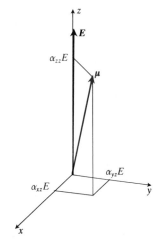

Fig. T.2 When an electric field is applied to a molecule parallel to the z direction, the induced electric dipole moment might not be parallel to it. The three components of the induced dipole moment are given by three components of the polarizability tensor α.

Further information

The physical aspects of tensors are described in Lovett (1989) and, in more advanced presentations, in Nye (1985) and Wooster (1973), all of whom describe

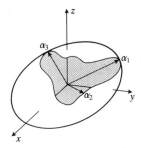

Fig. T.3 The three principal components of the polarizability tensor define the polarizability ellipsoid. The distance from the centre to a point on the surface gives the polarizability of the molecule in that direction.

their application to the description of the properties of solids. Tensors are invaluable in the theory of the behaviour of molecules in electromagnetic fields, and for an illustration of their use see Buckingham (1967) and Barron (1983). Certain special types of tensor that have particular rotational symmetries are described by Brink and Satchler (1968), Rose (1957), Edmonds (1974), and Fano and Racah (1959). The vector product $a \wedge b$ is often said to be a vector; it is in fact a second rank antisymmetrical tensor. The distinction is made very nicely in Altmann (1991), who deals, in an engaging way, with some of the muddles that people made of physics until this point was appreciated.

Term symbol

A spectroscopic **term** is a set of states that have the same or closely similar energies (see °atomic spectra), and transitions between terms are responsible for spectroscopic lines (see the °Bohr condition). A **term symbol** is a label used to specify a term. An atomic term is specified by its angular momentum quantum numbers S (the total spin), L (the total orbital angular momentum), and J (the total angular momentum) written in a special way. The terms of polyatomic molecules are reported using the symmetry classifications of their states. A typical atomic term symbol looks like $^2S_{1/2}$ or $^3P_2$ (Box T.1) and is constructed as follows.

- The letter denotes the value of L according to the correspondence

$$L = 0 \quad 1 \quad 2 \quad 3 \quad 4 \quad 5 \quad 6 \quad 7 \quad ...$$
$$\quad\quad S \quad P \quad D \quad F \quad G \quad H \quad I \quad K \quad ...$$

- The left superscript is the value of $2S+1$.

That is, the left superscript gives the °multiplicity of the term. Thus, ^2S is a 'doublet S term', and ^3P is a 'triplet P term'. Note the roman S for the term symbol with $L=0$, which should not be confused with the italic S for the total spin.

- The right subscript is the value of J for the °level of the term.

For example, the ^2P term has the two levels $2P_{1/2}$ and $^2P_{3/2}$.

Term symbols for linear molecules are constructed similarly, but in place of L (which is not a good quantum number in a linear molecule) the value of Λ, the orbital angular momentum around the internuclear axis (see °Hund coupling cases) is given instead according to the correspondence

$$|\Lambda| = 0 \quad 1 \quad 2 \quad 3 \quad 4 \quad ...$$
$$\quad\quad \Sigma \quad \Pi \quad \Delta \quad \Phi \quad \Gamma \quad ...$$

Thus, the ground state of an NO molecule is $^2\Pi$. The projection of

Box T.1 Term symbols

Multiplicity → $2S+1$

Value of L → X $_{J ← \text{Level}}$

$L = 0$ 1 2 3 4 5 6 7...

X = S P D F G H I K...

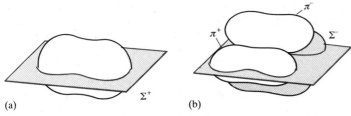

Fig. T.4 (a) A σ orbital (and any closed shell) has $+$ symmetry with respect to reflection in a plane containing the internuclear axis, and gives rise to a Σ^+ term. (b) One of the two π orbitals has $+$ symmetry and the other has $-$ symmetry with respect to reflection in the plane. If the configuration of the molecule is $(\pi^+)^1(\pi^-)^1$, the overall term has Σ^- symmetry because $(+1) \times (-1) = -1$.

the total orbital angular momentum J is denoted Ω and its absolute value is written as a subscript, as in $^2\Pi_{3/2}$ for one of the two levels of the ground term of NO (see °Hund coupling cases).

For centrosymmetric linear molecules the parity of the term is given by a subscript g or u (see °*gerade* and *ungerade*). Thus, CO_2 has a $^1\Sigma_g$ ground term. For Σ terms we need to distinguish between states that are respectively even and odd under reflection (Fig. T.4). This is achieved with a superscript $+$ or $-$, as in $^3\Sigma_g^-$ for the ground term arising from the π^2 configuration of O_2 (one π orbital is $+$ and the other is $-$, so their product is $-$). The term symbols of polyatomic molecules are expressed using the symmetry properties of the orbitals in the molecular point °group. Thus, the H_2O molecule belongs to the C_{2v} point group, is a singlet, and has a ground state wavefunction which is totally symmetric (symbolized A_{1g}): its term symbol is $^1A_{1g}$.

Further information

See MQM Chapter 9 for an account of atomic term symbols in slightly more detail. All texts that treat atomic spectra discuss the derivation and application of term symbols, so see Gerloch (1986) for an introduction and °atomic spectra for further references. The derivation of the values of the angular momentum quantum numbers that may arise from a given configuration is described in the entry on the °Clebsch–Gordan series. The derivation of the term symbols for molecules (both diatomic and polyatomic) is a task for °group theory (which is why it has been treated so lightly in this entry); therefore see Chapter 7 of MQM, Harris and Bertolucci (1978), and Kettle (1985).

Torsional barrier

A **torsional barrier** is a barrier to free rotation around an internuclear axis. An example is the barrier in ethane, which prevents one CH_3 group freely rotating relative to the other about the C—C bond on account of the interaction of the C—H bonds on the neighbouring groups.

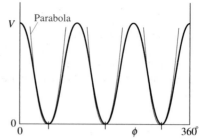

Fig. T.5 Although the potential energy of a molecule may vary harmonically as one group undergoes rotation through the angle ϕ relative to its neighbour, for small displacements the potential varies as ϕ^2 and the groups undergo simple harmonic torsional vibration.

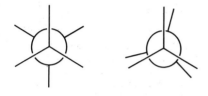

1   2

The staggered conformation (**1**) of ethane has the lowest energy and the eclipsed conformation (**2**) has the highest energy. For displacements close to the staggered conformation, the potential energy varies as ϕ^2 (Fig. T.5), and the group behaves like a °harmonic oscillator. It follows that the ground state of the ethane torsional motion is a zero-point rocking about the bond, with the difference that as the wavefunction extends through the barrier, °tunnelling may occur from one well to the next. The first excited state of the torsional mode resembles a harmonic oscillator in each of the wells, but the wavefunctions extend through the barriers even more than in the ground state (Fig. T.6) and so there is a greater chance that the rotor will tunnel into a new orientation. At high excitations, when the total energy greatly exceeds the peaks of the barrier potential, the angular variation in potential energy is negligible, and in this limit the group rotates essentially freely. The correlation of energy levels is illustrated in Fig. T.7.

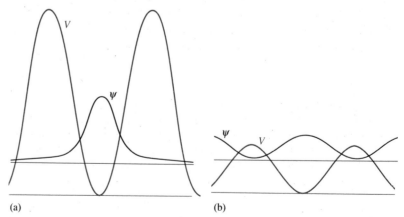

(a)   (b)

Fig. T.6 (a) The ground state of a torsional oscillator resembles a harmonic oscillator wavefunction, but there is a non-zero probability that the oscillator will be found in a neighbouring well. (b) For low barriers, the wavefunction resembles that of a free rotor.

Further information

Helpful discussions of the hindered-rotor problem are given in Hollas (1983) and in the ever-impressive Townes and Schawlow (1955). The problem is introduced by Pitzer (1983) and treated in detail by Papoušek and Aliev (1982). The group theoretical discussion of fluxional molecules is a complicated problem: see Ezra (1982).

Transition rate

A **transition probability** at a specified time is the probability that a system has changed from one state to another under the influence of a perturbation. The **transition rate** is the rate of change of the transition probability. The latter's principal significance is that it determines the intensity of spectral lines, for the rate at which energy is absorbed from or emitted into the surrounding electromagnetic field is proportional to the rate at which transitions occur.

The calculation of transition rates is normally based on expressions obtained from °perturbation theory, for the distortion caused by an electromagnetic field is interpreted as an admixture of excited states into the original state. Since the perturbed wavefunction is a superposition that includes these other states, there is a non-zero probability of finding the system in them.

The **Fermi golden rule** for calculating the rate of stimulated transitions is set out in Box T.2. It equates the rate to the product of the square of the transition moment (see °electric dipole transition) and the density of radiation at the transition frequency. The latter is a measure of the extent of movement of the charge during the transition: if the motion is great and has the appropriate symmetry (e.g., is dipolar), then the field can interact strongly and the transition rate is high. Fermi's formula is the basis of the discussion of °selection rules, for if the transition moment is zero, the transition rate is zero.

Further information

See Chapter 8 of MQM for a quantitative discussion of transition rates and a derivation of Fermi's golden rule. Most texts on quantum mechanics show how transition rates are calculated, so see Davydov (1976), Craig and Thirunamachandran (1984), Bransden and Joachain (1989), and Das and Melissinos (1986). Loudon (1979) gives a nice discussion; see also Heitler (1954).

Tunnelling

Quantum mechanical **tunnelling** is the penetration of a particle into a region from which, according to classical mechanics, it is excluded

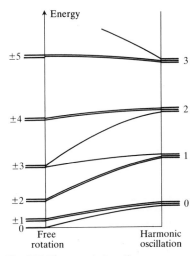

Fig. T.7 The correlation diagram showing how the states of a free rotor correlate with those of a torsional oscillator as the depth of the well is increased. Note that all the states of the free rotor (except the ground state) are doubly degenerate because rotation may occur in either direction.

Box T.2 Fermi's golden rule

The transmission rate, W, to a band of states where the number of states per unit energy is ρ, is given by

$$W = \frac{2\pi}{\hbar} |H^{(1)}|^2 \rho$$

For the evaluation of the density of states, see Davydov (1976).

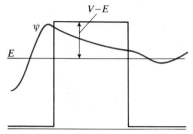

Fig. T.8 Tunnelling through a rectangular potential barrier (only the real part of the wavefunction is shown). The narrower the barrier and the smaller the difference between the total energy and the potential energy, the greater the tunnelling probability.

because it has insufficient energy to overcome the potential energy barrier.

Tunnelling is a consequence of the continuity of wavefunctions at barriers. A wavefunction that is not zero at the inside edge of the barrier decays towards zero (Fig. T.8) inside the barrier where its total energy, E, is less than the potential energy, V. If the amplitude has not reached zero at the outer edge of the barrier it will stop decaying and resume the oscillation it had on the other side of the barrier (but with a smaller amplitude). Since the wavefunction is non-zero on the right of the barrier, the particle may be found there. If it is, we conclude that it has tunnelled through the barrier and out of the well.

The wavefunction decays more rapidly inside the barrier the greater the mass of the particle and the greater the difference between its energy and the potential energy. Thus, electrons, which are very light, can tunnel much more efficiently than other particles, and protons can tunnel more efficiently than deuterons. Macroscopic particles tunnel negligibly. Tunnelling through a spatially finite barrier is most efficient when the energy of the particle is close to the top of the barrier.

The shape of the potential is also important. Sharply changing potentials reflect the particle more effectively than slowly varying potentials. This behaviour is similar to the propagation of light, which is reflected most strongly from regions where the refractive index changes abruptly.

The fact that the kinetic energy is locally negative wherever E is less than V does not imply that the kinetic energy of the particle could ever be measured as negative. The outcome of a measurement of the kinetic energy is determined (through the °expectation value) by the entire wavefunction, and the expectation value is always positive even if there are local negative contributions to it. If we tried to confine the particle into the region of the barrier where its kinetic energy is negative, we would find that the very act of confinement introduces an uncertainty into its energy that is sufficient to prevent us from concluding that the particle has a negative kinetic energy.

Further information

Tunnelling problems are extensively treated in introductory quantum mechanics texts, so see Davydov (1976), Bransden and Joachain (1989), and Das and Melissinos (1986). Bell (1980) and Khairutdinov, Zamaraev, and Zhdanov (1989) have given thorough accounts of the role of tunnelling in chemistry, and discuss the penetration of barriers of a wide variety of shapes. See Chalk (1988) for an interesting and reasonably full discussion, and Washburn (1988) for a comment.

Uncertainty principle

Heisenberg's **uncertainty principle** states that:

> The product of the uncertainties in the simultaneous specifications of two °complementary observables can never be less than a small quantity of the order of h.

The precise form of the principle is given in Box U.1. It implies that as one observable is constrained to have a more precise value, so its partner becomes more ill-defined. An experiment designed to measure the two observables simultaneously will be able to determine one precisely only at the expense of losing information about the other (see °duality). The principle was established by Werner Heisenberg in 1927, as a consequence of his formulation of °matrix mechanics.

The most basic example of the role of the uncertainty principle is the restriction it places on the simultaneous specification of the position of a particle with respect to a coordinate axis q and its linear momentum p parallel to the same axis. These two observables are complementary (their operators do not °commute), and if we require the position of the particle to be within a range Δq, then the range of uncertainty in the value of the linear momentum, Δp, must be such that $\Delta p \Delta q \geqslant \frac{1}{2}\hbar$. As the position is specified more precisely (and Δq is decreased), the spread in the possible range of p increases in such a way that the uncertainty product does not fall below $\frac{1}{2}\hbar$. Conversely, if the momentum is specified to within the range Δp, the range of locations of positions must be such that $\Delta p \Delta q \geqslant \frac{1}{2}\hbar$. Hence, the more closely we specify the position, the less we can say about momentum (parallel to the same axis), and vice versa. The uncertainty principle puts no restriction on the simultaneous values of position and momentum along perpendicular axes (the corresponding operators commute). Consequently, the position along x may be measured simultaneously with the momentum along y with arbitrary precision.

Box U.1 The uncertainty relations

Let A and B be two operators corresponding to the observables of interest, and ΔA and ΔB be the root mean square deviations from the mean:

$$\Delta A = \{\langle A^2 \rangle - \langle A \rangle^2\}^{1/2} \qquad \Delta B = \{\langle B^2 \rangle - \langle B \rangle^2\}^{1/2}$$

Then the uncertainty principle states that

$$\Delta A \Delta B \geqslant \tfrac{1}{2}|\langle [A, B] \rangle|$$

Typical uncertainty relations include

$$\Delta q \Delta p \geqslant \tfrac{1}{2}\hbar$$

$$\Delta l_x \Delta l_y \geqslant \tfrac{1}{2}|\langle l_z \rangle|\hbar$$

The angular location of a particle and its angular momentum are complementary and hence are also related by an uncertainty relation. However, cyclic systems must be treated with care because an uncertainty of 2π in an angle is equivalent to complete uncertainty in location. Special forms of the uncertainty principle are used in these cases (see *Further information*). The relation $\tau\Delta E \approx \hbar$ between the lifetime of a state, τ, and the imprecision, ΔE, with which its energy may be specified, is often treated as an aspect of the uncertainty principle. However, as there is no operator for time, the general principle does not apply to lifetime and energy directly (see the entry on °lifetime broadening).

An insight into the uncertainty principle comes from the form of the wavefunction of a particle with different degrees of localization. The wavefunction of a particle with a precise linear momentum is a plane wave of definite wavelength, and $\psi^*\psi$ is uniform. Hence, since the probability of finding the particle at any point is the same, its position is totally unpredictable. We can localize the particle partially by superimposing a number of waves with different wavelengths (Fig. U.1) and forming a °wavepacket. However, since the particle's wavefunction now contains components of several different wavelengths, its linear momentum is imprecise. To achieve complete localization we must superimpose an infinite number of plane waves of different wavelengths. Now the location is precise but the linear momentum is completely unpredictable.

Heisenberg devised a number of **gedanken** experiments (thought experiments) to show that the world was consistent with his principle. For example, he considered the use of a microscope to measure the position of an electron. He argued that to achieve high precision it was necessary to use very short wavelength (γ-ray) radiation. However, the shorter the wavelength the greater the momentum of each photon (by the °de Broglie relation, $p = h/\lambda$). Since at least one photon must be scattered into the microscope aperture for the position to be determined, the act of observation imparts a momentum to the particle, and that momentum is very large if γ radiation is used to locate the electron very precisely. If we seek to minimize the disturbance to the momentum by using very long wavelength radiation, diffraction effects are so great that the electron cannot be located.

The implication of the position–momentum uncertainty relation is that it is impossible to specify the trajectory of a particle. Thus, although we may know the position of the particle with arbitrary precision, we can know nothing of the particle's momentum. As a result, it is not possible to predict where the particle will be at the next instant. The general implication of the uncertainty principle is that there are two complementary modes of describing the physical world: we can

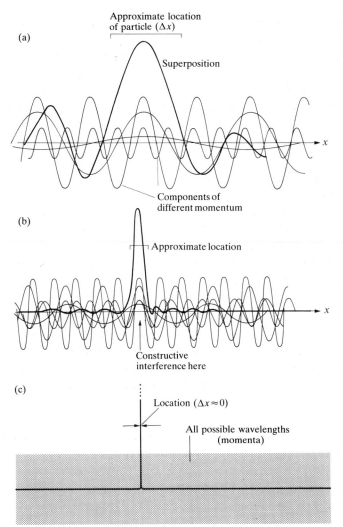

(a)

Approximate location
of particle (Δx)

Superposition

x

Components of
different momentum

(b)

Approximate location

x

Constructive
interference here

(c)

Location ($\Delta x \approx 0$)

All possible wavelengths
(momenta)

Fig. U.1 The complementarity of position and momentum expressed by the uncertainty principle can be interpreted in terms of the wavefunctions used to describe particles with positions and momenta defined to varying degrees of precision. (a) When several wavelengths contribute to the wavefunction both position and momentum are moderately well defined. (b) When more waves contribute the position is better defined, but at the expense of a less well defined momentum. (c) In the limit of a perfectly well defined location, there are an infinite number of waves, so the linear momentum is completely indeterminate.

elect to discuss it in terms of the locations of the particles or in terms of their momenta. It was an error of classical physics to mix these descriptions. Unfortunately, the conditioning of classical physics has been so great that many people still feel that the uncertainty principle denies them complete knowledge of the world. I consider the correct attitude to be that classical physics unconsciously sought to be *over-*

complete. The uncertainty principle reveals that there are alternative *complete* descriptions of the world; we should choose one description or the other, and not seek to mix them.

Further information

The uncertainty principle is derived in Chapter 5 of MQM. Heisenberg's own account, and a number of related papers, appear in an accessible form in Heisenberg (1930) and Price and Chissick (1977); the original paper is Heisenberg (1927). The emergence of the principle is described by Jammer (1966). An approach to the energy–time uncertainty relation is described in Salam and Wigner (1972), the book edited by Price and Chissick (1977), and Shalitin (1984), but see the entry on °lifetime broadening. The uncertainty principle for cyclic systems is described by Carruthers and Nieto (1968). For a survey of the philosophical implications of the principle see Jammer (1974), Hughes (1989), and Redhead (1987).

United atom

The **united atom method**, which in the early days of quantum chemistry was quite widely used in discussions of the structure of molecules, is an example of the use of a °correlation diagram, in this case between

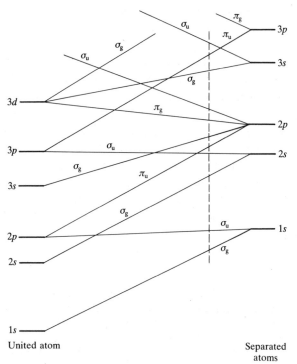

Fig. U.2 The correlation diagram between the states of two free atoms and a united atom. The energy levels of a diatomic molecule lie between the two extremes.

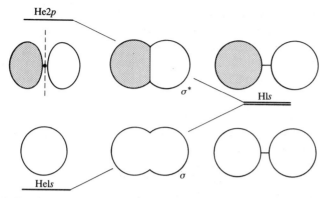

He2p

He1s

σ^*

H1s

σ

Fig. U.3 The correlation between the wavefunctions of two widely separated hydrogen atoms (right) and the *s* and *p* orbitals of a united helium-like atom (left). The out-of-phase combination correlates via the antibonding σ orbital with the He2p orbital.

the states of two widely separated atoms and the states formed when they have merged into a single atom (Fig. U.2). The energy levels of the separated atoms are known, and so are those of the united atom formed when the internuclear separation has become zero. The diatomic molecule of interest has energy levels that are represented by an intermediate stage in the correlation diagram.

The simplest illustration is obtained with two H atoms that are thought of as merging to give a He atom (that the united atom is not a true nuclide is irrelevant). The right hand side of Fig. U.3 shows the 1s orbitals of two H atoms, and the left hand side shows two orbitals of a single He atom. If the H1s orbitals come together with the same sign, they correlate with the He1s orbital of the united atom. If they come together with opposite signs, the °node half-way between the two nuclei is preserved as the united atom is formed. Therefore, that combination correlates with a He2p orbital of the united atom. Since a He2p orbital has a higher energy than a He1s orbital, we can conclude that at an intermediate stage the H_2 molecule has two orbitals, with the orbital without the node having a lower energy than the orbital with the node. These two orbitals are the σ and σ^* orbitals of °molecular orbital theory.

Further information

The concept of the united atom was introduced by Mulliken (1932), and some of the ways it was used are described thoroughly (as always) by Herzberg (1950). Coulson (1979) touches on it. The importance of the united atom is now really that it was an early example of the use of a °correlation diagram and the precursor of °Walsh diagrams and the °Woodward–Hoffmann approach to orbital correlations.

Valence bond theory

The **valence bond theory** (VB theory) was the first quantum mechanical theory of the chemical °bond when it was proposed by Walter Heitler and Fritz London in 1927, and was substantially developed by John Slater and Linus Pauling in the 1930s. It drew heavily on °Lewis's concept of a covalent bond as a shared electron pair. The theory makes the **perfect pairing approximation**, in which it is supposed that structures with electrons paired in all possible ways dominate the wavefunction of the molecule. The overall wavefunction is a °superposition of all possible perfectly paired **canonical structures**, with those of lowest energy making the main contribution. In the context of valence bond theory, the superposition is interpreted as a °resonance of the contributing structures.

We shall illustrate the approach first with H_2, in which there is only one perfectly paired canonical structure, the one corresponding to the Lewis structure H—H. At large internuclear separations the °wavefunction for the species is the wavefunction for the two separated atoms. It is written $A(1)B(2)$, where A and B are $1s$ orbitals on atoms A and B. When the atoms are at a bonding distance, it is supposed that the electrons can exchange their host orbital, and that the wavefunctions $A(1)B(2)$ and $A(2)B(1)$ make equal contributions. Since both distributions of electrons are equally likely, the overall wavefunction is written as the superposition $\psi = A(1)B(2) \pm A(2)B(1)$.

The ambiguity in sign is resolved by the °Pauli principle, which requires the total wavefunction to change sign under electron interchange. Since the spin state of two paired electrons is $\sigma = \alpha(1)\beta(2) - \alpha(2)\beta(1)$ (see °singlet and triplet states), the product $\psi\sigma$ changes sign when the labels 1 and 2 are interchanged only if we choose the $+$ combination of the orbital states. Hence, the VB wavefunction of the covalent bond in the H_2 molecule is $\{A(1)B(2) + A(2)B(1)\}\sigma$.

The energy of the H_2 molecule is now calculated using the superposition ψ as the molecular wavefunction and evaluating the °expectation value of the °hamiltonian. The expectation value is the sum of several integrals specified in Box V.1. Each integral depends on the internuclear distance as illustrated in Fig. V.1, and their sum passes through a minimum at the equilibrium bond distance.

Each of the integrals in Box V.1 has a physical interpretation, and an interpretation of the stability of the bond in terms of valence bond theory must take them all into account. The integral j is the repulsive interaction between the electron clouds on the two nuclei (Fig. V.2a). The integral $-j'$ is the attractive interaction between an electron on one nucleus and the other nucleus. The two integrals k and k' are

Box V.1 The valence bond treatment of H_2

The **energy** for the internuclear distance R is

$$E = 2E(H1s) + \frac{J \pm K}{1 \pm S^2}$$

where the **Coulomb integral** is

$$J = \frac{\kappa}{R} + j - 2j'$$

and the **exchange integral** is

$$K = \frac{\kappa S^2}{R} + k - 2Sk'$$

where $\kappa = 1/4\pi\varepsilon_0$. The constituent integrals are

$$S = \int A(1)B(1)\,d\tau$$

$$j = \kappa \int A(1)^2 \frac{1}{r_{12}} B(2)^2\,d\tau_1\,d\tau_2$$

$$j' = \kappa \int \frac{A(1)^2}{r_{1b}}\,d\tau_1 = \kappa \int \frac{B(1)^2}{r_{1a}}\,d\tau_1$$

$$k = \kappa \int A(1)B(1)\frac{1}{r_{12}}A(2)B(2)\,d\tau_1\,d\tau_2$$

$$k' = \kappa \int \frac{A(1)B(1)}{r_{1b}}\,d\tau_1 = \kappa \int \frac{A(1)B(1)}{r_{1a}}\,d\tau_1$$

specifically quantum mechanical in origin, and arise from interference between the $A(1)B(2)$ and $A(2)B(1)$ components of the total wavefunction ψ. An analysis of the electron distribution represented by ψ (by forming ψ^2) shows that there is a significant accumulation of electron density in the internuclear region where A and B interfere constructively. This extra accumulation of density is represented by the oval regions in Fig. V.2b. The integral k is the repulsive interaction k between the two electrons confined to the oval regions, and the integral $-k'$ is the attractive interaction between these electron-rich regions and the two nuclei.

The most important contribution to the stability of the H—H bond is the one represented by the integral k'. Hence, in the valence bond theory, the lowering of energy of the molecule below that of the separated atoms is in large measure due to the accumulation of electrons in the internuclear region where they are able to interact attractively with both nuclei.

One deficiency in this description of H_2 lies in the form of ψ, which implies that if one electron occupies the orbital A, then the other electron must be in B. In practice there is a significant probability that

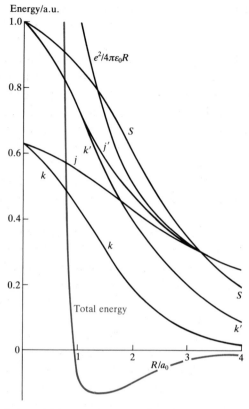

Energy/a.u.

$e^2/4\pi\varepsilon_0 R$

Total energy

R/a_0

Fig. V.1 The variation with internuclear separation of the integrals that occur in the valence bond description of the hydrogen molecule. The total energy is also shown: its minimum corresponds to the equilibrium bond length of the molecule.

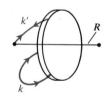

(a)

(b)

Fig. V.2 A diagrammatic portrayal of the integrals that occur in the valence bond description of a diatomic molecule (specifically H_2, Box V.1): (a) contributions to the Coulomb integral J and (b) contributions to the exchange integral K.

both electrons will be found on the same atom. Therefore, the wavefunction can be improved by including in ψ contributions from the ionic term $A(1)A(2)$, corresponding to H^-H^+, and from $B(1)B(2)$, corresponding to H^+H^-. The inclusion of these additional terms is called **ionic–covalent resonance**. According to the °variation principle, an improvement of the wavefunction leads to a lower energy, and hence the inclusion of resonance is an improved description of the bond.

When the valence bond method is applied to polyatomic molecules, each canonical structure is formed by pairing all the available electrons. Thus each canonical structure corresponds to a possible °Lewis structure of the molecule, and each structure has its characteristic energy. The overall structure is then regarded as a resonance hybrid of these canonical structures (see °benzene for an example). The central importance of pairwise interactions in the formulation of canonical structures led Pauling to introduce the concept of orbital °hybridization in order to construct local orbitals in molecules of the appropriate geometry. For instance, using hybridization, each C—H bond in CH_4 is represented by a perfect pairing structure in which A and B in the wave-function $A(1)B(2) + A(2)B(1)$ now represent the Csp^3 and $H1s$ orbitals respectively.

Although VB theory grew out of a chemist's perception of a bond, its original formulation had certain disadvantages that largely offset the advantage of chemical plausibility that inspired the theory. One disadvantage was that the number of canonical structures to be included increased dramatically with the number of atoms in the molecule. For example, there are five non-ionic structures for °benzene, but more than 10^5 for coronene (**1**). Since the structure is a superposition of all these canonical forms, including ionic structures, it can be appreciated that the calculation of the structure of a moderately large molecule was regarded as an enormous task.

1 Coronene

Recent years, however, have seen valence bond theory returning to the arena of serious computation in the form of **spin-coupled valence bond theory**. In this formulation, all N electrons occupy N distinct but not °orthogonal orbitals with their spins coupled into a resultant S in a coupling scheme. The wavefunction of the molecule is then expressed as a linear combination of all the coupling schemes that result in the same resultant spin. This approach has a number of advantages over conventional valence bond theory. For example, the ionic structures that are normally needed to achieve the correct dissociation into atoms are not required, and so a great deal of the labour of the conventional theory is eliminated. A special consequence of the use of non-orthogonal orbitals in the formulation of a structure is that bonding is *localized*, and this localization cannot be removed by a simple transformation.

Further information

Good accounts of conventional valence bond theory are given in Coulson (1979), DeKock and Gray (1980), and Levine (1983). Pauling (1960) is almost exclusively an account of the valence bond description of molecules, and is an unparalleled example of the power of quantum chemical reasoning within the format of that approach. For more modern assessments of the method and the introduction of spin-coupled valence bond theory see Cooper, Gerratt, and Raimondi (1986) and Klein and Trinajstić (1990). A comparison of °molecular orbital and valence bond theories is in the entry of that title.

Valence-shell electron pair repulsion theory

The **valence-shell electron pair repulsion theory**, which is almost universally referred to as **VSEPR theory**, is an attempt to account for the shapes of molecules in terms of the repulsions between the electron pairs belonging to each atom. It is usually applied to a molecule or

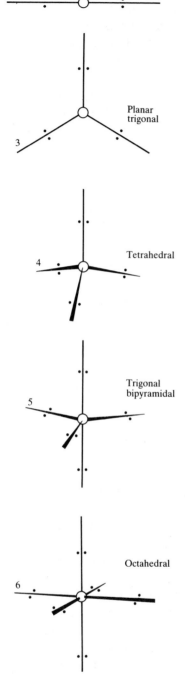

2   Linear

3   Planar trigonal

4   Tetrahedral

5   Trigonal bipyramidal

6   Octahedral

Fig. V.3 The arrangements of electron pairs around a central atom that minimize their repulsions and which are used to predict molecular shapes using VSEPR theory.

molecular fragment in which there is a recognizable 'central' atom to which other atoms are attached, such as the O atom in H_2O and the N atom in NH_3. The theory originated from the work of N.V. Sidgwick and H.M. Powell in the 1930s, but its current formulation is largely due to R. Nyholm and R.J. Gillespie.

The theory supposes that the shape of a molecule is such as to minimize repulsions between the bonding and lone electron pairs on the central atom A of an AB_n molecule. Thus, we think of the bonding electrons and lone pairs as free to move on the surface of a sphere centred on the atom until the distances between them are as large as possible. If there are four atoms attached to the central atom (and no lone pairs) the greatest separation of the bonding electrons is achieved if they occupy the corners of a tetrahedron; if there are six atoms attached to the central atom, they occupy the corners of an octahedron. If there are two atoms attached to the central atom and the latter has two lone pairs of electrons (as in H_2O), the two atoms and the two lone pairs adopt a tetrahedral arrangement. These basic shapes are summarized in Fig. V.3.

At this stage we do not distinguish between atoms that are singly or multiply bonded to the central atom: even if the atoms are bound by double bonds (as they may be in certain °Lewis structures of SO_4^{2-}, for instance), we treat the two electron pairs in each bond as bound together and behaving as a single 'superpair'. The shape of SO_4^{2-} is achieved by the four superpairs minimizing the repulsions between them, and the ion is tetrahedral.

In the second stage, we allow for differences between the strengths of repulsions between bonding and nonbonding pairs. The generally accepted view (accepted because it works, not because it is fully understood) is that lone pairs repel more strongly than bonding pairs, so the order of repulsions is

$$(LP,LP) > (LP,BP) > (BP,BP)$$

where LP stands for a nonbonding lone pair and BP stands for a bonding pair (or superpair). The molecule adjusts its shape so that the repulsions are minimized in the light of this ordering. Thus, bonds can afford to move together slightly if that relieves the repulsion arising from a lone pair.

As an example, consider the NH_3 molecule. The N atom has four electron pairs, three of which are bonding (involved in N—H bonds), and one of which is a lone pair. The basic shape of the molecule is therefore pyramidal, with an angle of 109.5° (the tetrahedral angle) between all the bonds and between the axis of the lone pair and the bonds. However, the molecule relieves repulsion from the lone pair by distorting slightly, with the N—H bonds closing together and moving away from the lone pair. This distortion is consistent with the experimentally observed HNH angle of 107°.

VSEPR theory is a reasonably successful rule of thumb for predicting molecular shapes qualitatively. It does not try to predict bond angles quantitatively. Thus, although it successfully accounts for the slight closing of the HNH angles in NH_3, it cannot predict that the molecule will adopt the angle 107° rather than some other value. The theory is also limited to closed-shell, main-group atoms, and it can fail disastrously when applied to compounds that contain d-block elements. The principal theoretical criticism of the theory is that it concentrates on the potential energy arising from repulsions between electron pairs and ignores contributions to the energy from changes in the kinetic energies of electrons and the composition of the orbitals that electrons occupy.

Further information

The VSEPR approach to chemical bonding is covered thoroughly in introductory chemistry courses, so see Atkins (1989). A good general introduction to the procedure is Gillespie (1972); see also Burdett (1980) for a broader perspective. The birth of the approach can be found in Sidgwick and Powell (1940). There has been considerable controversy about the origin of the repulsions between electron pairs. Something of the flavour of the dispute can be traced through Edmiston, Bartleson, and Jarvie (1986), who resolve the conflicting views in favour of Sidgwick and Powell's original suggestions. However, they also emphasize that the °Walsh approach is more reliable.

Valence state

The **valence state** is the hypothetical state of an atom when it is a component of a molecule. For example, a °valence bond description of a CH_4 molecule might regard it as an sp^3 °hybridized C atom with each of its four tetrahedrally orientated lobes overlapping one of the four surrounding H atoms. In this case we would picture the valence state as a C atom with one electron in each of its 2s and 2p orbitals. However, because electrons in neighbouring bonds have random spin orientations, the $2s^12p^3$ configuration of the valence state is a non-spectroscopic state without a definite orbital or spin angular momentum.

A valence state is a superposition of spectroscopic states with definite values of L, S, and J, and its energy is an average of these contributing terms. The energy of the valence state above the ground state is the energy needed to **promote** the atom to its valence state. The promotional energy is about 7 eV for a C atom in a tetrahedral environment. The ionization energy from the valence state, and its electron affinity, are similar averages of the ionization energies and electron affinities of the contributing spectroscopic states, and are the quantities used to calculate Mulliken °electronegativities.

In order to picture a valence state we can imagine the formation of a CH_4 molecule by bringing the four H atoms towards the central C atom in its ground state. As the atoms approach tetrahedrally, the surface of the C atom begins to stir, and the electron density undergoes transient fluctuations with accumulations in the tetrahedral directions. As the atoms get closer, the fluctuations get stronger and longer lasting. When the atoms are at their equilibrium bonding distance the fluctuations are essentially frozen into the four tetrahedral σ bonds. Only at this point would it be true to say that the central atom is in its valence state, for it has been drawn out of the ground state by the H atoms.

Further information

Helpful accounts of valence states will be found in McGlynn *et al.* (1972), Bratsch (1988*a*), and Huheey (1983). The concept was introduced by Mulliken (1934), and valence state energies are listed in Bratsch (1988*a*), who also gives a number of further references.

Variation principle

The **variation principle** states that:

> The change in energy for an arbitrary variation of the ground state wavefunction is always positive (so long as the hamiltonian is exact).

That is, if we calculate the energy using an approximate wavefunction (but the exact hamiltonian), which in this context is called a **trial function**, we shall obtain a value bigger than the true energy, and every improvement we make to that wavefunction will lower the energy we calculate. It follows that if we make a series of guesses about the form of the trial function, the one that gives the lowest energy will most closely resemble the true wavefunction of the system.

One systematic procedure is to express the trial function as a function of one or more parameters, and then to vary the parameters and look for an energy minimum. Thus, if the trial function depends on the parameter P, the expectation value of the hamiltonian will be a function of P. Then finding the value of P for which dE/dP is zero gives the optimum value of P to use in a trial function of that kind.

In the **Ritz procedure**, the trial function is written as a sum of fixed functions called the **basis set**, with each basis function multiplied by a variable coefficient. (Exactly the same principle is used in the °linear combination of atomic orbital construction of molecular orbitals.) The energy is calculated in terms of the coefficients, and all of them are varied until the energy has its lowest value.

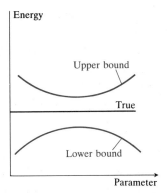

Fig. V.4 The variational minimum energy is an upper bound to the true energy. There are also techniques that give a lower bound, thus helping to bracket the energy within a range.

The variational minimum energy is an **upper bound** to the true energy (Fig. V.4). There are techniques of finding a **lower bound** beneath which the true energy cannot lie, and in principle this gives some indication of the accuracy of a variational calculation. However, the technique is difficult and has not been widely used.

Further information

For a simple derivation of the variation principle and determination of the minimum conditions, see Chapter 8 of MQM. For further discussion see Wilcox (1966). The determination of lower bounds is described in Löwdin (1966) and Lee (1987).

Vector model

The **vector model** is a pictorial representation of quantized °angular momenta in terms of vectors. Any given angular momentum with quantum numbers j and m_j is represented by a vector j of length $\{j(j+1)\}^{1/2}$ units and projection m_j units on an axis. The vector itself is taken to lie at an indeterminate position on a cone of °precession (Fig. V.5).

The coupled angular momenta permitted by the °Clebsch–Gordan series are represented by combining the vectors to give the specified resultant. Each component vector precesses around the resultant (Fig. V.6). The resultant itself precesses around the z axis, which is arbitrary if no external field is present but along the field if one is present.

The overall angular momentum of a many-electron atom is built up by combining the angular momenta of all the electrons. Fortunately this formidable exercise is simplified by noting that the **core electrons**

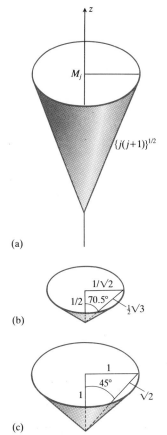

Fig. V.5 (a) The fundamental structure used in the vector model of angular momentum. An angular momentum with quantum numbers j and m_j is represented by a cone of side $\{j(j+1)\}^{1/2}$ and height m_j. The angular momentum vector lies at an indeterminate position on the surface of the cone. Scale drawing of the cone for (b) $j=\tfrac{1}{2}$, $m_j=+\tfrac{1}{2}$ and (c) $j=1$, $m_j=+1$.

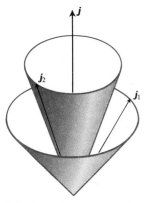

Fig. V.6 The vector representation of the coupling of two angular momenta j_1 and j_2 into a resultant j. The two component vectors lie at an indeterminate angle on their cones, but they lie at a definite angle to each other (so their resultant is well defined).

of the atom (the electrons in shells other than the valence shell) have zero angular momentum because all the spins are paired and the shell is complete. Then an approximation is normally made that is based on the relative strengths of Coulombic and spin–orbit interactions. There are two extreme cases:

- **Russell–Saunders coupling**, when the Coulombic interactions dominate,

- *jj*-**coupling**, when spin–orbit coupling dominates.

In Russell–Saunders coupling, the spin–orbit interaction is so small that it is effective only after all the orbital momenta have been combined into a resultant L, and all the spins have been combined into a resultant S. The electronic orbital motions are dominated by the electron–electron Coulombic interactions. For two electrons the coupling of the momenta would be represented by a diagram of the type in Fig. V.7a. The two spins also couple to give a definite resultant S: the coupling energy for this interaction arises from the spin-dependent °exchange energy, and so it too, although quantum mechanical in origin, is a Coulombic interaction. At this stage the two resultant momenta L and S couple together to form a resultant J, the total angular momentum, and the strength of this interaction depends on the strength of the spin–orbit coupling.

The Russell–Saunders scheme breaks down when the spin–orbit coupling is stronger than the Coulombic interaction because the spin and orbital angular momenta of individual electrons couple together. The *jj*-coupling scheme describes the extreme situation of this kind. In it (Fig. V.7b), each electron's spin is allowed to couple to its orbital momentum; thus s_1 and l_1 couple to form j_1. The two components °precess strongly around their resultant in the manner characteristic of a strongly interacting pair of momenta. This j_1 is now coupled to another j_2 and the total angular momentum J constructed: the latter coupling is relatively weak, for it depends on the Coulombic interactions of the electronic orbitals. We see that, although the total angular momentum obtained in this way might be the same as the total in the Russell–Saunders scheme, the states of the atoms are different, and hence their energies also differ.

Neither scheme is an exact representation of the true state of affairs because there is always some competition between the different types of interaction, and indeed it is quite possible for some electrons in the same atom to be coupled by one scheme and the remainder by the other. Nevertheless, for light atoms, which have small spin–orbit coupling constants, the Russell–Saunders scheme is often a good description of the valence electrons. Heavy atoms, which have large spin–orbit coupling constants, are often predominantly *jj*-coupled. It follows that

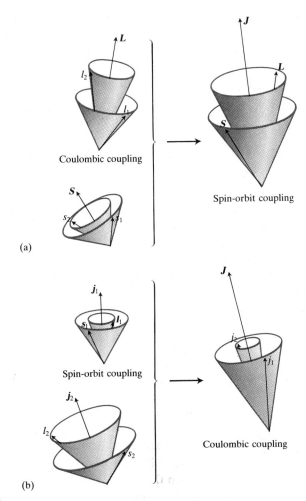

Coulombic coupling

Spin-orbit coupling

(a)

Spin-orbit coupling

Coulombic coupling

(b)

Fig. V.7 (a) Russell–Saunders coupling. The spins of the electrons couple into a resultant **S**, the orbital angular momenta of the electrons couple into a resultant **L**, and then the two combined angular momenta couple into an overall resultant **J**. (b) *jj*-coupling. The spin and orbital angular momenta of each electron couple together individually, and then the total angular momenta j_1 and j_2 of each electron couple together into an overall resultant **J**.

the wavefunctions corresponding to the Russell–Saunders scheme are a good starting point for more elaborate calculations on light atoms.

Further information

See Chapters 6 and 9 of MQM for further information about the vector model and the two main coupling schemes. Good accounts are to be found in Herzberg (1944), King (1964), and Kuhn (1962). Candler (1964) deals explicitly with the vector model of the atom, and Munowitz (1988) indicates some of the properties of vector representations of nuclear spins and the basis of the technique in terms of the °density matrix. See also the °Hund coupling cases and °singlet and triplet states for further concepts and references.

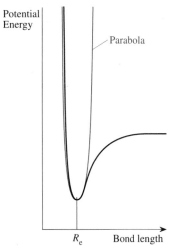

Potential Energy

Parabola

R_e Bond length

Fig. V.8 A typical molecular potential energy curve and the parabolic approximation to it that is valid at small displacements from equilibrium.

Vibrational spectroscopy

A typical molecular potential energy curve is shown in Fig. V.8. For small displacements the potential energy is parabolic, and oscillations confined to these displacements are °harmonic. Their frequency depends on both the °force constant, k, and the effective mass, μ, of the vibrational mode (see °reduced and effective mass), and is given by $\omega = (k/\mu)^{1/2}$. The quantized energy levels are those of a °harmonic oscillator of this frequency, and the °selection rule $\Delta v = \pm 1$ results in transitions of frequency ω and wavenumber $\omega/2\pi c$.

Molecular force constants and masses are such that vibrational transitions fall in the infrared region of the spectrum, and so vibrational spectroscopy is an infrared technique. Vibrational structure also appears in an °electronic spectrum (see the °Franck–Condon principle). As a rough guide, weak bonds between heavy atoms have vibrational transitions in the region of several 10^2 cm^{-1} (the I—I bond in I_2 vibrates at a wavenumber of 214 cm^{-1}, corresponding to a frequency of 6.4×10^{12} Hz, or a wavelength of 4.67×10^4 nm) and stiff bonds between light atoms vibrate in the region of several 10^3 cm^{-1} (the H—H bond in H_2 vibrates at 4395 cm^{-1}, corresponding to 1.3×10^{14} Hz, or 2280 nm). Bond stretches tend to be at higher frequencies than bond bends. See Table V.1 for the vibrational wavenumbers of some diatomic molecules, and Table V.2 for the typical wavenumbers of groups of atoms in molecules.

A vibration is active (observable) in the infrared if during it the dipole moment of the molecule changes. Diatomic molecules absorb in the infrared only if they are polar. In polyatomic molecules it is necessary to assess (using °group theory) whether each °normal mode is accompanied by a change in dipole moment. At normal temperatures the Boltzmann distribution ensures that essentially all molecules are in their ground vibrational state. As a result, the spectrum consists of

Table V.1 Vibrational wavenumbers and frequencies of diatomic molecules

	$\tilde{\nu}/\text{cm}^{-1}$	$\nu/(10^{13}\ \text{Hz})$
$^1\text{H}_2^+$	2322	6.961
$^1\text{H}_2$	4395	13.19
$^2\text{H}_2$	3118	9.349
$^1\text{H}^{19}\text{F}$	4138	12.41
$^1\text{H}^{35}\text{Cl}$	2991	8.967
$^1\text{H}^{81}\text{Br}$	2649	7.941
$^1\text{H}^{127}\text{I}$	2308	6.919
$^{14}\text{N}_2$	2358	7.069
$^{16}\text{O}_2$	1580	4.738
$^{19}\text{F}_2$	892	2.67
$^{35}\text{Cl}_2$	560	1.67

Table V.2 Typical vibrational wavenumbers, $\tilde{v}/(\mathrm{cm}^{-1})$

C–H stretch	2850–2960
C–H bend	1240–1465
C–C stretch, bend	700–1250
C=C stretch	1620–1680
C≡C stretch	2100–2260
O–H stretch	3590–3650
H-bonds	3200–3570
C=O stretch	1640–1780
C≡N stretch	2215–2275
N–H stretch	3200–3500
C–F stretch	1000–1400
C–Cl stretch	600– 800
C–Br stretch	500– 600
C–I stretch	500
CO_3^{2-}	1410–1450
NO_3^-	1350–1420
NO_2^-	1230–1250
SO_4^{2-}	1080–1130
Silicates	900–1100

Data: Bellamy (1975, 1980).

a single line for each vibrational mode of the molecule, and corresponds to the excitation of the normal mode from its ground state to its first excited state. Such a transition is called the **fundamental** of the mode.

The vibration of a molecule is not strictly harmonic because the potential in which the atoms move is only approximately parabolic (Fig. V.8). This mechanical °anharmonicity has several consequences. First, the selection rule $\Delta v = \pm 1$ is no longer strict, and other transitions become weakly allowed. Then instead of a single line for each mode, the fundamental is accompanied by weaker **overtones** or higher **harmonics** with $\Delta v = \pm 2$, and so on. The overtones do not appear at exactly 2ω, 3ω,... for the anharmonicity also affects the energy levels slightly.

It is possible for a single photon to excite two modes simultaneously and to give rise to a **combination band**. A combination band has zero intensity in the harmonic approximation, but it may steal intensity from a nearby active mode by the mixing process known as **Fermi resonance**. In Fermi resonance, the wavefunction for the jointly excited state is mixed with the wavefunction of an excited active mode that matches it in energy. The combination band thereby acquires some of the character of the active mode, and hence becomes allowed.

When a vibrational transition occurs in a gas phase molecule it may be accompanied by a °rotational transition and give rise to a °branch structure in the spectrum. Other features that affect a vibrational spectrum include °inversion doubling (for example, in NH_3) and λ-type °doubling in linear triatomic molecules (see °Coriolis interaction).

The main kinds of chemical information that may be obtained from vibrational spectra include the elementary but important one of the identification of a species by using its vibrational spectrum as a fingerprint. The major quantitative information that may be obtained is the rigidity of bonds under the stresses of stretching and bending: the force constant is an important feature of a chemical bond. The anharmonicities show how far the true potential differs from an ideal parabola. The rotational structure of vibrational transitions enables the molecular geometry to be determined in different vibrational states (bond angle and bond length dependence on vibrational state), and the vibrational and rotational structure of electronic transitions enables the same kind of information to be obtained about electronically excited states. This information contributes to a picture of the potential energy curves of molecules in different electronic states.

We have concentrated on electric dipole absorption spectra: vibrational transitions may also be observed in °Raman spectroscopy.

Further information

See MQM Chapter 11 for a further introduction to vibrational spectroscopy; see also the entry on °normal modes. For detailed information see Banwell (1983), Bunker (1979), Hollas (1983, 1987), Steinfeld (1985), Struve (1989), Graybeal (1988), and the authoritative monograph by Wilson, Decius, and Cross (1955). Vibrational Raman spectroscopy is treated in detail in Long (1977).

Vibronically allowed transition

A **vibronic transition** is a simultaneous vibrational and electronic transition of a molecule, such as is responsible for the band spectra of electronic spectra. A **vibronically allowed transition** is an electronic transition that is made possible by the vibrational motion of a molecule.

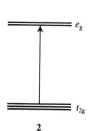

2

As an example of a vibronically allowed transition, consider a d–d transition in an octahedral complex, the type of transition that is often responsible for the colour of such complexes. In a d–d transition, a d electron of the central ion is excited from a t_{2g} orbital to an e_g orbital (**2**); because the separation of the two sets of orbitals is small the absorption occurs in the visible region of the spectrum. However, the transition is forbidden by the Laporte °selection rule, which excludes transitions that are not accompanied by a change of parity (that is, g–g and u–u transitions are forbidden).

The Laporte selection rule is valid if the complex has a centre of symmetry, for only then is the parity well defined; hence it applies only if the complex is strictly octahedral. Since the complex may vibrate, it may lose its centre of symmetry briefly if the vibrational mode is asymmetric (Fig. V.9): the transition then becomes weakly allowed.

To see the nature of the coupling in more detail, we concentrate on the electron distribution of the central ion, not the complex as a whole. Because the environment of the ion is briefly asymmetrical, the ion's electron distribution is briefly distorted in one direction or another. This distortion is equivalent to the admixture of metal p orbital character into the d orbitals (Fig. V.10). Since p orbitals have u parity, the ion can be excited by an electric dipole transition by making use of the allowed p–d transition. The intensity of the transition is proportional to the extent to which p character is mixed into the d orbitals.

Further information

For vibronic transitions, see the entry on the °Franck–Condon principle and the references therein. Vibronically allowed transitions are treated in Chapter 12 of MQM. For more detailed information see Griffith (1964), who gives a very good, straightforward analysis, and Hollas (1983, 1987) for one that is more wide ranging but also more involved. For a detailed treatment of the coupling between electronic and nuclear motion see Fischer (1984).

Virial theorem

In its simplest form the **virial theorem** states that:

> If the potential energy is proportional to r^n, the mean kinetic and potential energies are related by $T = \frac{1}{2}nV$.

In quantum mechanics, we interpret the mean value as the expectation value. Thus, for a harmonic oscillator in which the potential energy is parabolic ($n = +2$), the mean potential and kinetic energies are equal. For a Coulombic system (such as the H atom), for which $n = -1$, the kinetic and potential energies are related by $T = -\frac{1}{2}V$. In each case, the total energy can be expressed in terms of the mean potential or mean kinetic energy alone.

The **hypervirial theorems** are a set of theorems based on the vanishing of the average value of the °commutator of an °operator with the °hamiltonian of the system when the system is in an °eigenstate of the hamiltonian (Box V.2). A special case of the hypervirial theorems is the **quantum mechanical virial theorem**, which is given in the Box together with the form the theorem takes for a diatomic molecule. The latter is useful for discussing the changes in kinetic energy that accompany bond formation. For instance, we can use the theorem to conclude that, at the equilibrium bond length, the molecular potential energy is less than the sum of the potential energies of the separated atoms but that the total kinetic energy is greater.

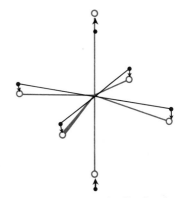

Fig. V.9 An asymmetric vibrational mode of an octahedral complex may eliminate the centre of inversion briefly and the Laporte rule may then be broken.

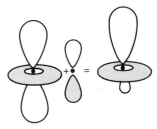

Fig. V.10 Because the environment of an ion is briefly asymmetrical if it undergoes a vibration like that shown in Fig. V.9, the ion's electron distribution is briefly distorted. The distortion is equivalent to the admixture of metal p orbital character into the d orbitals.

Box V.2 The virial theorems

The **hypervirial theorem** states that, for an eigenstate of the hamiltonian,

$$\langle [H, \Omega] \rangle = 0$$

Taking Ω as the sum over the products $q_i p_i$ for each particle leads to the **quantum mechanical virial theorem**:

$$\left\langle \sum_i q_i \frac{\partial V}{\partial q_i} \right\rangle = 2\langle T \rangle$$

For a diatomic molecule,

$$2\langle T \rangle_{\text{electronic}} = -\langle V \rangle - R\frac{dE}{dR}$$

where R is the internuclear separation, V is the sum of the electronic and nuclear potential energies, and E is the molecular potential energy (the total energy less the nuclear kinetic enegy).

Further information

The virial theorem is described in Appendix 5 of MQM. A good source on the classical virial theorem, including its proof, limitations, and applications, is Goldstein (1982). The quantum mechanical version is described by Hirschfelder (1960), who also deduces and describes the hypervirial theorems; see also Zhu (1986) and Crawford (1989b) for proofs. Sivardiere (1986) describes a number of applications of the theorem. For applications to molecular structure, see Deb (1972, 1981) and Feinberg, Ruedenberg, and Mehler (1970). A helpful introductory account, particularly with reference to chemical bonding, is provided by Levine (1983).

Virtual transition

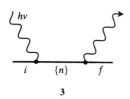

3

The distortion of a °perturbed system can be described by a °superposition of the °wavefunctions of the original system. Hence, the system behaves as though it contains features of the excited states. This admixture of other states is expressed by saying that the system has made a **virtual transition** to the excited states.

When electromagnetic radiation scatters from a molecule, the distortion it induces can be interpreted as a series of virtual transitions to the excited molecular states (**3**). The distortion is immediately released, and the photon flies off leaving the molecule either it its original state (Rayleigh scattering) or in one of the excited states populated by the initial virtual transition (°Raman scattering). As the frequency of the incident light approaches one of the transition frequencies of the molecule, the transition gradually loses its virtual character and becomes real: the molecule is really excited and the photon is really absorbed.

A virtual transition can be interpreted as occurring instantaneously, and hence the virtually excited state has zero lifetime. That being so, its energy is completely uncertain, and energy is not conserved during a virtual transition. Another way of expressing the same feature is to note that a virtual transition is just a name for a distortion and a means of taking into account the effect of a perturbation; hence, the question of energy conservation does not arise because the transition does not really take place.

Further information

See the discussion of °perturbation theory for the context in which virtual transitions appear. The transformation of virtual transitions into real transitions as a resonance is approached (that is, as the frequency of the perturbation approaches a transition frequency of the system) is described in Craig and Thirunamachandran (1984) and Heitler (1954).

Walsh diagram

A **Walsh diagram** is a °correlation diagram that portrays the variation in the orbital energies of a molecule as its shape is changed (typically as a bond angle is changed). By selecting the bond angle that results in the lowest total energy (which is approximated by the sum of the orbital energies) it is possible to predict the likely shape of the molecule from the occupation of its orbitals. It is also possible, from the changes in the populations of the orbitals, to predict how the shape will change when the molecule is electronically excited. The diagrams were first presented by A. D. Walsh in a classic series of papers that occupied a large part of one issue of the *Journal of the Chemical Society* in 1953.

The H_2A molecule will serve as an illustration of the general case. The atomic orbital basis set consists of the $1s$ orbitals of the two H atoms and the four valence As and Ap orbitals of the central A atom. Their relative energies are judged by considering how their overlap with each other varies as the bond angle is changed. Six molecular orbitals can be built from these six atomic orbitals. The plot of the energies of the orbitals as the bond angle changes from 180° in the linear molecule to 90° in the angular molecule gives the Walsh diagram of the H_2A molecule (Fig. W.1).

The lowest energy orbital in 90° H_2A is the one labelled $1a_1$, which is built from the overlap of the As and Ap_z orbitals with one (°symmetry adapted) combination of the H$1s$ orbitals. As the bond angle is changed to 180° the two H$1s$ orbitals overlap less, so on those grounds we expect the energy to rise. However, in the linear molecule only the As orbital contributes to the bond, and the elimination of the higher-energy Ap component of the orbital *lowers* the overall energy of the molecular orbital. (In some elementary treatments, the rehybridization of the central atom is ignored and it is concluded that the energy of the orbital falls as the molecule becomes linear. That would be valid in an H_3 molecule, where there is no Ap orbital and in molecules where the As and Ap orbitals are very far apart in energy.) When A orbital mixing is significant, the net effect is that the energy of the orbital falls as the molecule becomes linear.

The energy of the orbital labelled $1b_2$ is lowered because the H$1s$ orbitals move into a better position for overlap with the Ap_y orbital. The biggest change occurs for the $2a_1$ orbital. First, it contains considerable As character in the 90° molecule, but is pure Ap in the 180° molecule. Second, although it is largely AH nonbonding and slightly HH bonding in the 90° molecule, it is purely nonbonding in the 180° molecule.

The $1b_1$ orbital is a nonbonding Ap_x orbital perpendicular to the molecular plane in the 90° molecule and it remains nonbonding in the

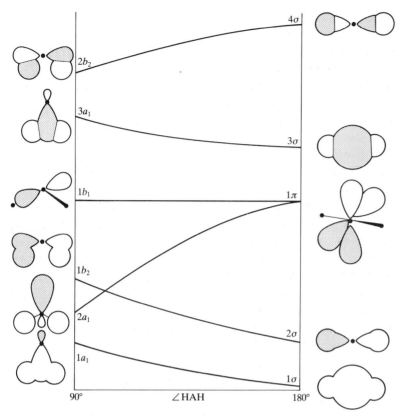

Fig. W.1 The Walsh diagram for H_2A molecules and the compositions of the orbitals in the 90° and 180° cases. For H_3 species (such as H_3^+) and some other H_2A species, the lowest orbital rises in energy from left to right. H_3^+ itself is angular (specifically, an equilateral triangle).

linear molecule. Hence, its energy barely changes with angle. The $3a_1$ orbital is antibonding in the 90° molecule and has both As and Ap character; in the linear molecule it is pure As on the central atom, and possibly less antibonding; hence its energy decreases as the molecule becomes linear. The uppermost orbital, $2b_2$, is the antibonding partner of the $1b_2$ orbital, and increases in energy as the H1s orbitals achieve better overlap in the linear molecule.

The principal feature that determines whether or not the H_2A molecule is bent is whether the $2a_1$ orbital is occupied. This $2a_1$ orbital has a much lower energy in the angular molecule than in the linear molecule. Hence, a lower total energy is achieved if, when it is occupied, the molecule is angular. The shape adopted by an H_2A molecule therefore depends on the number of electrons that occupy the orbitals. In H_2O, which has eight valence electrons to accommodate, the $2a_1$ orbital must be occupied, and so we can predict that the molecule will be bent. The transient BeH_2 molecule, which has four valence electrons, is predicted to be linear since the $2a_1$ orbital is not occupied, and the occupied orbitals have only a weak effect on the molecular geometry.

A selection of Walsh diagrams molecules is given in Figs. W.1 to W.4. (The B atoms of AB_n molecules have both p and s valence orbitals, and so the diagrams are correspondingly more complicated than those for AH_n molecules.)

The advantage of the Walsh approach over °VSEPR theory is that it is expressed in terms of the occupation of delocalized molecular orbitals, and the energies of these orbitals incorporate electron–electron and nucleus–nucleus repulsions, electron–nucleus attractions, and the kinetic energies of the electrons. Thus, in principle, the Walsh approach is richer than VSEPR theory, for the latter is expressed entirely in terms of repulsions between localized pairs of electrons. However, to some extent the advantage is illusory, because the VSEPR theory can be expressed in more sophisticated terms, and there is some uncertainty about the precise significance of the 'orbital energy' that is plotted in a Walsh diagram.

The two approaches are both qualitative attempts to identify dominant bonding influences, and both are helpful. The Walsh approach is particularly helpful for the discussion of excited states of molecules, where the VSEPR approach is almost completely silent.

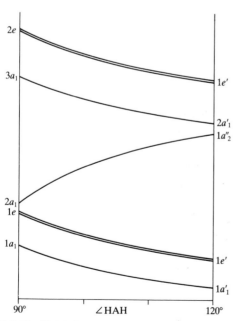

Fig. W.2 The Walsh diagram for an AH_3 molecule for bending from trigonal pyramidal to trigonal planar.

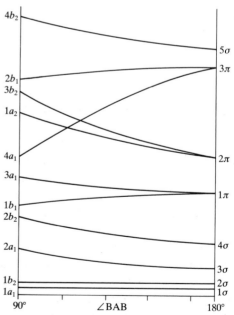

Fig. W.3 The Walsh diagram for an AB_2 molecule for bending from angular to linear.

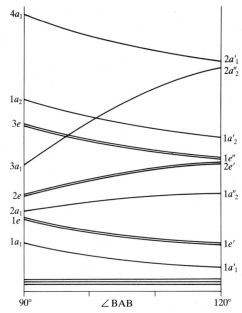

Fig. W.4 The Walsh diagram for an AB_3 molecule for bending from trigonal pyramidal to trigonal planar.

Further information

Walsh diagrams are introduced in Shriver, Atkins, and Langford (1990) and Atkins (1990). They are used in much the same spirit as Walsh first used them in Hollas (1983). There are good introductory discussions in DeKock and Gray (1980), Gimarc (1979), and Burdett (1980), who treat them in some detail. For the original papers see Walsh (1953).

Wavefunction

The **wavefunction** of a system is the function that contains all the information about its dynamical properties. It is found by solving the °Schrödinger equation of the system with the appropriate boundary conditions. If the wavefunction of the system is known, all the observable properties of the system may be deduced by performing the appropriate mathematical °operations.

The wavefunction may be a function of time, and is then often written Ψ. When it is not a function of time (or when the time-dependence has been factored out) it is usually denoted ψ and is a function of all the coordinates of all particles in the system. The wavefunction is often labelled with the °quantum numbers that distinguish the state of the system it describes. If the particles are independent of each other (in the sense that the potential energy is independent of their relative locations), the N-particle wavefunction is the product of N one-particle wavefunctions. The wavefunctions of a collection of identical particles must obey the °Pauli principle and be either symmetric (bosons) or antisymmetric (fermions) under particle interchange.

Time-independent wavefunctions may be °complex, in which case they represent a particle with net motion in a specific direction. A purely real (or purely imaginary) time-independent wavefunction represents a system with no net motion. Two examples of one-particle wavefunctions are e^{ikx} (a complex wavefunction) for a particle travelling freely in the positive x direction with °momentum $k\hbar$, and e^{-r/a_0} (a real wavefunction) for an electron in the ground state of the °hydrogen atom.

Some important features of wavefunctions (which are dealt with in other entries) are as follows:

● The °Born interpretation.

The °Born interpretation of the wavefunction is that $\psi^*\psi$ is the probability density of the particle at each point in space; hence, ψ is the **probability amplitude** of finding the particle at each point. For example, the probability density of the electron in the ground state of a °hydrogen atom decays exponentially with distance from the nucleus. For a free particle in a state of definite linear momentum, $\psi^*\psi$ is a constant and the probability density is the same at all locations in space.

● The acceptability of only certain wavefunctions implies °quantization.

For a survey of the limitations on the wavefunction, see the entries on the °Schrödinger equation and °quantization.

● Information is extracted from a wavefunction by acting on it with an °operator and calculating an °expectation value.

Broadly speaking, the amplitude, ψ, of the wavefunction at a point gives the probability of finding the electron there, the slope, ψ', of the wavefunction indicates the contribution to the °linear momentum of that part of the wavefunction, and the curvature, ψ'', likewise indicates the contribution to the °kinetic energy (Fig. W.5). The *observed* values of the momentum and the kinetic energy are the sums of these local contributions (see °expectation value).

● The wavefunction of a mixed state is a °superposition of the component states.

That is, to calculate the outcome of a joint process, we add the amplitudes of the probabilities (the wavefunctions), and deal with probabilities only at the end of the calculation.

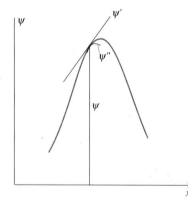

Fig. W.5 The amplitude (ψ) of a wavefunction gives the probability density of a particle at a point; the slope (ψ') gives the contribution to the linear momentum at the point; the curvature (ψ'') likewise gives the local contribution to the kinetic energy.

● The wavefunction of a state with definite energy is proportional to $e^{-iEt/\hbar}$.

Such a state is a **stationary state** (its square modulus does not depend on time) and persists for ever. If the state decays with a time constant τ, its energy is imprecise (see °lifetime broadening).

Further information

Wavefunctions pervade all texts on quantum mechanics (with a few exceptions), so see any standard text. For the emergence of an understanding of the significance of the wavefunction see Jammer (1966, 1974), Hughes (1989), and Moore (1989).

Wavepacket

A **wavepacket** is a °superposition of wavefunctions that is usually strongly peaked in one region of space and virtually zero elsewhere (Fig. W.6). The peak of the wavepacket denotes the most likely location of the particle; it occurs where the contributing wavefunctions are in phase and interfere constructively. Elsewhere the wavefunctions interfere destructively, and the net amplitude of the wavepacket is small or zero.

A wavepacket moves because all the component functions change at different rates, and at different times the point of maximum constructive interference is in different locations. The motion of the wavepacket corresponds very closely to the motion predicted for a classical particle in the same potential. An important difference from classical physics is that the wavepacket spreads with time, but this tendency is very small for massive, slow particles.

Further information

The motion of wavepackets is described in Appendix 3 of MQM. Thorough, quantitative discussions appear in Goldberger and Watson (1964) and Bransden and Joachain (1989).

Wien's displacement law

In 1893 the German physicist Wilhelm Wien (1864–1928) observed that:

> The wavelength corresponding to the maximum intensity of radiation emitted from a black body is displaced to shorter values as the temperature is raised.

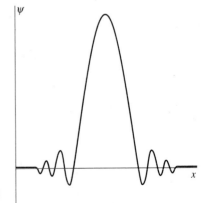

Fig. W.6 A wavepacket is a superposition of wavefunctions that interfere constructively in a localized region and destructively everywhere else.

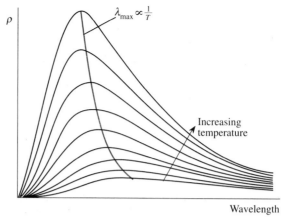

Fig. W.7 The migration of the wavelength of maximum energy density of black body radiation to shorter wavelength as the temperature is raised follows the crest of the Planck distribution.

The quantitative form of this **displacement law** is that at all temperatures

$$T\lambda_{max} = \tfrac{1}{5}c_2 \qquad c_2 = 1.439 \text{ cm K}$$

The constant c_2 is called the **second radiation constant**; its value was first measured by Otto Lummer and Ernst Pringsheim in 1899. It gradually became clear through the work of Lummer and Pringsheim that Wien's law is valid only for short wavelengths and high temperatures.

The theoretical justification of Wien's law is found in the Planck distribution law for °black body radiation, for it is quite easy to differentiate that function with respect to wavelength and to locate its maximum. The theoretical value of the second radiation constant is $c_2 = hc/k$, in agreement with the observed value. The physical picture for the migration of the maximum to shorter wavelengths (higher frequencies) is that as the temperature is raised, oscillators that could not previously acquire the necessary energy to oscillate can now do so, and so the crest of the Planck distribution rolls towards shorter wavelengths as these oscillators become active and grow in intensity (Fig. W.7).

Further information

The development and later derivation of Wien's displacement law are covered well in both Jammer (1966) and Kuhn (1978).

Wigner coefficient

A **Wigner coefficient** is the coefficient in the expansion of a state of total °angular momentum in terms of the contributing individual angular

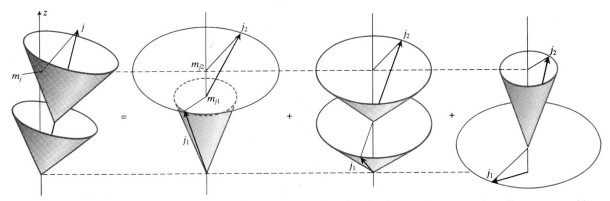

Fig. W.8 A state of coupled angular momenta (left) with quantum numbers j and m_j can be expressed as a linear superposition of states of the individual angular momenta for which $m_1 + m_2 = m$. The coefficients in the sum are the Wigner coefficients.

momenta (Fig. W.8). Wigner coefficients are also called **Clebsch–Gordan coefficients**, **vector-coupling coefficients**, or (with minor modifications to make them more symmetrical), **3j-symbols**. The coefficients were extensively used by the Hungarian–American theoretical physicist Eugene Wigner in his application of group theoretical methods to quantum mechanics.

As a specific example, consider the coupling of an α and a β spin into a state with $S=0$. As explained in the entry on °singlet and triplet states, the (normalized) wavefunction of the coupled state is $(1/\sqrt{2})(\alpha\beta - \beta\alpha)$. Therefore, the Wigner coefficient of the uncoupled state $\alpha\beta$ is $1/\sqrt{2}$ and that of $\beta\alpha$ is $-1/\sqrt{2}$. Since the $M_S=0$ state of a spin triplet is $(1/\sqrt{2})(\alpha\beta + \beta\alpha)$, both coefficients are $+1/\sqrt{2}$ for this combined state. Complete tables of Wigner coefficients are available, and they enable any coupled state to be expressed in terms of the contributing individual angular momentum states.

Racah coefficients are coefficients that arise when a total angular momentum state obtained by one coupling procedure is expressed in terms of total angular momentum states obtained by an alternative procedure. Thus, they arise when a state obtained by jj-coupling is expressed as a linear combination of the states of the same system that are obtained by Russell–Saunders coupling (see °vector model). The Racah coefficients are often expressed as the more symmetrical **6j-symbols**.

Further information

Wigner coefficients are discussed in detail in books on angular momentum, so see Brink and Satchler (1968), Zare (1987), Edmonds (1974), and Biedenharn and Louck (1981, 1984). An extensive and useful table of 3j-symbols and 6j-symbols has been combined by Rotenberg *et al.* (1959).

Woodward–Hoffmann rules

The **Woodward–Hoffmann rules** account for the products of certain concerted organic reactions, particularly **pericyclic reactions** in which the reaction occurs by the reorganization of electron pairs within a closed chain of interacting atomic orbitals. The rules are expressed in terms of the **conservation of orbital symmetry**, which is essentially °adiabatic passage through a °correlation diagram linking the electronic states of the reactants with those of the products. The rules were formulated by the American chemists Robert Woodward and Roald Hoffmann in 1965.

An example is the conversion of butadiene (**4**) to cyclobutene (**5**). The molecular orbitals of the two species are shown in Fig. W.9. However, although both molecules have a C_2 axis and two mirror planes in common, only some of these °symmetry elements are present at all stages of the reaction, and which ones are always present depends on the path. We can distinguish two orbital °correlation diagrams according to whether the pathway is **conrotatory** or **disrotatory**; that is, whether two CH_2 groups rotate in the same sense or opposite senses (Fig. W.10). To construct the correlation diagram we use the °noncrossing rule for states of the same symmetry with respect to the symmetry elements that survive during the passage from reactants to products. Thus, the conrotatory path preserves the C_2 axis at all stages (but not the mirror planes) and the disrotatory path preserves only one of the mirror planes.

It is clear from the correlation diagram in Fig. W.11 that only the conrotatory path is viable when the conditions are such that only small activation energies can be overcome. That is generally the case in a

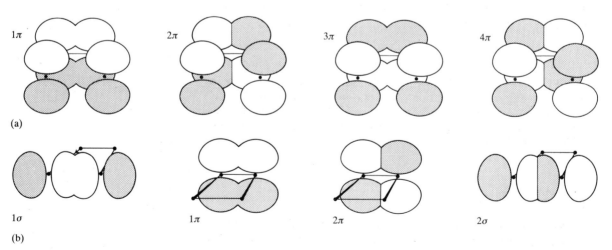

Fig. W.9 (a) The four π orbitals of butadiene and (b) the four orbitals (two σ and two π) that they become in cyclobutene.

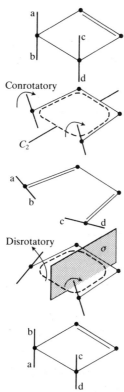

Conrotatory

C_2

Disrotatory

σ

Fig. W.10 The conrotatory and disrotatory motions that lead to the isomerization of butadiene to cyclobutene. Note that in the conrotatory mode the C_2 axis is preserved but the vertical mir or plane is not; in the disrotatory mode the opposite is true, for the C_2 axis is lost but the mirror plane is preserved.

thermal reaction pathway. Hence, we can predict that for such reactions the activation energy will be quite small and the products will have a definite (conrotatory) stereochemical relationship to the reactants.

Further information

The above entry has given only the slightest glimpse of the full rationalizing and predictive power of the rules. They are introduced in Chapter 12 of MQM. For more details see Woodward and Hoffmann (1970) and Gilchrist and Storr (1979). Useful accounts are also given in Salem (1982), Jones (1984), and Lowry and Richardson (1987).

Work function

The **work function** of a metal is the minimum energy required to remove an electron from it. Metals with low work functions can more easily lose their electrons than metals with high work functions (Table W.1).

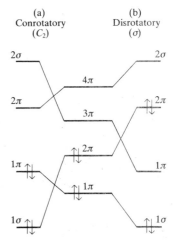

(a) Conrotatory (C_2)

(b) Disrotatory (σ)

Fig. W.11 The correlation diagram for the isomerization of butadiene (centre) to cyclobutene by conrotatory (left) and disrotatory (right) motion. If only a small amount of energy is available, as in a thermal isomerization, only the lower energy conrotatory path is accessible.

Table W.1 Photoelectric and (thermionic) work functions, Φ/eV

Li	Be		
2.42	3.9		
	(3.67)		
Na	Mg		
2.3	3.7		
K	Ca	Cu	Zn
2.25	3.2	4.8	4.3
		(4.5)	(2.2)
Rb		Ag	Cd
2.09		4.3	4.1
		(4.3)	
Cs	Au		
2.14	5.4		
(1.86)	(4.3)		

Data: Gray (1972).

The work function plays a role in the °photoelectric effect and in thermionic emission.

The analogy with the °ionization energy should be noticed. However, the work function includes a contribution from the work that must be done against the attraction between the electron and the **mirror charge** it induces in the metal.

Further information

For further information, including an analysis of the contributions to the work function, see Solymar and Walsh (1970), Anderson (1984), Kittel (1986, 1987), and Moruzzi.

X-ray spectra

X-rays are electromagnetic radiation with wavelengths of the order of 100 pm and less (see °electromagnetic spectrum). A principal terrestrial source of X-rays is the bombardment of metals with high-energy electrons. X-rays are also a component of **synchrotron radiation**, the radiation emitted by electrons that are constrained to travel in a circular path at high speed. The radiation produced by electron bombardment consists of two components (Fig. X.1). The continuous background, which is known as **Bremsstrahlung** (from the German words for 'brake' and 'ray'), is formed as the electrons decelerate as they plunge into the metal. Superimposed on the Bremsstrahlung are the sharp lines of the metal's X-ray spectrum.

The discrete spectral lines arise from transitions within the core of the atoms that constitute the material. The incoming electron ejects a core electron, either completely or into some unoccupied upper level. One of the remaining core electrons falls into the hole left by the ejected electron, and the difference in energy is radiated (Fig. X.2). High-energy ('hard') X-rays are formed when the electron is ejected from the K shell ($n = 1$); an electron falling from the L shell ($n = 2$) into the vacancy gives rise to the K_α line, one falling from the M shell ($n = 3$) gives the K_β line, and so on. Softer X-rays (longer wavelength radiation) are formed when the electron is ejected from the L shell, and the lines L_α, L_β, and so on are formed as electrons fall into the L shell from the M and N shells.

As a first approximation, the K radiation can be treated as resulting from a transition in a hydrogenic atom of nuclear charge $(Z-1)e$, where the reduction from Ze takes into account the °shielding by the remaining $1s$ electron. It follows from the information in Box H.6 that

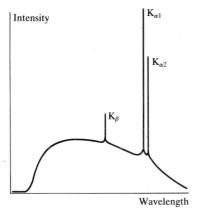

Fig. X.1 A typical X-ray spectrum of a metal (copper). The broad featureless background is the Bremsstrahlung.

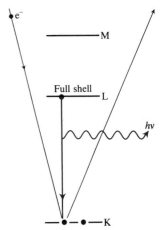

Fig. X.2 The mechanism of production of X-rays. The incoming electron ejects an electron from an inner shell (here the K shell), and an electron from a higher shell falls into the vacancy. The energy difference is carried away as an X-ray photon.

the frequencies of K radiation are $(Z-1)^2 cR(1-1/n^2)$ with $n=2,3,\ldots$. Similar expressions for other lines can be written, but with different screening constants. The hydrogenic expression is the justification of **Moseley's law**:

> The square root of the X-ray frequency is proportional to the atomic number Z.

The law was established by the British physicist H.G.J. Moseley (1887–1915) in 1913, shortly before he was killed by a sniper's bullet at Gallipoli. The importance of the law was that it enabled the elements to be put in an unambiguous order (according to atomic number) for the first time, and it showed that some elements had not yet been identified.

Further information

For an account of X-ray spectra see Kuhn (1962). A table of X-ray transition frequencies will be found in Gray (1972). For an introduction to synchrotron sources see Hollas (1983).

Zeeman effect

The **Zeeman effect** is the splitting of spectral lines into several components by a strong magnetic field. In the **normal Zeeman effect**, which is shown by atoms without electron spin, each line is split into three. In the more common **anomalous Zeeman effect**, which is shown by atoms with non-zero spin, the line structure is more complicated. The effect was first observed by the Dutch physicist Pieter Zeeman (1865–1943) in 1896.

The only source of °magnetic moment in the absence of spin is the orbital angular momentum of the electrons. The applied field interacts with the orbital moment and the energy of the state with projection M_L is changed by $\mu_B B M_L$, where μ_B is the °Bohr magneton. For example, a ^{1}P term is split into three equally spaced components and a ^{1}D term is split into five components (Fig. Z.1). Since the °selection rule for an °electric dipole transition is $\Delta M_L = 0, \pm 1$, all transitions fall into three groups and appear in the spectrum as three sets of lines. The splitting is about 1.5 cm^{-1} with fields of the order of 3 T, which is easily detectable.

The transitions are polarized. When viewed parallel to the direction of the magnetic field, the $\Delta M_L = 0$ line is absent and the $\Delta M_L = \pm 1$ lines are circularly polarized in opposite senses. When viewed perpendicular to the field, the $\Delta M_L = 0$ line is present and polarized parallel to the field; it is called a **π line**. The $\Delta M_L = \pm 1$ lines are plane polarized perpendicular to the field, and are called **σ lines** (*senkrecht* is the German word for perpendicular).

The anomalous Zeeman effect arises when the atom has non-zero electron spin. It is then necessary to take into account the different effective magnetic moments of each °level of the term (see the Landé °g factor). Since the g-factor g_J depends on L, S, and J, the splitting $g_J \mu_B B M_J$ may be different for different levels, and as a result the transitions no longer fall into three neat groups.

As an example, consider the transitions ^{3}P→^{3}S. The magnetic moment of the ^{3}S term arises solely from its spin angular momentum, and $g_J = 2$. The ^{3}P term has the three levels ^{3}P$_0$, ^{3}P$_1$, and ^{3}P$_2$ with $g_J = 0$, $\frac{3}{2}$, and $\frac{3}{2}$ respectively. The ^{3}P$_0$ level has only one state and is unaffected by the applied field (Fig. Z.2). The other two levels are split by the field into three and five states respectively, with energies $\frac{3}{2}\mu_B B M_J$. The selection rules $\Delta M_J = 0, \pm 1$ then result in the complicated ('anomalous') spectrum shown in Fig. Z.2. The polarizations of the lines can be used to assign the transitions. At very high fields an anomalous Zeeman spectrum becomes normal on account of the °Paschen–Back effect, which effectively eliminates the contribution of the spins.

One application of the Zeeman effect is to the determination of the

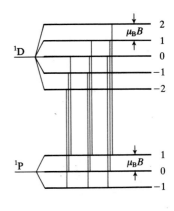

Fig. Z.1 The normal Zeeman effect for atoms without electron spin. Because the Landé *g*-factors are the same in the upper and lower terms, the allowed transitions fall into three groups, giving three lines where one is observed in the absence of a field.

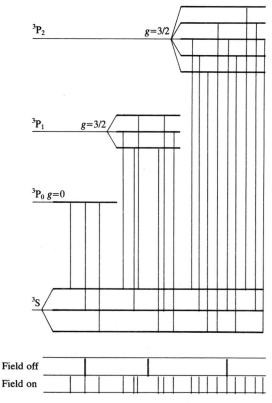

Fig. Z.2 The anomalous Zeeman effect is observed for atoms with non-zero electron spin. The Landé *g*-factors are different for the levels of the upper and lower terms, and consequently the allowed transitions do not fall into three neat groups.

multiplicity of terms. The splitting of energy levels by a magnetic field is also the basis of magnetic resonance techniques: see °electron spin resonance and °nuclear magnetic resonance.

Further information

The Zeeman effect is described in Kuhn (1962) and in the article in the encyclopaedia edited by Besançon (1985).

Zero point energy

The **zero point energy** of a system is the energy of its lowest permitted state. For example, the zero point energy of a °harmonic oscillator is the energy of the state with $v=0$ (the lowest permitted value of v), and is $\frac{1}{2}\hbar\omega$. This zero point energy accounts for the difference between the minimum of the molecular potential energy curve (D_e) and the dissociation energy (D_0) of a molecule. The zero point energy of a °particle

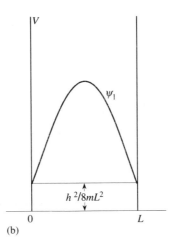
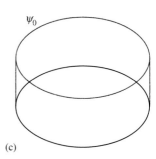

(a)   (b)   (c)

Fig. Z.3 (a) A harmonic oscillator has a zero point energy because its wavefunction penetrates into regions of non-zero potential energy, and it has non-zero kinetic energy on account of its curvature. (b) A particle in a box has a zero point energy because the wavefunction has a curvature and hence corresponds to non-zero kinetic energy. (c) A particle on a ring has no zero point energy because its wavefunction can satisfy the periodic boundary conditions without acquiring curvature.

in a box is the energy corresponding to $n = 1$ (the lowest permitted value of n), and is $h^2/8mL^2$.

The zero point energy arises from the boundary conditions that the wavefunctions of the system must satisfy. In a harmonic oscillator, the wavefunction must approach zero smoothly at large displacements. Consequently it samples regions of non-zero potential energy (Fig. Z.3a) even in its ground state; moreover, because the wavefunction has curvature, it also represents a state of non-zero kinetic energy. For a particle in a box, the wavefunction must vanish at the walls, not be zero everywhere, and be continuous. Hence it must curve from one wall to the other (Fig. Z.3b) and therefore correspond to a state of non-zero kinetic energy. A particle on a ring (or, equivalently, a rotating molecule) does not have a rotational zero point energy because the wavefunction need not have curvature to satisfy the cyclic boundary conditions (that it should repeat itself after a rotation by 2π, Fig. Z.3c).

In general, the greater the mass of a particle and the greater its spatial freedom, the smaller its zero point energy. Hence, replacing H by D in a molecule decreases the vibrational zero point energy of the X—H bond. Macroscopic objects and particles free to move over large regions of space have negligible zero point energies.

Further information

The zero point energy of °harmonic oscillators and °particles in boxes are described in those entries, and in more detail in Chapter 3 of MQM.

Bibliography

The reference MQM is to P. W. Atkins (1983). Molecular quantum mechanics. Oxford University Press.

Abragam, A. (1961). *The principles of nuclear magnetism.* Clarendon Press, Oxford.

Abragam, A. and Bleaney, B. (1970). *Electron paramagnetic resonance of transition ions.* Clarendon Press, Oxford.

Abramowitz, M. and Stegun, I. A. (1965). *Handbook of mathematical functions.* Dover, New York.

Agmon, N. (1988). Ionization potentials for isoelectronic series. *J. chem. Educ.,* **65**, 43.

Albeverio, S. (1988). *Solvable models in quantum mechanics.* Springer, New York.

Albright, T. A., Burdett, J. K., and Whangbo, M.-H. (1985). *Orbital interactions in chemistry.* Wiley, New York.

Allen, H. C. and Cross, P. C. (1963). *Molecular vib-rotors.* Wiley, New York.

Allen, L. C. (1989). Electronegativity is the average one-electron energy of the valence shell electrons in ground-state free atoms. *J. Amer. chem. Soc.,* **111**, 9003.

Allen, M. P. and Tildesley, D. J. (1987). *Computer simulation of liquids.* Clarendon Press, Oxford.

Allred, A. L. and Rochow, E. G. (1958). A scale of electronegativity based on electrostatic force. *J. inorg. nucl. Chem.,* **5**, 264.

Altmann, S. L. (1986), *Rotations, quaternions, and double groups.* Clarendon Press, Oxford.

Altmann, S. L. (1991*a*). *Icons and symmetries.* Oxford University Press.

Altmann, S. L. (1991*b*). *Band theory of solids: An introduction from the point of view of symmetry.* Oxford University Press.

Anderson, P. W. (1984). *Basic notions of condensed matter physics.* Benjamin/Cummings, Menlo Park.

Andrade e Silva, L. and Lochak, G. (1969). *Quanta.* Weidenfeld and Nicholson, London.

Arfken, G. B. (1985). *Mathematical methods for physicists.* Academic Press, Orlando.

Arons, A. B. and Peppard, M. B. (1965). Einstein's proposal of the photon concept—a translation of the *Annalen der Physik* paper of 1905. *Am. J. Phys.,* **33**, 367.

Ashcroft, N. W. and Mermin, N. D. (1976). *Solid state physics.* Holt, Reinhart, and Winston, New York.

Atkins, P. W. (1983). *Molecular quantum mechanics.* Oxford University Press.

Atkins, P. W. (1989). *General chemistry.* Scientific American Books, New York.

Atkins, P. W. (1990). *Physical chemistry*. Oxford University Press and W. H. Freeman and Co., New York.

Atkins, P. W., Child, M. S., and Phillips, C. S. G. (1970). *Tables for group theory*. Oxford University Press.

Atkins, P. W. and Symons, M. C. R. (1967). *The structure of inorganic radicals*. Elsevier, Amsterdam.

Ayscough, P. B. (1990). *Library of physical chemistry software*, Vol. 2. Oxford University Press and W. H. Freeman and Co., New York.

Bacry, H. (1988). *Localizability and space in quantum physics*. Springer, Berlin.

Bader, R. F. W. (1985). Atoms in molecules. *Acc. chem. Res.*, **18**, 9.

Bader, R. F. W. (1990). *Atoms in molecules: A quantum theory*. Clarendon Press, Oxford.

Baird, N. C. (1986). The chemical bond revisited. *J. chem. Educ.*, **63**, 660.

Baker, A. D. and Betteridge, D. (1977). *Photoelectron spectroscopy*. Pergamon Press, Oxford.

Baker, J. W. (1952). *Hyperconjugation*. Clarendon Press, Oxford.

Baker, J. W. (1958). *Conference on hyperconjugation*. Pergamon Press, Oxford.

Ballard, R. E. (1978). *Photoelectron spectroscopy and molecular orbital theory*. Adam Hilger, Bristol.

Ballhausen, C. J. (1962). *Introduction to ligand field theory*. McGraw-Hill, New York.

Ballhausen, C. J. (1979). *Molecular electronic structures of transition metal complexes*. McGwaw-Hill, New York.

Bander, M. and Itzykson, C. (1966). Group theory and the hydrogen atom. *Rev. mod. Phys.*, **38**, 330 (Part 1), 346 (Part 2).

Banwell, C. N. (1983). *Fundamentals of molecular spectroscopy*. McGraw-Hill, New York.

Barltrop, J. A. and Coyle, J. D. (1975). *Excited states in organic chemistry*. Wiley, Chichester

Barltrop, J. A. and Coyle, J. D. (1988). *Principles of photochemistry*. University Microfilms International, Ann Arbor.

Barron, L. D. (1983). *Molecular light scattering and optical activity*. Cambridge University Press.

Barrow, G. M. (1962). *Introduction to molecular spectroscopy*. McGraw-Hill, New York.

Barrow, J. D. and Tipler, F. J. (1986). *The anthropic cosmological principle*. Clarendon Press, Oxford.

Bashkin, S. and Stoner, J. O. (1976 *et seq.*). *Atomic energy levels and Grotrian diagrams*. North Holland, Amsterdam.

Batt, R. and Moore, J. W. (1987). *Iterations* and *Iterations II*. Published by *J. chem. Educ.*, Easton.

Bell, J. S. (1987). *Speakable and unspeakable in quantum mechanics. Collected papers on quantum philosophy*. Cambridge University Press.

Bell, R. P. (1980). *The tunnel effect in chemistry*. Chapman and Hall, London.

Bellamy, L. J. (1975). *The infrared spectra of complex molecules*, Vol. 1. Chapman and Hall, London.

Bellamy, L. J. (1980). *The infrared spectra of complex molecules*, Vol. 2. Chapman and Hall, London.

Berestetskii, V. B., Lifshitz, E. M., and Pitaevskii, L. P. (1971). *Relativistic quantum theory*. Pergamon Press, Oxford.

Bergmann, E. D. and Pullman, B. (1971). *Aromaticity, pseudoaromaticity, and anti-aromaticity*. Israel Academy of Sciences, Israel.

Besançon, R. M. (ed.) (1985). *The encyclopedia of physics*. Van Nostrand Reinhold, New York.

Bethe, H. A. (1964). *Intermediate quantum mechanics*. Benjamin, New York.

Bethe, H. A. and Salpetre, E. E. (1957). *Quantum mechanics of one and two electron atoms*. Academic Press, New York.

Biedenharn, L. C. and Louck, J. D. (1981). *The Racah-Wigner algebra in quantum theory*. Addison-Wesley, London.

Biedenharn, L. C. and Louck, J. D. (1984). *Angular momentum in quantum physics: Theory and application*. Addison-Wesley, London.

Biedenharn, L. C. and van Dam, H. (ed.) (1965). *Quantum theory of angular momentum*. Academic Press, New York.

Bingham, R. C., Dewar, M. J. S., and Lo, D. H. (1975). Ground states of molecules 25: MINDO/3. An improved version of the MINDO semiempirical SCF-MO method. *J. Amer. chem. Soc.*, **97**, 1285 (*et seq.*).

Bjorken, J. D. and Drell, S. D. (1964). *Relativistic quantum mechanics*. McGraw-Hill, New York.

Bohm, D. (1951). *Quantum theory*. Prentice-Hall, Englewood Cliffs.

Bohr, N. (1913). On the constitution of atoms and molecules. *Phil. Mag.*, **26**, 1, 476, and 857.

Born, M. (1970). *The Born–Einstein letters*. Walker, New York.

Born, M. and Beim, W. (1968). Dualism in quantum theory. *Phys. Today*, August, 51.

Born, M., Heisenberg, W., and Jordan, P. (1926). Zur Quantenmechanik. *Z. Phys.*, **35**, 557.

Born, M. and Jordan, P. (1925). Zur Quantenmechanik. *Z. Phys.*, **34**, 858.

Born, M. and Oppenheimer, J.R. (1927). Zur Quantumtheorie der Molekülen. *Annln. Phys.*, **84**, 457 (Sec. 1-2, Appendix 2).

Born, M. and Wolf, E. (1989). *Principles of optics*. Pergamon Press, Oxford.

Böttcher, C. J. F. and Bordewijk, P. (1978). *Theory of electric polarization*. Elsevier, Amsterdam.

Boulil, B., Deumié, M., and Henri-Rousseau, O. (1987). Where is the link between quantum mechanical and thermodynamic adiabaticity? *J. chem. Educ.*, **64**, 311.

Boys, S. F. (1950). Electronic wavefunctions. 1. A general method of calculation for the stationary states of any molecular system. *Proc. Roy. Soc.*, **A200**, 542.

Boys, S. F. (1960). Construction of some molecular orbitals to be approximately invariant for changes from one molecule to another. *Rev. mod. Phys.*, **32**, 296.

Bradley, C. J. and Cracknell, A. P. (1972). *The mathematical theory of symmetry in solids*. Clarendon Press, Oxford.

Brand, D. J. and Fisher, J. (1987). Molecular structure and chirality. *J. chem. Educ.*, **64**, 1035.

Brandt, S. and Dahmen, H. D. (1985). *The picture book of quantum mechanics*. Wiley, New York.

Bransden, B. H. and Joachain, C. J. (1989). *Introduction to quantum mechanics*. Longman, London and Wiley, New York.

Bratsch, S. G. (1988*a*). Revised Mulliken electronegativities: I. Conversion to Pauling units. *J. chem. Educ.*, **65**, 34.

Bratsch, S. G. (1988*b*). Revised Mulliken electronegativities: II. Applications and limitations. *J. chem. Educ.*, **65**, 34.

Bratsch, S. G. (1988c). Electronegativity and the acid–base character of binary oxides. *J. chem. Educ.*, **65**, 877.

Breneman, G. L. (1988). Order out of chaos: shapes of hydrogen orbitals. *J. chem. Educ.*, **65**, 31.

Brink, D. M. and Satchler, G. R. (1968). *Angular momentum*. Clarendon Press, Oxford.

Brown, C. A., Gerloch, M., and McMeaking, R. F. (1988). A cellular ligand-field model for '*l-l*' spectral intensities. I. Formalism and parameters. *Molec. Phys.*, **64**, 771 *et seq.*

Brumer, P. and Shapiro, M. (1989). Coherence chemistry: controlling chemical reactions with lasers. *Acc. chem. Res.*, **22**, 407.

Brundle, C. R. and Baker, A. D. (ed.) (1977). *Electron spectroscopy: theory, techniques, and applications*. Vols. 1 and 2. Academic Press, London.

Buckingham, A. D. (1965). Experiments with oriented molecules. *Chem. Brit.*, **1**, 54.

Buckingham, A. D. (1967). Permanent induced molecular moments and long range intermolecular forces. *Adv. Chem.*, **12**, 107.

Buckingham, A. D. and Orr, B. J. (1967). Molecular hyperpolarizabilities. *Q. Rev. chem. Soc.*, **21**, 195.

Bunker, P. R. (1979). *Molecular symmetry and structure*. Academic Press, New York.

Burdett, J. K. (1980). *Molecular shapes: Theoretical models of inorganic strereo-chemistry*. Wiley, New York.

Cagnac, B. and Pebay-Peyroula, J. C. (1975). *Modern atomic physics: Fundamental principles*. Macmillan, London.

Caldwell, D. J. and Eyring, H. (1971). *The theory of optical activity*. Wiley, New York.

Calvert, J. G. and Pitts, J. N. (1966). *Photochemistry*. Wiley, New York.

Candler, C. (1964). *Atomic spectra and the vector model*. Hilger and Watts, London.

Carlin, R. L. (1986). *Magnetochemistry*. Springer, Berlin.

Carrington, A. and McLachlan, A. D. (1967). *Introduction to magnetic resonance*. Harper and Row, New York.

Carruthers, P. and Nieto, M. M. (1968). Phase and angle variables in quantum mechanics. *Rev. mod. Phys.*, **40**, 411.

Chalk, J. D. (1988). A study of barrier penetration in quantum mechanics. *Am. J. Phys.*, **56**, 29.

Chandler, D. (1987). *Introduction to modern statistical mechanics*. Oxford University Press.

Charney, E. (1979). *The molecular basis of optical activity: Optical rotatory dispersion and circular dichroism*. Wiley, New York.

Cheetham, A. K. and Day, P. (ed.) (1987). *Solid state chemistry: Techniques*. Oxford University Press.

Chen, E. C. M. and Wentworth, W. E. (1975). The experimental values of atomic electron affinities. *J. chem. Educ.*, **52**, 486.

Child, M. S. (1974). *Molecular collision theory*. Academic Press, London.

Child, M. S. (1991). *Semiclassical mechanics*. Clarendon Press, Oxford.

Clark, T. (1985). *A handbook of computational chemistry: A practical guide to chemical structure and energy calculations*. Wiley, New York.

Clementi, E. and Raimondi, D. L. (1963). Atomic screening constants from SCF functions. *IBM Research Note* NJ-27.

Cochran, W. (1973). *The dynamics of atoms in crystals*. Edward Arnold, London.

Colpa, J. P. and Islip, M. F. J. (1973). Hund's rule and the Z expansion for the energy, electron repulsion, and electron-nuclear attraction. *Mol. Phys.*, **25**, 701.

Compton, A. H. (1923). A quantum theory of the scattering of X-rays by light elements. *Phys. Rev.*, **21**, 483.

Condon, E. U. (1928). Nuclear motions associated with electron transitions in diatomic molecules. *Phys. Rev.*, **32**, 858.

Condon, E. U. and Odabaşi, H. (1980). *Atomic structure*. Cambridge University Press.

Cook, D. B. (1978). *Structures and approximations in molecules*. Ellis Horwood, New York.

Cooper, D. L., Gerratt, J., and Raimondi, M. (1986). The electronic structure of the benzene molecule. *Nature*, **323**, 699.

Cooper, D. L., Gerratt, J., and Raimondi, M. (1989). The electronic structure of 1,3-dipoles: Hypervalent atoms. *J. chem. Soc. Perkin Trans. 2*, 1187.

Cooper, D. L., Gerratt, J., and Raimondi, M. (1990). The spin-coupled valence bond description of benzenoid aromatic hydrocarbons. *Topics in current chemistry*, **153**, 41.

Cooper, D. L., Gerratt, J., Raimondi, M., and Wright, S.C. (1989). The electronic structure of heteroatomic molecules. 1. Six-membered rings. *J. chem. Soc. Perkin Trans. 2*, 255.

Cooper, D. L., Gerratt, J., and Raimondi, M., Wright, S. C. and Hyams, P. A. (1989). The electronic structure of heteroatomic molecules. 3. A comparison of benzene, borazine, and boroxine. *J. chem. Soc. Perkin Trans. 2*, 719.

Corney, A. (1977). *Atomic and laser spectroscopy*. Clarendon Press, Oxford.

Cotton, F. A. (1990). *Chemical applications of group theory*. Wiley, New York.

Cotton, F. A. and Walton, R. A. (1983). *Multiple bonds between ligand atoms*. Wiley, New York.

Cotton, F. A. and Wilkinson, G. (1988). *Comprehensive inorganic chemistry*. Wiley, New York.

Coulson, C. A. (1970). π bonds. In *Physical chemistry, an advanced treatise*, (ed. H. Eyring, D. Henderson, and W. Jost), Vol. 5, p. 369. Academic Press, New York.

Coulson, C. A. (1979). *Valence* (revised by Roy McWeeny). Oxford University Press.

Coulson, C. A. (1982). *The shape and structure of molecules* (revised by Roy McWeeny). Oxford University Press.

Coulson, C. A., O'Leary, B., and Mallion, R. B. (1978). *Hückel theory for organic chemists*. Academic Press, London.

Coulson, C. A. and Rushbrooke, G.S. (1940). Note on the method of molecular orbitals. *Proc. Camb. Phil. Soc.*, **36**, 193.

Coulter, B. L. and Adler, C. G. (1971). The relativistic one-dimensional square potential. *Am. J. Phys.*, **39**, 305.

Cowan, R. D. and Duane, R. (1981). *The theory of atomic structure and spectra*. University of California Press, Berkeley.

Cox, P. A. (1987). *The electronic structure and chemistry of solids*. Oxford University Press.

Craig, D. P. and Thirunamachandran, T. (1984). *Molecular quantum electrodynamics*. Academic Press, London.

Craig, D. P. and Walmsley, S.H. (1968). *Excitons in molecular crystals*. Benjamin, New York.

Crawford, F. S. (1989*a*). Applications of Bohr's correspondence principle. *Am. J. Phys.*, **57**, 621.

Crawford, F. S. (1989*b*). Footnote to the quantum mechanical virial theorem. *Am. J. Phys.*, **57**, 555.

Das, A. and Melissinos, A. C. (1986). *Quantum mechanics: A modern introduction*. Gordon and Breach, New York.

Dasent, W. A. (1982). *Inorganic energetics*. Cambridge University Press.

Davidson, N. (1962). *Statistical mechanics*. McGraw-Hill, New York.

Davies, D. W. (1967). *The theory of the electric and magnetic properties of molecules*. Wiley, London.

Davydov, A. S. (1976). *Quantum mechanics*. Pergamon Press, Oxford.

Davytan, L. S., Pogosyan, G. S., Sissakian, A. N., and Ter-Antonyan, V. M. (1987). On the hidden symmetry of a one-dimensional hydrogen atom. *J. Phys. A.*, **20**, 2765.

Dean, J. A. (1987). *Handbook of organic chemistry*. McGraw-Hill, New York.

Deb, B. M. (1972). The force concept in chemistry. *Rev. mod. Phys.*, **45**, 22.

Deb, B. M. (1981). *The force concept in chemistry*. Van Nostrand Reinhold, New York.

DeKock, R. L. (1987). The chemical bond. *J. chem. Educ.*, **65**, 934.

DeKock, R. L. and Gray, H. B. (1980). *Chemical structure and bonding*. Benjamin/Cummings, Menlo Park.

Derome, A. E. (1987). *Modern NMR techniques for chemistry research*. Pergamon Press, Oxford.

d'Espagnat, B. (1976). *Conceptual foundations of quantum mechanics*. Benjamin, Menlo Park.

DeVault, D. (1984). *Quantum-mechanical tunnelling in biological systems*. Cambridge University Press.

Dewar, M. J. S. (1969). *The molecular orbital theory of organic molecules*. McGraw-Hill, New York.

Dewar, M. J. S. and Thiel, W. (1977). Ground states of molecules. 38. The MNDO method. Approximations and parameters. *J. Amer. chem. Soc.*, **99**, 4899.

Dewar, M. J. S., Zoebisch, E. G., Healy, E. F., and Stewart, J. J. P. (1985). AM1: A new general purpose quantum mechanical molecular model. *J. Amer. chem. Soc.*, **107**, 3902.

Dirac, P. A. M. (1958). *The principles of quantum mechanics*. Clarendon Press, Oxford.

Dirac, P. A. M. (1964). *Lectures on quantum mechanics*. Belfer Graduate School, New York.

Douglas, B. E. and Hollingsworth, C. A. (1985). *Symmetry in bonding and spectra*. Academic Press, New York.

Duffy, J. A. (1990). *Bonding, energy levels, and bands in inorganic solids*. Longman, London.

Duncan, A. B. F. (1971). *Rydberg series in atoms and molecules*. Academic Press, New York.

Earnshaw, A. (1968). *Introduction to magnetochemistry*. Academic Press, New York.

Ebsworth, E. A. V., Rankin, D. W. H., and Cradock, S. (1987). *Structural methods in inorganic chemistry*. Blackwell Scientific, Oxford.

Edmiston, C. and Ruedenberg, K. (1963). Localized atomic and molecular orbitals. I. *J. chem. Phys.*, **35**, 457.

Edmiston, C. and Ruedenberg, K. (1965). Localized atomic and molecular orbitals. II. *J. chem. Phys.*, **43**, 597.

Edmiston, C., Bartleson, J., and Jarvie, J. (1986). Pauli forces and the valence shell electron pair repulsion model for H_2O and NH_3. *J. Am. chem. Soc.*, **108**, 3593.

Edmonds, A.R. (1974). *Angular momentum in quantum mechanics*. Princeton University Press.

Einstein, A. (1905). Uber einen die Erzeugung und Verwandlung des Lichtes betreffenden heuristischen Gesichtspunkt. *Ann. Phys.*, **17**, 132.

Eisberg, R. N. and Resnik, R. (1985). *Quantum physics of atoms, molecules, solids, nuclei, and particles*. Wiley, New York.

Eland, J. H. D. (1983). *Photoelectron spectra*. Open University Press, Milton Keynes.

Ellis, A. B. (1987). Superconductors: Better levitation through chemistry. *J. chem. Educ.*, **64**, 836.

Emsley, J. (1989). *The elements*. Clarendon Press, Oxford.

Englefield, M. J. (1972). *Group theory and the Coulomb problem*. Wiley, New York.

Englman, R. (1972). *The Jahn-Teller effect in molecules and crystals*. Wiley, New York.

Englman, R. (1976). *Non-radiative decay of ions and molecules in solids*. North Holland, Amsterdam.

Ezra, G. S. (1982). *Symmetry properties of molecules*. Springer-Verlag, Berlin.

Fano, U. (1970). Spin–orbit coupling: A weak force with conspicuous effects. *At. mol. Phys.*, **2**, 30.

Fano, U. and Racah, G. (1959). *Irreducible tensorial sets*. Academic Press, New York.

Feinberg, M. J., Ruedenberg, K., and Mehler, E. L. (1970). The origin of bonding and antibonding in the hydrogen molecule-ion. *Adv. quantum Chem.*, **5**, 21.

Feynman, R. P. (1939). Forces in molecules. *Phys. Rev.*, **56**, 340.

Feynman, R. P. (1949). Space-time approach to quantum electrodynamics. *Phys. Rev.*, **76**, 769.

Feynman, R. P. and Hibbs, A. R. (1965). *Quantum mechanics and path integrals*. McGraw-Hill, New York.

Feynman, R. P., Leighton, R. B., and Sands, M. (1963). *Lectures in physics*. Addison-Wesley, Reading.

Figgis, B. (1966). *Introduction to ligand fields*. Wiley, New York.

Fischer, G. (1984). *Vibronic coupling: The interaction between the electronic and nuclear motions*. Academic Press, London.

Fischer, H. and Hellwege, K.-H. (1977). *Magnetic properties of free radicals*. Landolt-Börnstein Vol. 9, Springer Verlag, Berlin.

Flygare, W. H. (1978). *Molecular structure and dynamics*. Prentice-Hall, Englewood Cliffs.

Fowler, R. H. and Guggenheim, E. A. (1965). *Statistical thermodynamics*. Cambridge University Press.

Frago, S., Karwowski, J., and Saxena, K. M. S. (1976). *Handbook of atomic data*. Elsevier, Amsterdam.

Franck, J. (1925). Elementary processes of photochemical reactions. *Trans. Faraday Soc.*, **21**, 536.

Franck, J. and Hertz, G. (1914). Uber Zusammenstosse zwischen Elektronen und den Molekülen des Quecksilberdampfes und die Ionisierungsspannung desselben. *Verh. dt. phys. Ges.*, **16**, 457.

Freed, K. (1983). Is there a bridge between *ab initio* and semiempirical theories of valence? *Acc. chem. Res.*, **16**, 137.

Freeman, A. J. and Frankel, R. B. (1967). *Hyperfine interactions.* Academic Press, New York.

Freeman, R. (1987). *A handbook of nuclear magnetic resonance.* Longman, London and Wiley, New York.

Galindo, A. and Sanchez del Rio, C. (1961). Intrinsic magnetic moment as a nonrelativistic phenomenon. *Am. J. Phys.*, **29**, 582.

Gallup, G. A. (1988). The Lewis electron-pair model, spectroscopy, and the role of the orbital picture in describing the electronic structures of molecules. *J. chem. Educ.*, **65**, 671.

Gaydon, A. G. (1968). *Dissociation energies and spectra of diatomic molecules.* Chapman and Hall, London.

Gerloch, M. (1986). *Orbitals, terms, and states.* Wiley, Chichester.

Gerratt, J. (1986). *Aromaticity.* Wiley, New York.

Gilbert, R. G. and Smith, S. C. (1990). *Theory of unimolecular and recombination reactions.* Blackwell Scientific Publications, Oxford.

Gilchrist, T. L. and Storr, R. C. (1979). *Orbital reactions and orbital symmetry.* Cambridge University Press.

Gillespie, R. (1972). *Molecular geometry.* Van Nostrand Reinhold, London.

Gimarc, B. M. (1979). *Molecular structure and bonding: The qualitative molecular orbital approach.* Academic Press, New York.

Glidewell, C. and Lloyd, D. (1986). The arithmetic of aromaticity. *J. chem. Educ.*, **63**, 306.

Goldberger, M. L. and Watson, K. M. (1964). *Collision theory.* Wiley, New York.

Goldin, E. (1982). *Waves and photons: An introduction to quantum optics.* Wiley, New York.

Goldstein, H. (1982). *Classical mechanics.* Addison-Wesley, Reading.

Goodenough, J. B. (1963). *Magnetism and the chemical bond.* Wiley, New York.

Gordy, W. and Cook, R. L. (1984). *Microwave molecular spectra.* Wiley, New York.

Gray, D. E. (ed.) (1972). *American Institute of Physics handbook.* McGraw-Hill, New York.

Graybeal, J. D. (1988). *Molecular spectroscopy.* McGraw-Hill, New York.

Green, H. S. (1965). *Matrix mechanics.* Noordhoff, Groningen.

Greenwood, N. N. and Earnshaw, A. (1984). *Chemistry of the elements.* Pergamon Press, Oxford.

Gribov, L. A. and Orville-Thomas, W. J. (1988). *Theory and methods of calculation of molecular spectra.* Wiley, Chichester.

Griffith, J. S. (1964). *The theory of transition metal ions.* Cambridge University Press.

Griffiths, J. (1976). *Colour and constitution of organic molecules.* Academic Press, New York.

Gumbs, G. and Kiang, G. (1986). The relativistic one-dimensional square potential. *Am. J. Phys.*, **54**, 462.

Haddon, R. C. (1988). π-electrons in three dimensions. *Acc. chem. Res.*, **21**, 243.

Hall, H. E. (1974). *Solid state physics.* Wiley, Chichester.

Hameka, H. F. (1965). *Advanced quantum chemistry.* Wiley, New York.

Hamermesh, M. (1962). *Group theory and its applications to physical problems.* Addison-Wesley, Reading.

Hamilton, W. C. and Ibers, J. A. (1968). *Hydrogen bonding in solids: Methods of molecular structure determination.* Benjamin, New York.

Hankins, T. L. (1980). *Sir William Rowan Hamilton.* Johns Hopkins University Press, Baltimore.

Harris, D. C. and Bertolucci, M. D. (1978). *Symmetry and spectroscopy.* Oxford University Press.

Harris, D. C., Hills, M. E., and Hewston, T. A. (1987). Preparation, iodometric analysis, and classroom demonstration of superconductivity in $YBa_2Cu_3O_{8-x}$. *J. chem. Educ.*, **64**, 847.

Harris, R. K. (1986). *Nuclear magnetic resonance spectroscopy.* Longman, London.

Harrison, W. A. (1980). *Electronic structure and the properties of solids.* W. H. Freeeman and Co., San Francisco.

Hatfield, W. E. (1987). Magnetic measurements. In *Solid state chemistry: Techniques.* (ed. Cheetham, A. K. and Day, P.) Clarendon Press, Oxford.

Hecht, C. E. (1990). *Statistical thermodynamics and kinetic theory.* W. H. Freeman and Co., New York.

Hecht, J. (1988). *Understanding lasers.* Howard W. Sams, Indianapolis.

Hegstrom, R. A., Chamberlain, J. P., Seto, K., and Watson, J. G. (1988). Mapping the weak chirality of atoms. *Am. J. Phys.*, **56**, 1086.

Hegstrom, R. A. and Kondepudi, D. K. (1990). The handedness of the universe. *Sci. Amer.*, **262**(1), 98.

Hehre, W.J., Radom, L. Schleyer, P. v.R., and Pople, J.A., (1986). *Ab initio molecular orbital theory.* Wiley, New York.

Heisenberg, W. (1925). Uber quantentheoretische Undentung kinematischer und mechanischen Beziehungen. *Z. Phys.*, **33**, 879.

Heisenberg, W. (1927). Uber den anschaulichen Inhalt der quantentheoretischen Kinematik und Mechanik. *Z. Phys.*, **43**, 172.

Heisenberg, W. (1930). *The physical principles of quantum theory.* Dover, New York.

Heisenberg, W. (1989). *Physics and philosophy: The revolution in modern science.* Penguin Books, Harmondsworth.

Heitler, W. (1954). *The quantum theory of radiation.* Clarendon Press, Oxford.

Hellmann, H. (1933). Zur rolle der kinetischen elektronenergie für die zwischenatomaren Kräfte. *Z. Phys.*, **85**, 180.

Herman, F. and Skillman, S. (1963). *Atomic structure calculations.* Prentice Hall, Englewood Cliffs.

Herzberg, G. (1944). *Atomic spectra and atomic structure.* Dover, New York.

Herzberg, G. (1945). *Molecular spectra and molecular structure. II. Infrared and Raman spectra of polyatomic molecules.* Van Nostrand, New York.

Herzberg, G. (1950). *Molecular spectra and molecular structure. I. Spectra of diatomic molecules.* Van Nostrand, New York.

Herzberg, G. (1966). *Molecular spectra and molecular structure. III. Electronic spectra and electronic structure of polyatomic molecules.* Van Nostrand, New York.

Hey, A. J. G. (1987). *The quantum universe.* Cambridge University Press

Hinchcliffe, A. (1987). *Ab initio determination of molecular properties.* Adam Hilger, Bristol.

Hinchcliffe, A. (1988). *Computational quantum chemistry.* Wiley, Chichester.

Hinchcliffe, A. and Munn, R.W. (1985). *Molecular electromagnetism.* Wiley, Chichester.

Hirschfelder, J. O. (1960). Classical and quantum mechanical hypervirial theorems. *J. chem. Phys.*, **33**, 1462.

Hirschfelder, J. O. (1967). Intermolecular forces. *Adv. chem. Phys.* **12**.

Hirschfelder, J. O., Byers Brown, W., and Epstein, S. T. (1964). Recent developments in perturbation theory. *Adv. quantum Chem.* **1**, 256.

Hirschfelder, J. O., Curtiss, C. F., and Bird, R. B. (1954). *Molecular theory of gases and liquids.* Wiley, New York.

Hirst, D. M. (1985). *Potential energy surfaces: Molecular structure and reaction dynamics.* Taylor and Francis, London.

Hirst, D. M. (1990). *A computational approach to quantum chemistry.* Blackwell Scientific Publications.

Hoff, A. J. (ed.) (1989). *Advanced EPR.* Elsevier, Amsterdam.

Hoffmann, R. (1989). *Solids and surfaces: A chemist's view of bonding in extended structures.* VCH, Weinheim.

Hollas, J. M. (1983). *High resolution spectroscopy.* Butterworth, London.

Hollas, J. M. (1987). *Modern spectroscopy.* Wiley, Chichester. Homans, S. W. (1989). *A dictionary of concepts in NMR.* Oxford University Press.

Hotop, H. and Lineberger, W. C. (1975). Binding energies in atomic negative ions. *J. phys. chem. Ref. Data*, **4**, 539.

Hotop, H. and Lineberger, W. C. (1985), Binding energies in atomic negative ions 2. *J. phys. chem. Ref. Data*, **14**, 731.

Hout, R. F., Pietro, W. J., and Hehre, W. J. (1984). *A pictorial approach to molecular structure and reactivity.* Wiley, New York.

Hu, B. Y.-K. (1989). Kramers–Krönig in two lines. *Am. J. Phys.*, **57**, 821.

Hughes, R. I. G. (1989). *The structure and interpretation of quantum mechanics.* Harvard University Press.

Huheey, J. E. (1986). A novel method for assigning R,S labels to enantiomers. *J. chem. Educ.*, **63**, 474.

Huheey, J. E. (1983). *Inorganic chemistry.* Harper and Row, New York.

Hunt, R. W. G. (1987). *The reproduction of colour.* Fountain Press, Tolworth.

Infeld, L. and Hull, T. E. (1951). The factorization method. *Rev. mod. Phys.*, **23**, 21.

Jablonski, A. (1935). Uber den mechanismus der photolumieszenz von Farbstoffephosphoren. *Z. Phys.*, **94**, 38.

Jackson, J. D. (1962). *Classical electrodynamics.* Wiley, New York.

Jahn, H. E. and Teller, E. (1937). Stability of polyatomic molecules in degenerate electronic states. 1. Orbital degeneracy. *Proc. Roy. Soc.*, **161A**, 220.

Jahn, H. E. and Teller, E. (1938). Stability of polyatomic molecules in degenerate electronic states. 2. Spin degeneracy. *Proc. Roy. Soc.*, **164A**, 117.

Jammer, M. (1966). *The conceptual development of quantum mechanics.* McGraw-Hill, New York.

Jammer, M. (1974). *The philosophy of quantum mechanics: The interpretation of quantum mechanics in historical perspective.* Wiley, New York.

Jauch, J. M. (1968). *Foundations of quantum mechanics.* Addison-Wesley, Reading.

Jauch, J. M. and Rohrlich, F. (1955). *The theory of photons and electrons.* Addison-Wesley, Reading.

Jeener, J. (1982). Superoperators in magnetic resonance. *Adv. mag. Reson.*, **10**, 1.

Jelley, N. A. (1990). *Fundamentals of nuclear physics*. Cambridge University Press.

Jennings, B. R. and Morris, V. J. (1974). *Atoms in contact*. Oxford University Press.

Jensen, W. B. (1980). *The Lewis acid-base concepts*. Wiley, New York.

Jensen, W. B. (1982). The positions of lanthanum (actinium) and lutetium (lawrencium) in the periodic table. *J. chem. Educ.*, **59**, 635.

Jensen, W. B. (1984). Abegg, Lewis, Langmuir, and the octet rule. *J. chem. Educ.*, **61**, 191.

Johnson, C. S. and Pedersen, L. G. (1973). *Problems and solutions in quantum chemistry and physics*. Addison-Wesley, New York.

Johnson, D. A. (1982). *Some thermodynamic aspects of inorganic chemistry*. Cambridge University Press.

Jones, R. A. Y. (1984). *Physical and mechanistic organic chemistry*. Cambridge University Press.

Jordan, T. F. (1969). *Linear operators for quantum mechanics*. Wiley, New York.

Jordan, T. F. (1986). *Quantum mechanics in simple matrix form*. Wiley, New York.

Jorgensen, P. (1981). *Second quantization-based methods in quantum chemistry*. Academic Press, New York.

Jørgensen, W. L. and Salem, L. (1973). *The organic chemists' book of orbitals*. Academic Press, New York.

Judd, B. R. (1975). *Angular momentum theory for diatomic molecules*. Academic Press, New York.

Kasha, M. (1984). The triplet state: An example of G.N. Lewis' research style. *J. chem. Educ.*, **61**, 204.

Katriel, J. (1972). A study of the interpretation of Hund's rule. *Theor. Chim. Acta.*, **23**, 309.

Kauffman, J. M. (1986). Simple method for determination of oxidation numbers of atoms in compounds. *J. chem. Educ.*, **63**, 474.

Kauzmann, W. (1957). *Quantum chemistry: an introduction*. Academic Press, New York.

Kaye, G. W. C. and Laby, T. H. (1973). *Tables of physical and chemical constants*. Longman, London.

Kemble, E. C. (1958). *The fundamental principles of quantum mechanics*. Dover, New York.

Kettle, S. F. A. (1985). *Symmetry and structure*. Wiley, New York.

Kettle, S. F. A. (1989). Why is ammonia not optically active? *J. chem. Educ.*, **66**, 841.

Khairutdinov, R. F., Zamaraev, K. I., and Zhdanov, V. P. (1989). *Electron tunneling in chemistry*. Vol. 30 of *Comprehensive chemical kinetics* (ed. R. G. Compton). Elsevier, Amsterdam.

Kibble, T. W. B. (1985). *Classical mechanics*. Longman, London. Kidd, R., Ardini, J., and Anton, A. (1989). Evolution of the modern photon. *Am. J. Phys.*, **57**, 27.

Kihara, T. (1978). *Intermolecular forces*. Wiley, Chichester. King, G. W. (1964). *Spectroscopy and molecular structure*. Holt, Reinhardt, and Winston, New York.

Kitao, O. and Nakatsuji, H. (1988). Cluster expansion of the wave function. Valence and Rydberg excitation and ionizations of benzene. *J. chem. Phys.*, **87**, 1169.

Kittel, C. (1986). *Introduction to solid state physics*. Wiley, New York.

Kittel, C. (1987). *Quantum theory of solids*. Wiley, New York.

Kittel, C. and Kroemer, H. (1980). *Thermal physics*. W. H. Freeman and Co., San Francisco.

Klein, D. J. and Trinajstić, N. (1990). *Valence bond theory and chemical structure*. Elsevier, Amsterdam.

Koopmans, T. (1933). Uber die Zuordnung von Wellenfunktionen and Eigenwerten zu den Einzelnen Elektronen eines Atoms. *Physica*, **1**, 104.

Kovaćs, I. (1969). *Rotational structure in the spectra of diatomic molecules*. Adam Hilger, London.

Kramers, H. A. (1964). *Quantum mechanics*. Dover, New York.

Kuhn, H. G. (1962). *Atomic spectra*. Longmans, London.

Kuhn, T. S. (1978). *Black-body theory and the quantum discontinuity, 1894–1912*. Clarendon Press, Oxford.

Kyrala, A. (1967). *Theoretical physics: Applications of vectors, matrices, and quaternions*. Saunders, Philadelphia.

Ladd, M. F. C. (1979). *Structure and bonding in solid state chemistry*. Wiley, New York.

Ladd, M. F. C. (1989). *Symmetry in molecules and crystals*. Wiley, New York.

Landau, L. D. and Lifshitz, E. M. (1958*a*). *Quantum mechanics*. Pergamon Press, Oxford.

Landau, L. D. and Lifshitz, E. M. (1958*b*). *Statistical physics*. Pergamon Press, Oxford.

Landau, L. D. and Lifshitz, E. M. (1960). *Electrodynamics of continuous media*. Pergamon Press, Oxford.

Landsberg, G. and Mandelstam, L (1928). Eine neue Erscheinung bei der Lichtzerstreung in Krystallen. *Naturwiss.*, **16**, 557.

Langhoff, P. W., Epstein, S. T., and Karplus, M. (1972). Aspects of time-dependent perturbation theory. *Rev. mod. Phys.*, **44**, 602.

Lee, J. (1987). The upper and lower bounds of the ground state energies using the variational method. *Am. J. Phys.*, **55**, 1039.

Lemberger, A. and Pauncz, R. (1970). The theoretical interpretation of Hund's rule. *Acta Phys. Hung.*, **27**, 169.

Lever, A. B. P. (1984). *Inorganic electronic spectroscopy*. Elsevier, Amsterdam.

Levine, I. N. (1983). *Quantum chemistry*. Allyn and Bacon, Boston.

Levine, R. D. and Bernstein, R. B. (1987). *Molecular reaction dynamics and chemical reactivity*. Oxford University Press.

Lewis, D. and Peters, D. (1975). *Facts and theories of aromaticity*. Macmillan, London.

Lin, S. H. (ed.) (1980). *Radiationless transitions*. Academic Press, New York.

Lister, D., Macdonald, J. N., and Owen, N. L. (1978). *Internal rotation and inversion: An introduction to large amplitude motions in molecules*. Academic Press, London.

Lloyd, D. R. (1986). On the lanthanide and 'scandinide' contractions. *J. chem. Educ.*, **63**, 502.

Lockwood, E. H. and Macmillan, R. H. (1978). *Geometric symmetry*. Cambridge University Press.

Long, D. A. (1977). *Raman spectroscopy*. McGraw-Hill, New York.

Loudon, R. (1979). *The quantum theory of light*. Clarendon Press, Oxford.

Louisell, W. H. (1973). *Quantum statistical properties of radiation*. Wiley, New York.

Lovett, D. R. (1989). *Tensor properties of crystals*. Adam Hilger, Bristol.

Löwdin, P. O. (1966). The determination of upper and lower bounds of energy eigenvalues in perturbation theory by means of partitioning techniques. In *Perturbation theory and its applications* (ed. C. H. Wilcox). Wiley, New York.

Lowry, T. H. and Richardson, K. S. (1987). *Mechanism and theory in organic chemistry.* Harper and Row, New York.

Maitland, G. C., Rigby, M., Smith, E. B., and Wakeham, W. A. (1981). *Intermolecular forces: Their origin and determination.* Clarendon Press, Oxford.

Makerewitz, J. (1988). Coulomb and Fermi holes in a two-electron model atom. *Am. J. Phys.*, **56**, 1100.

Mandl, L. and Wolf, E. (ed.) (1970). *Coherence and fluctuations of light.* Dover, New York.

Maradudin, A. A. (1971). *Theory of lattice dynamics in the harmonic approximation.* Academic Press, New York.

Margenau, H. (1944). The exclusion principle and its philosphical importance. *Phil. Science,* **11**, 187.

Margenau, H. and Murphy, G. M. (1956). *The mathematics of physics and chemistry.* Vol. 1. Van Nostrand, New York.

Martin, W. C., Hagan, L., Reader, J., and Sugar, J. (1974). Ground levels and ionization potentials for lanthanides and actinide atoms and ions. *J. phys. chem. Ref. Data,* **3**, 771.

Martin, W. C., Zalubas, R., and Hagan, L. (1978). *Atomic energy levels: the rare earth elements.* Nat. Stand. Ref. Data Series, National Bureau of Standards, Washington.

Mason, J. (1988). Periodic contractions among the elements: Or, on being the right size. *J. chem. Educ.*, **65**, 17.

Mason, S. F. (1982). *Molecular optical activity and the chiral discriminations.* Cambridge University Press.

Matos, J. M. O, Roos, B. O., and Malmqvist, P. A. (1987). A CASSCF-CCI study of the valence and lower excited states of the benzene molecule. *J. chem. Phys.*, **86**, 1458.

Matsen, F. A. (1987). Three theories of superconductivity. *J. chem. Educ.* **64**, 842.

Mattuck, R. D. (1976). *A guide to Feynman diagrams in the many-body problem.* McGraw-Hill, New York.

McClellan, A. L. (1963). *Tables of experimental dipole moments.* W. H. Freeman and Co., San Francisco.

McDowell, C. A. (1969). Ionization potentials and electron affinities. In *Physical chemistry, an advanced treatise,* Vol 3 (ed. H. Eyring, D. Henderson, and W. Yost). Academic Press, New York.

McGlynn, S. P., Azumi, T., and Kinoshita, M. (1969). *Molecular spectroscopy of the triplet state.* Prentice-Hall, Englewood Cliffs.

McGlynn, S. P., Vanquickenborne, L. C., Kinoshita, M., and Carroll, D. G. (1972). *Introduction to applied quantum chemistry.* Holt, Reinhardt, and Winston, New York.

McIntosh, H. V. (1959). On accidental degeneracy in classical and quantum mechanics. *Am. J. Phys.*, **27**, 620.

McIntosh, H. V. (1971). Symmetry and degeneracy. In *Group theory and its applications,* Vol. 2, (ed. E. Loebl), p.75. Academic Press, New York.

McLachlan, A. D. (1960). Dangers of the 'average energy approximation' in perturbation theory. *J. chem. Phys.*, **32**, 1263.

McQuarrie, D. A. (1983). *Quantum chemistry.* University Science Books, Mill Valley.

McWeeny, R. (1970). *Spins in chemistry*. Academic Press, New York.

McWeeny, R. (1989). *Methods of molecular quantum mechanics*. Academic Press, New York.

Memory, J. D. (1968). *Quantum theory of magnetic resonance parameters*. McGraw-Hill, New York.

Miller F. A. and Kauffman, G. B. (1989). C. V. Raman and the discovery of the Raman effect. *J. chem. Educ.*, **66**, 795.

Mills, I. (ed.) (1988). *Quantities, units, and symbols in physical chemistry*. Blackwell Scientific Publications, Oxford.

Minkin, V. I., Osipov, O. A., and Zhdanov, Yu. A. (1970). *Dipole moments in organic chemistry*. Plenum, New York.

Møller, C. and Plesset, M. S. (1934). Note on an approximation treatment for many-electron systems. *Phys. Rev.*, **46**, 618.

Moore, C. E. (1965 *et seq.*). *Selected tables of atomic spectra*. Nat. Stand. Ref. Data Series, National Bureau of Standards, Washington.

Moore, C. E. (1971). *Atomic energy levels as derived from the analyses of optical spectra*. Nat. Stand. Ref. Data Series, National Bureau of Standards, Washington.

Moore, W. J. (1989). *Schrödinger: Life and thought*. Cambridge University Press.

Morse, P. M. (1929). Diatomic molecules according to the wave mechanics. II. Vibrational levels. *Phys. Rev.*, **34**, 57.

Morse, P. M. and Feshbach, H. (1953). *Methods of theoretical physics*. McGraw-Hill, New York.

Moruzzi, V. L., Janak, J. F., and Williams, A. R. (1978). *Calculated electronic properties of metals*. Pergamon Press, Oxford.

Moss, R. E. (1973). *Advanced molecular quantum mechanics*. Chapman and Hall, London.

Mott, N. F. and Davis, E. A. (1971). *Electronic processes in non-crystalline media*. Clarendon Press, Oxford.

Mulliken, R. S. (1932). The interpretation of band spectra. III. Electron quantum numbers and states of molecules and their atoms. *Rev. mod. Phys.*, **4**, 1.

Mulliken, R. S. (1934). A new electronegativity scale; together with data on valence states and on valence ionization potentials and electron affinities. *J. chem. Phys.*, **2**, 782.

Mulliken, R. S., Rieke, C. A., Orloff, D., and Orloff, H. (1949). Analytical expressions for overlap integrals. *J. chem. Phys.*, **17**, 1248.

Munowitz, M. (1988). *Coherence and NMR*. Wiley, New York.

Murrell, J. N. (1971). *The theory of the electronic spectra of organic molecules*. Chapman and Hall, London.

Murrell, J. N., Carter, S., Farantos, S. C., Huxley, P., and Varandas, A. J. C. (1984). *Molecular potential energy functions*. Wiley, Chichester.

Murrell, J. N. and Harget, A. J. (1972). *Semi-empirical self-consistent field molecular orbital theory of moelcules*. Chapman and Hall, London.

Murrell, J. N., Kettle, S. F. A., and Tedder, J. M. (1970). *Valence theory*. Wiley, New York.

Murrell, J. N., Kettle, S. F. A., and Tedder, J. M. (1985). *The chemical bond*. Wiley, Chichester.

Muus, L. T. and Atkins, P. W. (ed.) (1970). *Electron spin relaxation in liquids*. Plenum Press, New York.

Naray-Szabo, G., Surjan, P. R., and Angyan, J. G. (1987). *Applied quantum chemistry*. Reidel, Dordrecht.

Nassau, K. (1983). *The physics and chemistry of color*. Wiley, New York.

Newman, J. R. (1954). Laplace. *Mathematics in the modern world*. In *Readings from Scientific American*, (ed. M. Kline). W. H. Freeman, San Francisco.

Newnham, R. E. (1974). *Structure–property relations*. Springer, Berlin.

Nye, J. F. (1985). *The physical properties of crystals*. Oxford University Press.

O'Donnell, S. (1983). *William Rowan Hamilton: Portrait of a prodigy*. Boole Press, Dublin.

Orchin, M. and Jaffé, H. H. (1967). *The importance of antibonding orbitals*. Houghton–Mifflin, Boston.

Overheim, R. D. and Wagner, D. L. (1982). *Light and color*. Wiley, New York.

Pais, A. (1982). *Subtle is the Lord*. Oxford University Press, New York.

Papoušek, D. and Aliev, M. R. (1982). *Molecular vibrational-rotational spectra*. Elsevier, Amsterdam.

Parker, S. B. (ed.) (1987). *Solid state physics source book*. McGraw-Hill, New York.

Parr, R. G., Craig, D. P., and Ross, I. G. (1950). Molecular orbital calculation of the lower excited electronic levels of benzene, configuration interaction included. *J. chem. Phys.*, **18**, 1561.

Parr, R. G. and Yang, W. (1989). *Density-functional theory of atoms and molecules*. Oxford University Press.

Pauli, W. (1940). The connection between spin and statistics. *Phys. Rev.*, **58**, 716. This paper is included in Schwinger (1958).

Pauling, L. (1960). *Nature of the chemical bond*. Cornell University Press.

Pauling, L. (1984). G. N. Lewis and the chemical bond. *J. chem. Educ.*, **61**, 201.

Pauling, L. and Wilson, E. B. (1935). *Introduction to quantum mechanics*. McGraw-Hill, New York.

Pauncz, R. (1967). *Alternant molecular orbital method*. Saunders, Philadelphia.

Pauncz, R. (1969). Electron correlation in atoms and molecules. *Physical chemistry, an advanced treatise* (ed. H. Eyring, D. Henderson, and W. Jost), **3**, 185. Academic Press, New York.

Pearson, R. G. (1987). Recent advances in the concept of hard and soft acids and bases. *J. chem. Educ.*, **64**, 561.

Pearson, R. G. (1988). Absolute electronegativity and hardness. *Inorg. Chem.*, **27**, 734.

Pearson, R. G. (1990). Electronegativity scales. *Acc. chem. Res.*, **23**, 1.

Penrose, R. (1989). *The emperor's new mind*. Oxford University Press.

Piepho, S. B. and Schatz, P. N. (1983). *Group theory in spectroscopy*. Wiley, New York.

Pilar, F. (1968). *Elementary quantum chemistry*. McGraw-Hill, New York.

Pitzer, R. M. (1983). The barrier to internal rotation in ethane. *Acc. chem. Res.*, **16**, 207.

Poirier, R., Kari, R., and Csizmadia, I. G. (1985). *Handbook of gaussian basis sets: A compendium for ab initio molecular orbital calculations*. Elsevier, Amsterdam.

Power, E. A. (1964). *Introductory quantum electrodynamics*. Longman, London.

Price, W. C. and Chissick, S. S. (ed.) (1977). *The uncertainty principle and foundations of quantum mechanics: A fifty years' survey*. Wiley, London.

Puddephatt, R. J. and Monaghan, P. K. (1986). *The periodic table of the elements*. Oxford University Press.

Racah, G. (1942). Theory of complex spectra. II. *Phys. Rev.*, **62**, 438.

Rae, A. I. M. (1986). *Quantum mechanics*. McGraw-Hill, Maidenhead.

Raman, C. V. and Krishnan, K. S. (1928). A new type of secondary radiation. *Nature*, **121**, 501.

Rao, C. N. R. (1975). *Ultraviolet and visible spectroscopy*. Butterworths, London.

Rao, C. N. R. and Gopalakrishnan, J. (1986). *New directions in solid state chemistry*. Cambridge University Press.

Redhead, M. (1987). *Incompleteness, nonlocality, and realism: A prolegomenon to the philosophy of quantum mechanics*. Clarendon Press, Oxford.

Reed, A. E. and Weinhold, F. (1986). On the role of d orbitals in SF_6. *J. Amer. chem. Soc.*, **108**, 3586.

Reif, F. (1965). *Fundamentals of statistical and thermal physics*. McGraw-Hill, New York.

Reissland, J.A. (1973). *The physics of phonons*. Wiley, New York.

Rich, R. L. and Suter, R. W. (1988). Periodicity and some graphical insights on the tendency toward empty, half-full, and full subshells. *J. chem. Educ.*, **65**, 702.

Richards, W. G. (1983). *Quantum pharmacology*. Butterworth, London.

Richards, W. G. (ed.) (1989). *Computer-aided molecular design*. IBC Technical Services, Byfleet.

Richards, W. G. and Cooper, D. L. (1983). Ab initio *molecular orbital calculations for chemists*. Oxford University Press.

Richards, W. G. and Scott, P. R. (1976). *Structure and spectra of atoms*. Wiley, Chichester.

Richards, W. G. and Scott, P. R. (1985). *Structure and spectra of molecules*. Wiley, Chichester.

Richards, W. G., Trivedi, H. P., and Cooper, D. L. (1981). *Spin-orbit coupling in molecules*. Clarendon Press, Oxford.

Ridley, B. K. (1988). *Quantum processes in semiconductors*. Clarendon Press, Oxford.

Rigby, M., Smith, E. B., Wakeham, W. A., and Maitland, G. C. (1986). *The forces between molecules*. Oxford University Press.

Riley, K. F. (1974). *Mathematical methods for the physical sciences: An informal treatment for students of physics and engineering*. Cambridge University Press.

Rivers, R. J. (1987). *Path integral methods in quantum field theory*. Cambridge University Press.

Roberts, M. W. and McKee, C. S. (1978). *Chemistry of the metal-gas interface*. Clarendon Press, Oxford.

Roman, P. (1965). *Advanced quantum theory*. Addison-Wesley, Reading.

Roman, P. (1969). *Introduction to quantum field theory*. Wiley, New York.

Rose, M. E. (1955). *Multipole fields*. Wiley, New York.

Rose, M. E. (1957). *Elementary theory of angular momentum*. Wiley, New York.

Rose-Innes, A. C. and Rhoderick, E. H. (1976). *Introduction to superconductivity*. Pergamon Press, Oxford.

Rotenberg, M., Bivins, R., Metropolis, N., and Wooten, J.K. (1959). *The 3j and 6j symbols*. MIT Press.

Ruedenberg, K. (1962). The physical nature of the chemical bond. *Rev. mod. Phys.*, **34**, 326.

Ruekberg, B. (1987). An astonishingly easy method for determining *R* and *S* for Fischer projections. *J. chem. Educ.*, **64**, 1034.

Rye, R. R. and Houston, J. E. (1984). Molecular Auger spectroscopy. *Acc. chem. Res.*, **17**, 41.

Sacks, L. J. (1986). Coulombic models in chemical bonding. *J. chem. Educ.*, **63**, 288.

Salahub, D. R. and Zerner, M. C. (1989). *The challenge of d and f electrons*. American Chemical Society Symposium No. 394.

Salam, A. and Wigner, E. P. (ed.) (1972). *Aspects of quantum theory*. Cambridge University Press.

Salem, L. (1966). *The molecular orbital theory of conjugated systems*. Benjamin, New York.

Salem, L. (1982). *Electrons in chemical reactions: First principles*. Wiley, New York.

Sanders, J. K. M. and Hunter, B. K. (1987). *Modern NMR spectroscopy*. Oxford University Press.

Sanderson, R. T. (1976). *Chemical bonds and bond energy*. Academic Press, New York.

Sanderson, R. T. (1988). Principles of electronegativity: II. Applications. *J. chem. Educ.*, **65**, 277.

Sanderson, R. T. (1989). *Simple inorganic substances: A new approach*. Robert E. Krieger Publishers, Malabar.

Sandorfy, C. (1964). *Electronic spectra and quantum chemistry*. Prentice-Hall, Englewood Cliffs.

Schaefer, H. F. (1972). *The electronic structure of atoms and molecules: A survey of rigorous quantum mechanical results*. Addison-Wesley, Reading.

Schaefer, H. F. (1984). *Quantum chemistry: The development of* ab initio *methods in molecular electronic structure theory*. Clarendon Press, Oxford.

Schlegel, R. (1980) *Superposition and interaction: Coherence in physics*. University of Chicago Press.

Schrödinger, E. (1926). Quantisierung als Eigenwertproblem. *Annln. Phys.*, **79**, 361 and 489; **80**, 437; **81**, 109.

Schweber, S. S. (1961). *An introduction to relativistic quantum field theory*. Harper and Row, New York.

Schwinger, J. (ed.) (1958). *Quantum electrodynamics*. Dover, New York.

Schwinger, J. (1965). On angular momentum. *Quantum theory of angular momentum* (ed. L.C. Biedenharn and H. van Dam). Academic Press, New York.

Segal, G. A. (ed.) (1977). *Semiempirical methods of electronic structure calculation. Part A: Techniques; Part B: Applications*. Plenum Press, New York.

Series, G. W. (1957). *The spectrum of atomic hydrogen*. Clarendon Press, Oxford.

Shalitin, D. (1984). On the time-energy uncertainty relation. *Am. J. Phys.*, **52**, 1111.

Shannon, R. D. (1976). Revised effective ionic radii in oxides and systematic studies of interatomic distances in halides and chalcogenides. *Acta Cryst.*, **A32**, 751.

Shannon, R. D. and Prewitt, C. T. (1969). Effective ionic radii in oxides and fluorides. *Acta Cryst.*, **B25**, 925

Shore, B. W. and Menzel, D. H. (1968). *Principles of atomic spectra*. Wiley, New York.

Shriver, D. F., Atkins, P. W., and Langford, C. H. (1990). *Inorganic chemistry.* Oxford University Press and W. H. Freeman and Co., New York.

Sidgwick, N. V. and Powell, H. M. (1940). Stereochemical types and valency groups. *Proc. R. Soc.*, **176A**, 153.

Siegman, A. E. (1986). *Lasers.* University Science Books, Mill Valley.

Sinanoğlu, O. and Brueckner, K. A. (1970). *Three approaches to electron correlation in atoms.* Yale University Press.

Singh, S. B. and Singh, C. A. (1989). Extension of the Hellmann-Feynman theorem and applications. *Am. J. Phys.*, **57**, 894.

Sivardiere, J. (1986). Using the virial theorem. *Am. J. Phys.*, **54**, 1100.

Slater, J. C. (1963). *Quantum theory of molecules and solids: I Electronic structure of molecules.* McGraw-Hill, New York.

Slichter, C. P. (1988). *Principles of magnetic resonance.* Springer-Verlag, New York.

Smyth, C. P. (1955). *Dielectric behavior and structure.* McGraw-Hill, New York.

Solymar, L. and Walsh, D. (1988). *Lectures on the electrical properties of materials.* Clarendon Press, Oxford.

Starzak, M. E. (1989). *Mathematical methods in chemistry and physics.* Plenum, New York.

Stebbings, R. F. and Dunning, F. B. (ed.) (1983). *Rydberg states of atoms and molecules.* Cambridge University Press.

Steiner, E. (1976). *The determination and interpretation of molecular wave functions.* Cambridge University Press.

Steinfeld, J. I. (1985). *Molecules and radiation: An introduction to modern molecular spectroscopy.* MIT Press, Cambridge, Mass.

Stephenson, G. (1973). *Mathematical methods for science students.* Longman, London.

Stern, O. (1921). Ein Weg zur experimentellen Prüfung der Richtungsquantelung im Magnetfeld. *Z. Phys.*, **7**, 249.

Stern, O. and Gerlach, W. (1922). Der experimentelle Nachweis des magnetischen Moments des Silberatoms. *Z. Phys.*, **8**, 110.

Stevenson, R. (1965). *Multiplet structure of atoms and molecules.* Saunders, Philadelphia.

Stewart, J. J. P. (1989). Optimization of parameters for semiempirical methods. 1. Method; 2. Applications. *J. comp. Chem.*, **10**, 207 and 221.

Stiddard, M. H. B. (1975). *The elementary language of solid state physics.* Academic Press, New York.

Stone, A. J. and Alderton, M. (1985). Distributed multipole analysis. *Molec. Phys.*, **56**, 1047.

Stranges, A. N. (1984). Reflections on the electron theory of the chemical bond: 1900-1925. *J. chem. Educ.*, **61**, 185.

Streitweiser, A. (1961). *Molecular orbital theory for organic chemists.* Wiley, New York.

Struve, W. S. (1989). *Fundamentals of molecular spectroscopy.* Wiley, New York.

Sugden, T. M. and Kenney, C. N. (1965). *Microwave spectroscopy.* Van Nostrand, London.

Symons, M. C. R. (1978). *Chemical and biological aspects of electron spin resonance spectroscopy.* Van Nostrand Reinhold, New York.

Szabo, A. and Ostlund, N. S. (1982). *Modern quantum chemistry: introduction to advanced electronic structure theory.* Macmillan, New York.

Tanabe, Y. and Sugano, S. (1954). On the absorption spectra of complex ions 2. *J. Phys. Soc. Japan*, **9**, 766.

Tedder, J. M. and Nechvatal, A. (1985). *Pictorial orbital theory*. Pitman, London.

Teller, E. (1937). The crossing of potential surfaces. *J. phys. Chem.*, **41**, 109.

Thirring, W. E. (1958). *Principles of quantum electrodynamics*. Academic Press, New York.

Tilley, D. R. and Tilley, J. (1986). *Superfluidity and superconductivity*. Adam Hilger, Bristol.

Tinkham, M. (1964). *Group theory and quantum mechanics*. McGraw-Hill, New York.

Tolman, R. C. (1938). *The principles of statistical mechanics*. Clarendon Press, Oxford.

Townes, C. H. and Schawlow, A. L. (1955). *Microwave spectroscopy*. McGraw-Hill, New York.

Uhlenbeck, G. E. and Goudsmit, S. (1925). Ersetzung der Hypothese vom unmechanischen Zwang durch eine Forderung bezüglich des inneren Verhaltens jedes einzelnen Elektrons. *Naturwiss.*, **13**, 953.

Uhlenbeck, G. E. and Goudsmit, S. (1926). Spinning electrons and the structure of spectra. *Nature*, **117**, 264.

van der Waerden, B. L. (ed.) (1967). *Sources of quantum mechanics*. North-Holland, Amsterdam.

van Siclen, C. (1988). The one-dimensional hydrogen atom. *Am. J. Phys.*, **56**, 9.

van Vleck, J.H. (1932). *The theory of electric and magnetic susceptibilities*. Clarendon Press, Oxford.

Vanquickenborne, L. G., Pierloot, K., and Devoghel, D. (1989). Electronic configuration and orbital energies: The 3*d*-4*s* problem. *Inorg. Chem.* **28**, 1805.

Verkade, J. G. (1986). *A pictorial approach to molecular bonding*. Springer, New York.

Volhardt, K. P. C. (1987). *Organic chemistry*. W. H. Freeman and Co., New York.

von Neumann, J. (1955). *Mathematical foundations of quantum mechanics*. Princeton University Press.

von Neumann, J. and Wigner, E. P. (1929). Uber das Verhalten von Eigenwerten bei adiabatischen Prozessen. *Phys. Z.*, **30**, 467.

Walsh, A. D. (1953). The electronic orbitals, shapes, and spectra of polyatomic molecules. *J. chem. Soc.*, 2260 *et seq.*

Wambaugh, J. (1974). *The delta star*. Bantam Books, New York.

Wangsness, R. K. (1979). *Electromagnetic fields*. Wiley, New York.

Washburn, S. (1988). A comment on 'A study of barrier penetration in quantum mechanics'. *Am. J. Phys.*, **56**, 679.

Wayne, R. P. (1988). *Principles and applications of photochemistry*. Oxford University Press.

Webster, B. (1990). *Chemical bonding theory*. Blackwell Scientific Publications, Oxford.

Wells, A. F. (1984). *Structural inorganic chemistry*. Clarendon Press, Oxford.

Wertz, J. E. and Bolton, J. R. (1972). *Electron spin resonance: elementary theory and practical applications*. McGraw-Hill, New York.

West, A. R. (1984). *Solid state chemistry and its applications*. Wiley, Chichester.

Wheatley, P.J. (1970). *The chemical consequences of nuclear spin.* North-Holland, Amsterdam.

Wheeler, J. A. and Zurek, W. H. (ed.) (1983). *Quantum theory and measurement.* Princeton University Press.

Whiffen, D. H. (1972). *Spectroscopy.* Longman, London.

White, H. E. (1935). *Introduction to atomic spectra.,* McGraw-Hill, New York.

White, R. M. (1983). *Quantum theory of magnetism.* Springer, Berlin.

Whittaker, E. (1954). William Rowan Hamilton. *Readings from Scientific American: Mathematics in the modern world* (ed. M. Kline), W. H. Freeman and Co., San Francisco.

Wigner, E. P. (1959). *Group theory.* Academic Press, New York.

Wilcox, D. H. (ed.) (1966). *Perturbation theory and its applications in quantum mechanics.* Wiley, New York.

Williams, A. F. (1979). *A theoretical approach to inorganic chemistry.* Springer, Berlin.

Williams, B. G. (1977). *Compton scattering.* McGraw-Hill, New York.

Wilson, E. B., Decius, J. C., and Cross, P. C. (1955). *Molecular vibrations.* McGraw-Hill, New York.

Wilson, J. and Hawkes, J. F. B. (1987). *Lasers: Principles and applications.* Prentice Hall, New York.

Wilson, S. (1984). *Electron correlation.* Oxford University Press.

Wilson, S. (ed.) (1987). *Methods in computational chemistry. 1. Electron correlation in atoms and molecules.* Plenum Press, New York.

Woodgate, G. K. (1980). *Elementary atomic spectra.* McGraw-Hill, New York.

Woodward, L. A. (1972). *Introduction to the theory of molecular vibrations and vibrational spectroscopy.* Clarendon Press, Oxford.

Woodward, R. B. and Hoffmann, R. (1970). *The conservation of orbital symmetry.* Academic Press, New York.

Woolf, A. A. (1988). Oxidation numbers and their limitations. *J. chem. Educ.,* **65**, 45.

Wooster, W. A. (1973). *Tensors and group theory for the physical properties of solids.* Oxford University Press.

Wybourne, B. G. (1974). *Classical groups for physicists.* Wiley, New York.

Yang, C. N. (1987). Square root of -1, complex phases, and Erwin Schrödinger. In *Schrödinger: Centenary celebration of a polymath* (ed. C.W. Kilmister). Cambridge University Press.

Yates, K. (1978). *Hückel molecular orbital theory.* Academic Press, New York.

Yourgrau, W. and van der Merwe, A. (1979). *Perspectives in quantum theory: Essays in honor of Alfred Landé.* Dover, New York.

Zare, R. N. (1987). *Angular momentum: Understanding spatial aspects in chemistry and physics.* Wiley, New York.

Zhou, Z. and Parr, R. G. (1989). New measures of aromaticity: Absolute hardness and relative hardness. *J. Amer. chem. Soc.,* **111**, 7371.

Zhu, D.-P. (1986). Proof of the quantum virial theorem. *Am. J. Phys.,* **54**, 267.

Zijlstra, H. (1967). *Experimental methods in magnetism.* North Holland, Amsterdam.

Ziman, J. M. (1969). *Elements of advanced quantum theory.* Cambridge University Press.

Index

The Periodic table

APR

Period	I (1)	II (2)	3	4	5	6	7	8	9	10	11	12	III (13)	IV (14)	V (15)	VI (16)	VII (17)	VIII (18)
1	1 H 1.008																	2 He 4.003
2	3 Li 6.94	4 Be 9.01											5 B 10.81	6 C 12.01	7 N 14.01	8 O 16.00	9 F 19.00	10 Ne 20.18
3	11 Na 22.99	12 Mg 24.31											13 Al 26.98	14 Si 28.09	15 P 30.97	16 S 32.06	17 Cl 35.45	18 Ar 39.95
4	19 K 39.10	20 Ca 40.08	21 Sc 44.96	22 Ti 47.90	23 V 50.94	24 Cr 52.01	25 Mn 54.94	26 Fe 55.85	27 Co 58.93	28 Ni 58.71	29 Cu 63.54	30 Zn 65.37	31 Ga 69.72	32 Ge 72.59	33 As 74.92	34 Se 78.96	35 Br 79.91	36 Kr 83.80
5	37 Rb 85.47	38 Sr 87.62	39 Y 88.91	40 Zr 91.22	41 Nb 92.91	42 Mo 95.94	43 Tc 98.91	44 Ru 101.07	45 Rh 102.91	46 Pd 106.4	47 Ag 107.87	48 Cd 112.40	49 In 114.82	50 Sn 118.69	51 Sb 121.75	52 Te 127.60	53 I 126.90	54 Xe 131.30
6	55 Cs 132.91	56 Ba 137.34	71 Lu 174.97	72 Hf 178.49	73 Ta 180.96	74 W 183.85	75 Re 186.2	76 Os 190.2	77 Ir 192.2	78 Pt 195.09	79 Au 196.97	80 Hg 200.59	81 Tl 204.37	82 Pb 207.19	83 Bi 208.98	84 Po 210	85 At 210	86 Rn 222
	87 Fr 223	88 Ra 226.03	103 Lr 257	104 Unq	105 Unp	106 Unh	107 Uns	108 Uno	109 Une									

Lanthanides

57 La 138.91	58 Ce 140.12	59 Pr 140.91	60 Nd 144.24	61 Pm 146.92	62 Sm 150.35	63 Eu 151.96	64 Gd 157.25	65 Tb 158.92	66 Dy 162.50	67 Ho 164.93	68 Er 167.26	69 Tm 168.93	70 Yb 173.04

Actinides

89 Ac 227.03	90 Th 232.04	91 Pa 231.04	92 U 238.03	93 Np 237.05	94 Pu 239.05	95 Am 241.06	96 Cm 247.07	97 Bk 249.08	98 Cf 251.08	99 Es 254.09	100 Fm 257.10	101 Md 258.10	102 No 255